# Parallel System Interconnections and Communications

Miltos D. Grammatikakis
Integrated Systems Development S. A.
Athens, Greece

D. Frank Hsu
Fordham University
New York, U.S.A.

Miro Kraetzl
Defence Science and Technology Organisation
Salisbury SA, Australia

CRC Press
Boca Raton London New York Washington, D.C.

**Library of Congress Cataloging-in-Publication Data**

Grammatikakis, Miltos D.
Parallel system interconnections and communications / Miltos D. Grammatikakis, D. Frank Hsu, Miro Kraetzl.
p. cm.
Includes bibliographical references and index.
ISBN 0-8493-3153-6 (alk. paper)
1. Parallel processing (Electronic computers) 2. Computer networks. I. Hsu, D. Frank (Derbiau Frank), 1948-II. Kraetzl, Miro. III. Title.

QA76.58 .G72 2000
004′.35—dc21 00-051908
CIP



Direct all inquiries to CRC Press LLC, 2000 N.W. Corporate Blvd., Boca Raton, Florida 33431.


International Standard Book Number 0-8493-3153-6
Library of Congress Card Number 00-051908
Printed in the United States of America 1 2 3 4 5 6 7 8 9 0
Printed on acid-free paper

# Preface

In this monograph, we describe a comprehensive environment of parallel interconnection networks and communications. The obvious growing trend to find solutions of complex computational tasks is making the need for fast and efficient parallel computing extremely important. Many areas of human activity require enormous computational power; computer vision, various medical applications, robotics, navigation, air traffic control, information networks management, weather prediction, stock market analysis, symbolic algebra, artificial intelligence, and numerous military applications are just few examples. The need to interconnect the processing elements in computers solving such huge tasks in such a way that will ensure optimal network performance is paramount. Of all parallel architectures implemented within new-generation computers, massively parallel systems containing hundreds or more processors are the most prevalent. They can be interconnected in a network of processors in static (multicomputers), or dynamic (multiprocessors) ways. We describe many of the possible topologies to build and assess the performance of multicomputers and multiprocessors. We give detailed descriptions of all important network architectures with emphasis on multistage interconnection networks, hypercubes, and meshes. We address all communication models, routing strategies, router performance, fault tolerance, reliability, and fault diagnosis in interconnection networks. Wherever possible, we outline the interconnection architecture and describe the performance of commercial systems.

We are aware of the fact that our monograph is far from being a complete exposition of ideas, research, and applications within the parallel interconnects environment. The amount of literature covering this topic within last few decades is vast, and sometimes it was hard to be up-to-date in such a fast-growing area of computer science research. However, we tried to cover a broad spectrum of relevant literature, at the expense of going deeply into details. That is why we provided very few proofs of outlined results. The idea of writing a monograph originated while working on the survey paper on packet routing in fixed-connection networks [235], when the authors realized that a generalization of results to the broader topic of communication in parallel interconnection networks would be an important resource of fundamental interest to students, designers, programmers, engineers, and researchers in the area. For the most advanced topics we assume few general prerequisites, while we try to develop an elegant narrative approach in discussing fundamental issues in parallel communication and algorithm design.

The book is organized in seven chapters. The first chapter describes network topologies, models, and applications in parallel computing. We concentrate on parallel systems, various programming paradigms, and theoretical models used in parallel computing. In the second chapter, we outline routing strategies, communication patterns, fault tolerance, and performance of packet routers. This chapter is devoted to general parallel interconnection communications on multicomputers and multiprocessors. Applications to multidimensional FFT and other parallel algorithms are given. The third chapter concerns multistage interconnection networks (MINs), with emphasis on blocking, rearrangeable, nonblocking MINs, and generalized connectors. Furthermore, fault tolerance, reliability, and performability of MINs are considered, together with special purpose interconnects, VLSI layout and customization, and commercial MINs. In the fourth chapter, most important topological, communication, and sorting properties of hypercubic networks are outlined. We start with topological properties of hypercubes, and then provide detailed exposition of hypercube communication, sorting and embedding algorithms, together with scheduling and allocation strategies. At the end of this chapter, a plethora of commercial hypercube systems is described. In the fifth chapter, we give a detailed exposition of topological properties and performance (sorting, routing) assessment of mesh (and mesh-like) interconnection networks. We discuss allocation, scheduling, VLSI layout and customization of meshes, and commercial mesh systems. The sixth chapter concerns additional important parallel interconnects like de Bruijn networks, shuffle-exchange, circulant, Cayley, and fat-tree networks. The seventh chapter addresses some important issues of future parallel communications in network design (realizability) and performance. As much as we tried to explore (in a comprehensive and exact way) the literature and results of the huge area of parallel communication in interconnection networks, we are sure that we might have omitted a number of important or relevant results.

The first author acknowledges the support from the University of Oldenburg and the University of Hildesheim where, during his service as Assistant Professor, most of his contribution towards this book was completed. He would also like to express his thanks to faculty members at these institutions for their continuous encouragement and support, in particular to Professor Eike Best (Department of Parallel Systems), Professor Günther Stiege (Department of Operating Systems and Networks), and also to Visiting Professor Alexander Lavrov (Kiev Polytechnic Institute). The second author would like to express his gratitude to Fordham University and to many members of the Fordham community, especially Dr. Robert Himmelberg; Rev. Jeffrey von Arx, S.J.; Rev. Robert R. Grimes, S.J.; and Dr. Nancy Busch for their continued support and encouragement. He would also like to thank Dr. Gen-Huey Chen of the National Taiwan University's Department of Computer Science and Information Engineering, and to all participants of the CAN (Combinatorics, Algorithms, and Networks) seminar conducted by the second author in Spring 1999. The third author would like to thank the many people from Communications Division of DSTO for their support while working on this book, in particular Chris Woodruff, Peter Shoubridge, and Nigel McGinty.

We would also like to thank Jop F. Sibeyn (Department of Computing Science, Umeå University) for the significant contribution to the exposition of the material in Chapter 5. Without his expertise in routing on meshes and his contribution to this book, this monograph would definitely lack the completeness in coverage of the important topics of meshes, arrays, and mesh-like interconnects.

Any views stated within this book are completely personal and are not related in any way to the Integrated Systems Development S.A., or to the Australian Department of Defence.

| M.D. Grammatikakis | D.F. Hsu | M. Kraetzl |
|---|---|---|
| Athens, Greece | New York, U.S.A. | Salisbury, South Australia |

# Contents

# List of Figures

# List of Tables

*We dedicate this book to Μάρω, Karen, and Leoni for their support and encouragement.*

# Chapter 1

# Network Topologies, Models, and Applications

## 1.1 Parallel Systems

Many examples of parallelism exist in life situations, such as open heart surgery, war crisis (coordination of Army, Navy, and Air Force), highway driving in lanes, and aircraft design. Parallel computing has evolved through: (a) the need for Science and Engineering applications, and (b) significant changes in the economy of scale (VLSI).

Many problems within the domain of *large scale scientific computation* require a vast number of computations. Thus, certain Science and Engineering problems are classified as *grand challenge problems.* In 1987, the U.S. Government Office of Science and Technology Policy defined grand challenge problems as fundamental applications to science or engineering, whose solution would be enabled by applying high-performance computing resources that could become available in the near future. These problems usually aim at predicting the effect of parameter changes on matter and energy. Some examples include,

- *Computational Fluid Dynamics:* long-range weather prediction, global climate change, computational ocean sciences, enhanced oil and gas recovery, nuclear reactor design, automobile and hypersonic aircraft design, quiet submarines,

- *Electronic structure calculations for the design of new materials:* chemical catalysts, immunological agents, drug design, human genome, semiconductors, and superconductors,

- *Plasma dynamics for fusion energy applications and military:* nuclear fusion, combustion systems, air, sea, and undersea surveillance for safe and efficient military technology,
- *Calculations to understand the fundamentals of matter:* quantum chromodynamics, astrophysics, structural analysis, seismology, and condensed matter theory,
- *Symbolic computations:* speech recognition, natural language processing, computer vision, image processing, automated reasoning, data mining for modeling business and financial processes, and discrete and continuous simulation of design, manufacturing, and production issues, e.g., in transportation systems,

Two solution approaches to these problems exist. In the *classical approach* one deals with the design of models which are used later in experiments. In some fields (e.g., particle physics) the cost of experimental facilities (supercolliders) is in the billions of dollars. An alternative approach consists of the *numerical simulation* of underlying physical processes. This method is cheaper, more versatile, and offers the possibility to obtain deeper knowledge of the problem. However, the mathematical modeling is not at all simple, and actual models are still needed for final verification. While simulation models for two-dimensional lattices are popular, three-dimensional ones are more relevant, but computationally intense. The time complexity is proportional to the number of lattice points, the number of variables per each lattice point, the number of operations per each variable, and the number of iterations needed for algorithmic convergence.

Consider an example of solving a partial differential equation using difference equations. For weather forecasting over a volume of 100 $km \times$ 100 $km \times 20$ $km$, with uniform segments of size 0.5 $km \times 0.5$ $km \times 0.5$ $km$, $k = 10^6$ lattice points must be used! Furthermore, for a single prediction a linear system $A_{k \times k}X = b$ must be solved, where matrix $A$ has only seven nonzero elements per row (corresponding to lattice point interactions). A numerical solution would require $\mathcal{O}(k^3/3) \sim 10^{18}/3$ arithmetic operations. For 2 days prediction with 30-minute evaluations, $10^{20}/3$ operations are needed. Even the fastest existing parallel computer (1 $T$ flops/sec) would take $\sim$1 year on this problem doing nothing else! Moreover, the prediction would be 1 year overdue!!

Parallel computing has been made possible by several scientific and technological breakthroughs which have decreased costs and increased memory space, switching speed, reliability, and computing power (whether expressed in MIPS/sec or FLOPS/sec). Several architectures for exploiting parallelism have been explored; the following classification is approximately from the least to the most general architecture, and from the fastest to the slowest system:

- *Systolic Arrays*
- *Associative Processors*
- *Vector Processors*
- *Data-flow Machines*
- *Functional Language Processors (e.g., Lisp)*
- *Massively Parallel Systems*
    - *Bus Systems (e.g., orthogonal buses)*
    - *Multiprocessors (e.g., Omega, butterfly)*
    - *Multicomputers (e.g., hypercube, mesh)*
    - *Hybrid Systems (e.g., banyan hypercube)*
    - *Distributed Systems (heterogeneous computing)*
    - *Reconfigurable Systems (e.g., NYU CHiP)*

In this book, we concentrate on *massively parallel processing systems* (MPP), the most prevalent of all parallel architectures. Most MPP systems are currently configured with 100 or more powerful processors. In particular, we consider *multicomputers* (called also static, point-to-point, or loosely coupled systems), and *multiprocessors* (named also dynamic, shared memory, or tightly coupled systems). Both multicomputers and multiprocessors allow processing elements (PEs) to execute separate instruction streams.

In multicomputers, PEs with local memories communicate with each other using send and receive operations. The success of this distributed memory hierarchy depends on *locality of references*, which is a function of the network topology, the parallel algorithm, the data allocation scheme, and *latency hiding techniques*.

In multiprocessors, PEs have access to a single address space. Most current multiprocessors feature a large distributed shared memory accessed through a switch-based multistage network. Each switch connects to other switches and possibly to incoming and outgoing network interfaces of processors. The use of a shared memory model helps in preserving the sequential paradigm of programming, by imposing sequential semantics on writable shared values.

In 1966, Flynn provided an interesting four-way classification of parallel architectures, based on the division between program control and data [208]. Single-instruction single-data stream (SISD) corresponds to the traditional Von Neumann "stored program" sequential computer, including pipelined vector machines, such as the Cray-1. Single-instruction multiple-data (SIMD) signifies a centralized data parallel approach, where all participating PEs execute the same instruction on different data, and synchronize before execution of the following instruction. SIMD is frequently called single-program multiple-data (SPMD). Notice that in SPMD, some PEs may be selected to *abstain* from a particular type of operation: this operation proceeds without input from that PE, and without delivering a result to that PE. SIMD systems are the ILLIAC IV, Goodyear MPP, and the Thinking Machines Corp. (TMC) CM-2. Multiple-instruction multiple-data (MIMD) corresponds to control parallelism, where PEs may execute different instructions. Typical MIMD systems are the Cray X-MP/4, Cray T3E, BBN Butterfly, iPSC/2, NCube, and Intel Paragon XP/S. Hybrid systems support both SIMD and MIMD capabilities; an example is the TMC CM-5.

Multiple-instruction single-data (MISD) machines were not practical to implement; however, in a broad sense, MISD corresponds to vector pipelining. A pipelined computation is described by a reservation table that describes usage through time of the different stages of a pipeline. Efficient pipelining depends on the ability to create a continuous stream of similar operations:

- By providing vector data from memory (called strip mining) and splitting the original vector into equal-size segments to fit the size of the vector registers; this is performed either by the compiler, or directly by hardware, as in the Fujitsu VP series.
- By *chaining* vectors in linked vector operations, without having to rely on storing intermediate results.
- By avoiding *collision hazards* due to data dependencies or instruction branches. To increase pipeline efficiency and avoid collision hazards one usually inserts (noncomputing) delay stages within the pipeline [344].

Both shared memory multiprocessors, and distributed memory multicomputers can be operated in SIMD or MIMD fashion. As shown in

Figure 1.1, the distinction between these two classes is based on the design of the interconnection network. For a multiprocessor, caches can be incorporated either together with the processors, or with the shared memory banks.

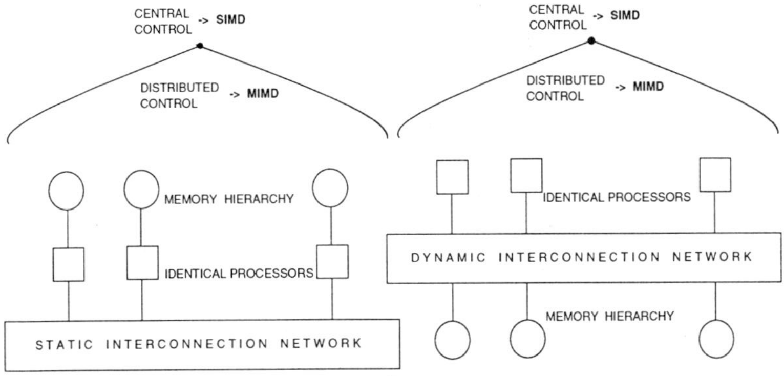


**FIGURE 1.1**
**Multicomputer and multiprocessor systems**

Some simple multicomputer and multiprocessor topologies are shown in Figure 1.2. The last two topologies shown are dynamic, connecting eight PEs either to memory, or back to themselves via wraparound links (connections not shown). We will encounter several more dynamic topologies in Chapter 3. Static topologies, including popular ones such as meshes and hypercubes, will be discussed in detail in Chapters 4, 5, and 6.

Depending on memory organization, parallel systems are further classified into shared memory, and distributed memory systems. Distributed systems communicate by exchanging messages, and there is no remote memory access (NORMA). Shared memory systems are further classified into three categories: uniform memory access (UMA), nonuniform memory access (NUMA), and cache-only memory access (COMA) systems.

UMA systems offer a central shared memory and allow any data element to be accessed (through shared read and write operations) in the same, constant time. Notice that UMA programs do not need to partition program data, thus providing flexibility and ease of program-

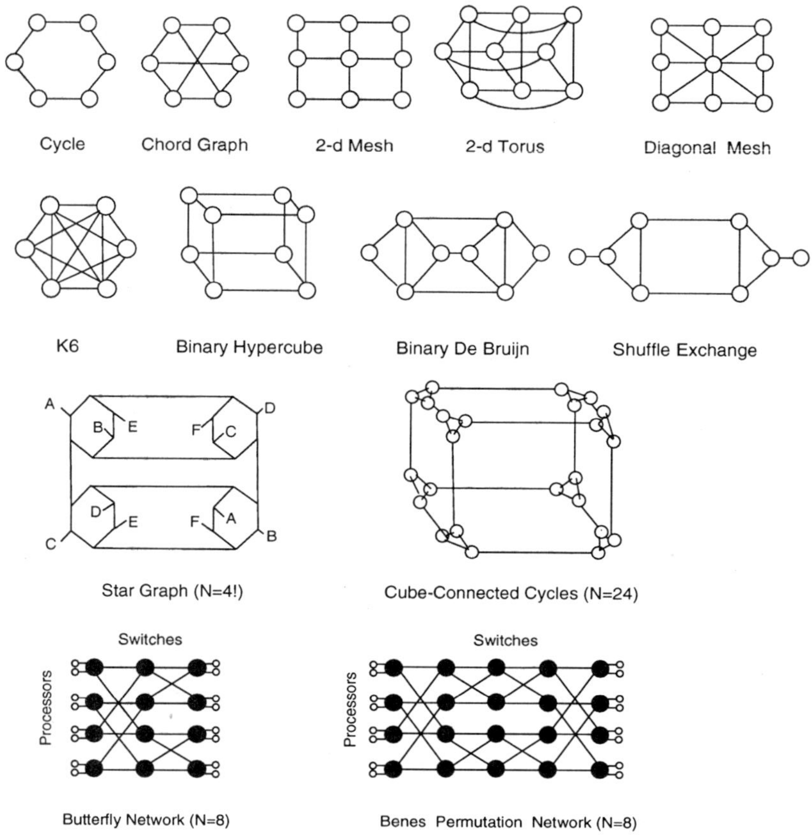


**FIGURE 1.2**
**Examples of static and dynamic networks**

ming. Examples of UMA systems include the Encore Multimax and the Sequent Symmetry.

NUMA systems do not support constant-time read and write operations, with remote memory access much slower than local memory access (typically two orders of magnitude slower). However, communication latency, which refers to the time interval between the instant that a message is issued until the instant that the message is completely received by the recipient, is kept low due to the exploitation of data locality, especially for common communication patterns. Communication latency consists of network latency, i.e., time spent on the interconnection network, and software latency introduced by the messaging layer and the network/processor interface. The value of exploiting locality is apparent, but the techniques for accomplishing this are not. For example, in most NUMA architectures, memory is organized hierarchically, so that node-to-neighbor memory access is faster. Examples are the Cray T3D/T3E series, the CM-5, the Paragon XP/S, and older systems such as the BBN TC2000, and Intel's N-cube.

Cache-only memory access (COMA) systems are NUMA machines with all physical memory distributed into very large caches. This reduces effective communication latency to maintain cache coherence, and thus requires less interconnection network bandwidth. Network bandwidth is defined as the amount of data transmitted in the network per unit of time, expressed in bits/sec. The KSR-1/2 was the first commercial COMA multiprocessor. Recent approaches improve the network interface, e.g., the University of Illinois IACOMA project [668].

Shared memory systems are further classified as either symmetric, or asymmetric. A symmetric multiprocessor provides all PEs "equal access" to system devices, such as disk and I/O interface.

Current massively parallel systems form a reasonable mix of UMA and NUMA architectures. Low level mechanisms may provide a view of shared memory on a message-passing based system or vice versa. The emulation of a message passing system on a shared memory machine is simple. We partition the shared memory among all PEs. Messages are exchanged between PEs by writing each other's memory partition. However, emulating a shared memory system on a message passing machine is very costly. This is usually called *virtual*, or *distributed shared memory* (VSM or DSM). The VSM idea stems from Kai Li [427], who also provided the first VSM implementation on a network of workstations [425]. Nowadays most commercial supercomputers are NUMA VSM systems; this includes the Cray T3E, NEC SX-4, and to a lesser degree the Hitachi SR-2201, and the Fujitsu VPP700.

Efficient parallel execution requires a delicate balance between *program granularity*, communication latency, and synchronization overhead between the different processes. Thus, if communication and synchronization latency is minimal, as in embarrassingly parallel or data parallel applications, *fine grain* partitioning will yield the best performance by allocating processes to a large number of less powerful PEs, e.g., a CM-2 MPP system. Fine grain parallelism is often exploited by a parallelizing compiler. Also, if communication and synchronization costs are large, as in a loosely coupled system, then *coarse grain* partitioning is more appropriate. Coarse grain partitioning works by executing essentially independent problems on systems with a small number of very powerful processors, e.g., a vectorizing Cray T90 parallel system. Coarse grain parallelism is exploited by an efficient vectorizing compiler, and the run time operating system.

## 1.2 Programming Paradigms

Two parallel programming techniques are ideally possible. *Explicit parallel programming* makes the user specify how sequential processes cooperate. In this case the compiler can be implemented in a straightforward manner. *Implicit parallel programming* uses sequential algorithms and a "smart" compiler for parallelizing an application. Implementations using implicit parallelism are usually not as efficient as explicit ones.

In explicit parallel programming several *paradigms* are used, which abstract implementation details allowing portability of software across different machine architectures. The *sequential programming paradigm*, in which the programmer has a simplified view of the system as a single processor which can access memory, is a familiar concept. Parallel programming paradigms must also provide abstract mechanisms for sharing information among PEs. The most common parallel programming paradigms are *message passing*, *data parallel*, and *(possibly virtual) shared memory*. These paradigms differ in many aspects, such as:

- The amount of effort programmers have to invest in coding parallel programs,

- The way code and data distribution are specified,
- The parallel architecture for which the paradigm fits best, and
- The amount of application concurrency which can be exploited by the paradigm.

Notice that *code distribution* refers to the assignment of various tasks to PEs, while *data allocation* refers to data scattering allocation to particular physical PEs, arranged by the compiler or through library calls. As shown in Figure 1.3, two forms of software interleaving are commonly used. With *fine interleaving*, each $n$-byte module contains a block of $n$ consecutive storage locations, while in *coarse interleaving*, each $n$-byte memory module contains a block of $n$ shifted, nonconsecutive storage locations. Data allocation with fine interleaving (data granularity) is better for crossbars or multistage networks, since simultaneous access to different modules is possible. Coarse interleaving is better when a broadcasting switch or bus is used, since remote reads from the same memory location may proceed simultaneously [218].

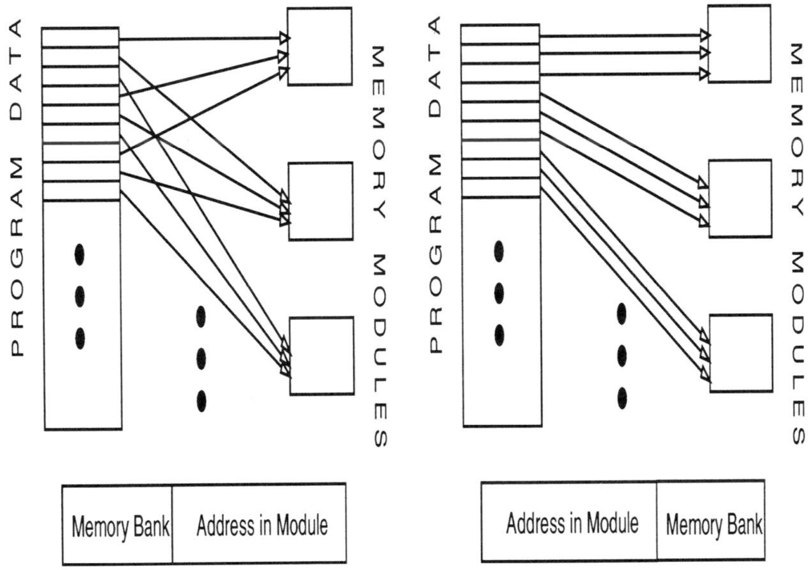


**FIGURE 1.3**
**Data allocation using (a) coarse and (b) fine interleaving**

### 1.2.1 Parallelizing Compilers

Parallelism inherent in loop structures but not explicitly specified by the programmer can be extracted by a *parallelizing compiler* capable of detecting parallelism automatically, and identifying, and exploiting data parallel operations [556]. The amount of parallelism that the compiler is able to detect depends on data, control, and shared resource, e.g., I/O dependencies, which impose a sequential order in the execution of statements. Data and control dependencies can be categorized into four classes:

- Flow (or write-read) dependence: statement S2 is reachable from statement S1, and statement S2 consumes the results of statement S1.
- Antidependence (or read-write dependence): statement S2 is reachable from statement S1, and statement S1 uses results that statement S2 may change.
- Output (or write-write) dependence: statements S1 and S2 may both result in changes to the same data.
- Control dependence: independence between statements S1 and S2 cannot be established, e.g., S1 and S2 manipulate, within an if or loop, an array variable which is indirectly addressed.

For optimizing data parallel compiling, several techniques, such as loop substitution through variable aliases, loop unrolling, loop redistribution (using strip mining), and nested loop reordering have been proposed [556].

### 1.2.2 General Run Time Optimizations

#### 1.2.2.1 Parallel Operating Systems

Early versions of parallel operating systems (OS), e.g., for Cosmic Cube and BBN Butterfly, provided a very rudimentary environment by compiling programs at a host machine.

Later, more general purpose symmetric OS appeared based on simple modifications of the uniprocessor Unix OS, e.g., CMU's Mach OS. The symmetry in the kernels make it easy to restart the OS, and also provide graceful degradation on faults. Thus, each PE runs a similar Unix microkernel, which supports several languages and interfaces for data parallel, shared memory, and message passing models, and provides support for *interactive and batch* execution, security, and accounting. It also includes libraries for interprocess communications, task management and interrupt handling (with routines to create, control, and destroy processes and threads), and virtual memory management, i.e., paging and shared memory. It provides interface to Ethernet, ATM, TCP/IP, FDDI, and through HIPPI high performance I/O storage management.

The latest parallel systems provide, along with the parallel processing nodes, dedicated OS and I/O nodes [607]. A new interesting research effort is directed towards fault tolerant distributed OS, such as Stanford's *Hive* Unix-like OS prototype [591].

#### 1.2.2.2 Dynamic Load Balancing

A common runtime optimization technique which improves PE utilization in parallel applications is *load balancing*. In its most common implementation, excess elements are locally diffused to neighbor PEs, which carry smaller loads. Thus, idle (or underutilized) PEs share the load with donor PEs. In general, donor PEs may be selected in various ways.

- With *asynchronous round-robin*, the donor PE is selected by accessing and incrementing a local variable initialized to: $(my\_rank + 1) \bmod N$, where $my\_rank$ is the distinct PE id number ($my\_rank \in [0, N-1]$).
- With *global round-robin*, the donor PE is selected by accessing a global variable using a *fetch and add* instruction. This method can cause contention, but it's not possible that PEs ask simultaneously for work from the same donor.
- With *random polling*, the donor PE is selected at random.

Load redistribution is also useful in connection to faults. With real-time computing, load redistribution in the presence of faults can be achieved using *spare PEs and links*, i.e., redundant elements which are fault-free prior to reconfiguration. However, in matrix problems, local redistribution presents a form of algorithmic fault tolerance, which involves PEs which hold: (a) the data to the left and/or right of the current failing PE, (b) the data (north, east, west, south) of the failing PE (NEWS approach), or (c) in global schemes, e.g., all PEs holding data in the same row as the current PE. Global schemes outperform local schemes for typical matrix-vector products and similar row dependent grid-based problems [186]. These efficient load balancing techniques are based on basic communication operations, such as semi-contraction, semi-expansion, broadcast, and parallel scan [301, 552, 580].

### 1.2.3 Message Passing Model

In the *message passing programming paradigm*, possibly heterogeneous, "sequential" processors, each consisting of a control unit (CU), an arithmetic and logical unit (ALU) and local memory (M), are connected together using an interconnection network (see Figure 1.4). Information is shared among processors by exchanging messages through the network. In a message transfer, data moves from local memory of one processor to local memory of another processor.

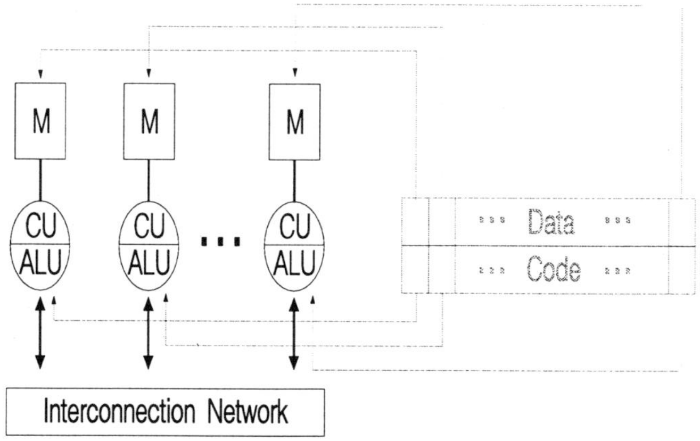


**FIGURE 1.4**
**The message passing programming paradigm**

Message passing programs consist of processes operating on private variables with the ability to exchange data by sending messages. Both data and code are distributed among processors. Efficient message passing programs strive towards minimizing network communication.

Hardware platforms which are naturally suited for this model are massively parallel processing systems with point-to-point links, and networks of workstations (NOW). Message passing can be easily emulated on shared memory platforms, e.g., by splitting the memory space into separate mailboxes.

Message passing programming languages are usually sequential languages, augmented with special libraries providing message passing subroutines. PARMACS, PVM, and MPI (up to version 1.2, and a subset of version 2.0) are currently used message passing libraries. Although PVM is an older communication standard, in most situations, increased efficiency and portability for heterogeneous distributed systems can be obtained by using the MPI standard together with C, C++, or Fortran languages. MPI offers more functionality than hardware message passing, since it provides over 140 communication, synchronization, and global reduction routines.

### 1.2.4 Data Parallel Model

In the *data parallel programming paradigm*, a single control unit (CU) supervises several identical arithmetic and logical units (ALUs) (see Figure 1.5). Each ALU has its own local memory (M). Local memories are connected together using an interconnection network, which usually is a synchronous bus or a point-to-point network, e.g., hypercube, mesh, or torus.

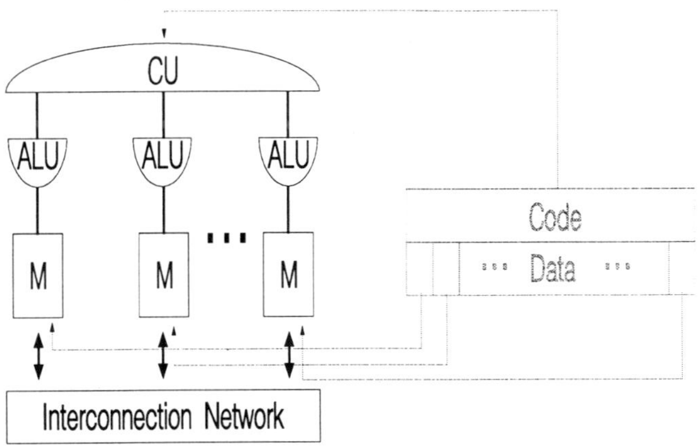


**FIGURE 1.5**
**The data parallel programming paradigm**

The paradigm supports a single instruction multiple data (SIMD), or single program multiple data (SPMD) abstraction. Some compilers, e.g., High Performance Fortran (HPF), provide data parallel behavior on various platforms. Programs consist of one main process executing identical operations in parallel on several local variables in a synchronous manner. Programs are implemented on a higher abstraction level. For example, *virtual processors* abstract from the available physical processors. Data parallelism is expressed by assigning data to virtual processors, which perform identical computation on its data. Since the compiler inserts synchronization constructs needed for parallel execution into the code, user-included deadlocks and data race hazards are completely avoided and program debugging is made easier.

Since data parallel programs have only one control flow, problems with inherent distributed control cannot be parallelized efficiently.

Data parallel programming languages are usually provided by special compilers which extend the set of operations of standard sequential languages. Fortran 90, HPF, Fortran D, ML, C*, and parallel C++ are data parallel languages. Fortran 90 and HPF are portable and widely accepted standards. In these languages, data exchanges are implicit, i.e., performed within (partial) permutation, replication, segmented *scan and reduction*, and broadcasting operations. Each time, either (a) PEs, or in initial implementations a host, execute(s) a scalar instruction, or (b) selected PEs execute the same parallel operation. Data allocation and communication details are left to the parallelizing compiler.

### 1.2.5 Shared Memory Model

As shown in Figure 1.6, in the *shared memory programming paradigm* each processor consisting of a control unit (CU), and an arithmetic and logical unit (ALU) is connected to memory (M) through an interconnection network which is usually a bus, a crossbar switch, or a multistage switching network. The memory access network can be equipped with caches, either between processors and the interconnection network, or between the interconnection network and memory banks. The shared memory paradigm occurs in two forms: a) physical shared memory multiprocessors providing uniform memory access (these systems are called symmetric multiprocessors - SMPs), and virtual shared memory multiprocessors providing nonuniform memory access. Notice that low level emulation mechanisms may provide a programmer's view of a virtual shared memory on message passing systems.

With shared memory programming it is essential to minimize memory contention. Although writing efficient concurrent (shared memory) programs has a reputation for being exotic and hard, it is neither when the system provides good primitives and suitable libraries, and the pro-

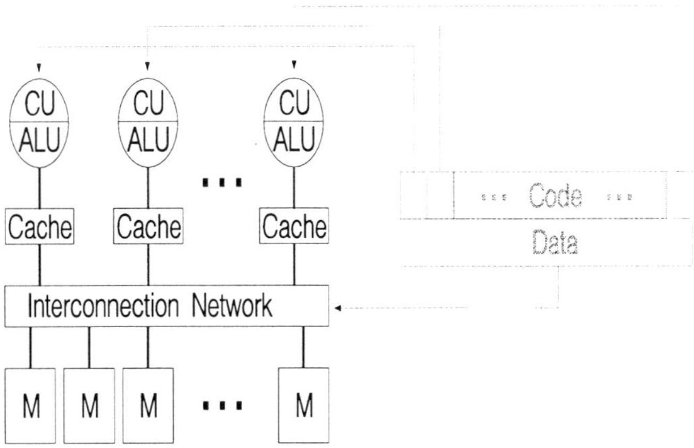


**FIGURE 1.6**
**The shared memory programming paradigm**

grammer exercises basic caution and alertness to avoid common pitfalls, such as data races and deadlocks. New program analysis tools, such as the Eraser, which detects program data races are major steps in this direction [599].

**Physical Shared Memory**

Programs are collections of processes accessing a central pool of shared variables. The programmer has to express code distribution only, most data are stored in global memory. Since data can be accessed concurrently by different processes, mutual exclusion operations are vital for shared memory programming. Although the global memory address space provides a simpler conceptual model for programming than message passing, shared memory programming is more error-prone due to synchronization problems involving data races and deadlocks.

Shared memory programming languages usually consist of sequential languages, together with special libraries providing memory access and synchronization routines. For example parts of MPI-2, Open-MP, and Cray's ShMem are shared memory programming utilities. MPI and Open-MP are portable and widely accepted standards, while Shmem may also undergo a standardization process. MPI-2 and Shmem implement one-way communications using put/get from remote memory, while Open-MP implements relatively high-level directives to specify parallel tasks or loops. For all three interfaces, the user must explicitly consider the memory consistency model, and apply synchronization mechanisms, such as release and acquire, for program consistency. HPF also provides extrinsic procedures to invoke other parallelism paradigms not directly supported in HPF (e.g., explicitly parallel code for branch-and-bound parallelism). Such procedures may have an independent memory consistency model, thus HPF requires all processors to enter and exit these procedures together.

**Virtual Shared Memory**

The idea behind the *virtual shared memory model* (VSM) is to implement coherent shared memory on a network of processors without any physically shared memory. The idea stems from Li [427], who in 1988 provided the first implementation (called IVY) on a network of workstations [425]. The VSM approach allows the flexibility of shared memory programming in a message passing environment. In addition, it improves portability and enhances system scalability through modular memory growth.

With VSM, the distributed memory forms a global virtual address space which is shared among loosely coupled processing nodes by exchanging messages. The programmer of a VSM environment has to bear in mind that remote memory access may significantly increase memory latency compared to local memory, thus the system behaves as a nonuniform memory access (NUMA) system.

#### 1.2.5.1 Cache Coherence

The existence of multiple cached copies of data creates the possibility of inconsistency between a cached copy and the shared memory, or between cached copies themselves. The cache coherence problem arises from the lack of knowledge where the correct data exists when writing shared variables, e.g., during process migration, or DMA I/O writes to memory.

One may certainly prevent data inconsistency by

- Prohibiting caching of writable shared variables, e.g., semaphores and shared data,
- Invalidating the entire cache directory upon a write to shared memory, or
- Permitting caching of variables only within a critical section.

In the former cases, accesses to shared variables are at cache-miss speeds, and nonlocal communication is reduced only by using data replication or recomputation. Thus, for large data size, performance decreases significantly. These schemes require OS, compiler, and programmer involvement in specifying coherent data. Although in the latter case access latency is somewhat reduced, all locks are accessed through atomic operations, thus sharing of writable data is effectively serialized.

When unlimited data caching and multiprocessing is required, more complicated schemes are needed to avoid coherence problems. For example, modern uniprocessors may reorder a read and an independent write upon a read-miss to improve performance. Obviously, this introduces no consistency problems. However, in a multiprocessor the read may be fetching a shared value, while the write may be releasing a lock on that value. Thus, another PE may actually be able to write to the value before the read completes, causing data inconsistency! Also notice, that unless all accesses are reads, programmers should avoid references by more than one PE to the same *cache line*; a cache line is a group of surrounding words which are loaded together upon a memory reference. Otherwise, the effect of *false sharing* may occur: cache lines alternatively ping-pong between different PEs that need only a small part of the cache line. Such effects are also common when accessing software synchronization (lock and barrier) routines which are needed for ensuring data coherence [471]. Depending on the size of the cache line in memory blocks, padding of data can help in avoiding this problem [385].

With a bus interconnect, cache coherence can be maintained by adopting a "snoopy protocol." Each cache controller "snoops" on the transactions of other caches and guarantees validity of cached data. Some systems using a write through policy, i.e., updating the main memory simultaneously with writing the cache, invalidate copies residing in remote caches. Thus, this protocol reduces latency to access new values. Other systems use a write back cache policy, updating the main memory either at block replacement (swap) time, i.e., when the cache block is swapped out, or when another PE needs the data item. In this case, any PE modifying data in its own cache, signals all PEs that their copies are invalid. Thus, this protocol reduces bus contention. Notice that the snooping logic must also watch for reads that access main memory locations corresponding to dirty (modified) locations in the cache, i.e., locations that have been changed but not yet written back. Several protocols implement cache coherence on write back caches, such as the Illinois write invalidate protocol, the Berkeley write invalidate protocol, and Dragon update protocols [137],

In a single stage or multistage network, the unavailability of either a system "bus", or efficient all-to-all broadcasting algorithms renders snoopy protocols useless. Instead, directory based schemes send consistency commands only to caches that keep a copy of a particular cache line (block), thus limiting network traffic to only essential updates, and providing scalability.

Three possible implementations of directory schemes exist. In a *full map* directory, a bit-vector at a central PE indicates which PEs share a given cache line, and which PE is the owner of this cache block. For a large number of PEs, a central full map directory is not only expensive in terms of memory, but also prone to contention causing long search time. Since at any time only a few PEs share data, a *limited map* protocol stores a small number of PE tags with each line in main memory. If more PEs must share the data than there are slots available in the directory, then broadcast is used. Fully distributed *chained* directories provide a main memory pointer to a linked list with PEs sharing the cache line. Thus, a write operation must first trace a chain of pointers from cache to cache, invalidating other copies along the chain; the actual write is stalled until the chain has been traversed.

A standard directory-based chained cache coherence protocol is SCI (Scalable Coherent Interface) [261]. SCI is designed for point to point topologies, with links operating at speeds of 1 Gbit/s or 1 Gbyte/s, depending on the medium in use. SCI is scalable to $2^{16}$ nodes. It supports interconnecting arbitrary heterogeneous machines, and provides 64-bit addresses, with 16 bits allocated for node selection, and the remaining 48 bits for internal addressing.

#### 1.2.5.2 Shared Memory Consistency

Consistency models specify the order by which the execution of instructions from one PE should be observed from other PEs. Four common memory consistency models exist: sequential (SC), weak (WC), processor, and release consistency.

Parallel program behavior depends on the way of interleaving instruction sequences. The SC model has been implemented on most parallel systems because of its simple semantics, which ease programming. However, in certain cases, it may lead to poor performance due to strict constraints on memory order, especially for a large number of PEs. Under Lamport's *sequential consistency* model, the result of any execution appears as an interleaving of individual processor operations, as if executed on some multithreaded sequential machine [386].

Some references to sequential consistency often attribute stronger conditions, such as a total order of all reads and writes (a property which is better called *dynamic atomicity*); this strict form of consistency requires the ability to determine the latest write, which implies a total ordering of requests and thus the existence of absolute global time. Although, some early uniprocessors have traditionally observed strict consistency, the total ordering of requests leads to inefficiency due to excessive data movement and synchronization requirements.

With sequential consistency, write back does not have to occur until another PE accesses data. However, if two PEs actively share data, then they must continually update the main memory copy. Weak consistency relaxes this constraint by providing the ability to abandon SC semantics when they are not needed, e.g., in updating finite elements for solving partial differential equations. However, WC may be more costly in terms of maintaining sequential consistency in software, since it requires direct programmer involvement [269].

Notice that in practice, we encounter various relaxed memory models, as in IBM 370, total store ordering model (TSO), partial store ordering model (PSO), processor consistency (PC), weak ordering (WO), release consistency (RC), PowerPC, relaxed memory order (RMO), and Alpha consistency [11]. Implementation details of these protocols are usually transparent to application programs. It is up to the programmer to define "well-behaved" applications which provide a sequential consistent memory view. For best performance, hardware would like to overlap as many operations as possible. To make sure that an application runs correctly we must restrict the overlaps. In all four relaxed models described next, we impose the following uniprocessor ordering restrictions:

- Flow dependence (write-read): If a PE reads variable $v$ after writing to $v$, then the read must wait for the write to be completed to avoid a uniprocessor read-write hazard.
- Antidependence (read-write): If a PE writes to variable $v$ after reading $v$, then the write must wait for the read to be completed to avoid a uniprocessor write-read hazard.
- Output Dependence (write-write): If a PE writes to variable $v$ after writing to $v$ previously, then the second write must wait for the first write to be completed to prevent a uniprocessor write-write hazard.

Under *processor consistency*, the memory model need not always honor a *write before read* instruction sequence; although many applications still operate correctly, a synchronization operation can be issued between the write and the read where this ordering is important. Thus, writes issued by the same PE are never seen out of order by any other PE in the system. However, the order of writes from two different PEs can be observed differently, since read operations may bypass a previous write operation, i.e., be observed by other PEs as happening before the write. This model allows the memory to hide some of the write latency: writes can be buffered, and bypassed by reads that don't read any location in the write buffer. Processor consistency is a weaker model than SC, and is sometimes called *total store ordering*.

Under *partial store ordering*, the memory model need not always honor *write before read*, and *write before write* ordering restrictions; when these orderings are important, a synchronization operation can be appropriately inserted. This model allows the memory to complete write operations out of order.

Under the *weak consistency model*, the memory model need not always honor *write before read*, *write before write*, *read before read*, and *read before write* ordering restrictions; when the application requires any of these ordering restrictions, appropriate synchronization operations must be inserted. Synchronization operations are guaranteed to be sequentially consistent, i.e., they are not allowed to perform until all preceding load and store operations have completed, and similarly load/store operations are not allowed to perform until all preceding synchronization operations have completed. To take advantage of this model, PEs must implement nonblocking reads, so that a PE does not stall on a read miss. However, the software must maintain consistency by enforcing synchronizations explicitly. This is also a weaker model than SC.

Under *release consistency*, memory accesses to shared data must always be preceded by an acquire (e.g., special read, or lock), and followed by a release (e.g., special write, or unlock) synchronization primitive [11]. There are two flavors of release consistency, RCsc and RCpc. RCsc maintains sequential consistency among synchronization operations, while RCpc maintains processor consistency among such operations. Thus, releases by the same PE are never seen out of order, while, for processor consistency, acquires (special reads) may bypass releases (special writes) on the same PE – also, releases by different PEs may be seen in different orders by different PEs. Adopting a processor or release consistency protocol may likely reduce the "idle time" due to write misses, since a following read or acquire does not have to wait for the completion of the write.

#### 1.2.5.3 Consistency Rules

It is interesting to examine in more detail possible reorders in a shared memory program. We describe consistency rules as computation rules (CMP), uniprocessor order (UPO and URR), processor order (PO = {RR, WW, WR, RW}), and cache coherence rules CC1 (or CC2), and CC3.

**Computation (CMP)**

Since the computation rules are intuitively obvious they are almost never stated explicitly. However, a formal definition is needed for reasoning about interactions between processes on shared data. An example of a failure to obey one of the rules of computation is shown below.

| | |
|---|---|
| Initially, | $A = 0.$ |
| | $\mathrm{PE}_0$ |
| | $A = 1;$ |
| Terminally, | $A = 2.$ |

Operand $A$ is initialized to zero. Then, $\mathrm{PE}_0$ sets $A$ to 1. However, at the end it has the value 2. Clearly, the memory system did not behave in the way one normally expects.

**Uniprocessor Order (UPO)**

Message reordering may violate *uniprocessor order* rules, when operations performed on the same shared operand appear to be reordered. If a machine obeys UPO, and if a first operation on a shared operand logically occurs in a program before a second operation on the *same shared operand*, then the first operation is executed before the second.

- **Read-Read order (RR).** If a machine obeys URR, and if a first read operation on shared data logically occurs in a program before a second read operation on shared data, then the first read operation is executed before the second read operation.

- **Write-Write order (WW).** If a machine obeys WW, and if a first write operation on shared data logically occurs in a program before a second write operation on shared data, then the first write operation is executed before the second write operation.

- **Write-Read order (WR).** If a machine obeys WR, and if a write operation on shared data logically occurs in a program before a read operation on shared data, then the write operation is executed before the read operation.

- **Read-Write order (RW).** If a machine obeys RW, and if a read operation on shared data logically occurs in a program before a write operation on shared data, then the read operation is executed before the write operation.

**Program order (PO)**

A machine obeys *program order* (PO), if all operations performed on *different shared operands* occur in the order defined by the program. Specifically, the following program order rules describe the constraints that a processor in a parallel or distributed system may be required to obey:

- **Read-Read order (RR)** requires that all read operations on shared data occur in the order defined by the program.
- **Write-Write order (WW)** requires that all write operations on shared data occur in the order defined by the program.
- **Write-Read order (WR)** requires that if a write instruction occurs before a read instruction in the logical order defined by a program (and if both instructions access shared data), then the write instruction will be executed before the read instruction.
- **Read-Write order (RW)** requires that if a read instruction occurs before a write instruction in the logical order defined by a program (and if both instructions access shared data), then the read instruction will be executed before the write instruction.

It is very hard to provide a test to detect UPO rule violations on a machine that violates PO rules [129].

**Cache Coherence (CC)**

Caching of shared data can lead to execution sequence reordering which may violate consistency rules. The term *cache coherence* is used to describe memory behavior when performing write operations. Several different levels of coherence have been proposed:

- **CC1:** All threads see each operand change value at exactly the same instant. CC1 is also called write atomicity (WA) [11].
- **CC2:** All threads see exactly the same sequence of changes of values for all operands.

- **CC3:** All threads see exactly the same sequence of changes of values for each separate operand.
- **CC4:** Different threads may see an operand assume different sequences of values.

No program can distinguish between CC1 and CC2 machines [129]. However, programs can distinguish between other types of coherence. A test program can demonstrate that a machine does not obey CC1 (and CC2) by detecting that the machine makes stale data visible to programs. For detecting CC1 or CC3 failures, shared memory programs must record value changes in at least one shared variable.

## 1.3 Theoretical Models

### 1.3.1 Parallel Communication Models

Communication in multicomputers is usually performed in *packet routing* mode. A message along with its destination address is placed in a packet that moves through the nodes of the interconnection network without establishing a physical connection path between the packet's source and destination. In contrast to packet routing, *circuit switching* allows the data to move through the network only after establishing a physical connection path between the source and destination. Routing along a network of buses is somewhat in-between these two extremes. Here a connection is not switched, but it is up to the algorithm designer to guarantee that no two packets are using the bus at the same time. However, since the bus is a shared resource and each PE must compete for it, bus performance quickly saturates when more than $\sim$30 active PEs are configured.

In the *SIMD model* every routing step is restricted to a single dimension; every node sends and receives at most one packet along a single direction. In the *MIMD model* (also called *all-port model*) every node can communicate with all its neighbors in a single step. While both models allow for simultaneous sends and receives (the network links operate in a full duplex fashion), the SIMD communication model assumes 1-port processors, i.e., this is a DMA-bound model.

Although packet routing models (also called switching techniques) are usually defined based on *unicasting, node to neighbor* communication, they are easily extended to more general situations involving one to many (*multicasting*), or one to all (*broadcasting*) communications.

Notice that the communication models presented in this book cover only part of the research on broadcasting and gossiping. Several other complex models lead to distinct formulations and analysis techniques [259, 277]. For example, parallel complexity is sometimes expressed in terms of the number of communication rounds. In this case, optimization corresponds to the (minimum) amount of messages that must be exchanged. This measure is appropriate if each communication is realized approximately in the same amount of time, e.g., when synchronization time is greater than communication time and messages sent at each step are of the same length.

The simplest packet routing model is *store-forward*. In this model, each node has a store/send capability, it sends a packet and waits for an acknowledge message from the receiver. Each such "store-forward" step is called a hop. With this strategy, a packet incurs a delay which is at least proportional to the distance it has to travel, since it fails to take advantage of the potential to traverse multiple nodes at once when there is no blockage. In the *constant model*, transmission of a packet from one network node to any of its neighbors takes one time step. In the *linear model*, this transmission takes $\alpha + \beta\, b$ time, where $\alpha$ is the startup time, $b$ the size of the packet transmitted, and $\beta$ the transmission rate.

In order to reduce the distance dependency, alternative packet routing models have been proposed. With *wormhole routing*, every node has a queue associated with each of its adjacent links which can hold a small number of bits. A packet may be spread over many nodes on the path from its source to its destination and progresses only if there is space in the queue at the next node. When the head moves, or when bits are consumed by the destination node, the entire packet can move by moving to the free space created [153]. The chunks into which packets are divided are called *flits* (typically around 100 bytes), while a phit (physical bit) refers to the number of packet bits moving over a link during a single clock cycle.[1] Notice that because of the way the routing is organized at the hardware level, flits should not stay in the same PE, except in their source and destination PEs. Furthermore, deadlock-free routing becomes more complicated, since flits belonging to different worms may share the same physical link. Wormhole routing was intro-

[1]Notice that unlike our definition, in the standard DS or HS parallel links a flit refers to how often and how much control information is sent *back* as credit information, when a given message moves forward.

duced in Ametec 2010, a two-dimensional mesh. It has also been used on NCube2 (hypercube), Intel Paragon, and Intel/DARPA Touchstone Delta (two-dimensional meshes), IWarp, Inmos, and T9000 transputers, and Cray T3D/T3E [607, 609].

*Virtual cut-through* is similar to wormhole routing, except that the entire packet can pile up at a node, since sufficient queuing is provided [336]. Assuming no congestion, the communication time for the various routing models is listed in Figure 1.7.

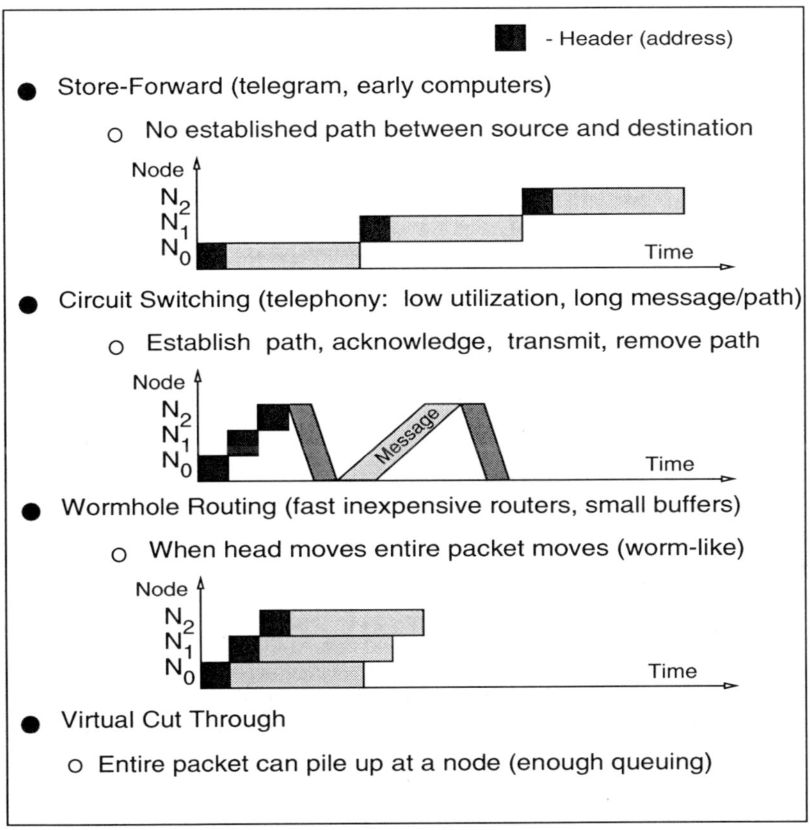


**FIGURE 1.7**
**Various routing models (with examples)**

Ignoring queuing delays arising from packets waiting for congested channels to become available, we may perform a simple comparison by considering packet propagation time for each routing model. Let $b$ be the packet size in flits, $w$ the flit size in bits, $\tau$ the time to transmit one bit to a neighbor, i.e., the inverse of the link bandwidth, $d$ the channels traversed, and $R$ the transmission rounds (for circuit switching). In Table 1.1 we compare the latency of a packet in the four different message switching techniques.

**Table 1.1** Performance of switching techniques (no congestion)

| *Switching Techniques* | *Latency* |
|---|---|
| Store-Forward | $b \cdot w \cdot \tau \cdot d$ |
| Circuit Switching | $\sim c_1 R \cdot \tau + c_2 \cdot b \cdot w \cdot d$ |
| Virtual Cut-Through | $w \cdot \tau \cdot (b + d)$ |
| Wormhole Routing | $w \cdot \tau \cdot (b + d)$ |

In general, one hopes to reduce network latency, and make it independent of the network load, communication pattern, and network size. Virtual cut-through achieves lower latency at the expense of greater consumption of communication resources, which lower effective network capacity for heavy traffic conditions (saturation rate) and increase the blocking probability. Notice that if $b \gg d$ the transfer time for wormhole routing (and virtual cut through) is almost distance independent.

In the presence of congestion, for efficient packet routing, the total communication delay (propagation and queuing) must be small, while the system maintains high reliability, i.e., multiple disjoint paths between nodes exist, and transmission errors are rare. Also, the queue size, defined as the maximum number of packets stored at any network node during communication, should be kept as small as possible. This is an important design consideration because of physical constraints arising from buffer space limitations and required fast access to buffer elements. Various models have been developed to specify more precisely these criteria. Some queuing methods from the literature, although not always realistic, are random queuing, round-robin, further-first (overall, or per-phase), oldest first, longest first (variant length packets), and also combinations of the above.

### 1.3.2 Parallel Computation Models

Parallel computation models focus on the parallel complexity of parallel algorithms, abstracting parallel system details.

In the Parallel Random Access Memory (PRAM) model a parallel system is viewed as an SIMD UMA machine consisting of a number of PEs with a globally addressable memory. Thus, PEs are always synchronized in their read, compute, and write memory cycles, and memory accesses take exactly the same time, i.e., there is no synchronization, or memory access overhead. Since more than one PE may read/write a particular memory address, a policy has to be adopted to determine the outcome for any concurrent access. The following are possible PRAM variants:

- The Exclusive Read Exclusive Write (EREW) PRAM model forbids concurrent reads or writes to the same memory address.
- The Concurrent Read Exclusive Write (CREW) PRAM model allows different processors to concurrently read the same memory address, but forbids concurrent writes.
- The Concurrent Read Concurrent Write (CRCW) PRAM model allows different PEs to concurrently read/write the same memory address. Conflicting writes are resolved by adopting a policy that either restricts all written values to be identical, or else chooses a value according to a predetermined policy (e.g., random, priority, or minimum). It can be shown, that any CRCW-PRAM can be simulated by an EREW-PRAM with $\mathcal{O}(\log N)$ slowdown, where $N$ is the number of PEs.

The simple definition of PRAM provides an abstract model of parallel computation, in which PEs share memory, but do not have communication links between each other. The advantage of this model is that it is easy for programmers. However, it is questionable whether a PRAM can be directly built. Other more recent, popular parallel computation models are PRAM extensions, such as asynchronous PRAM, local PRAM, block PRAM, and X-PRAM which includes modeling the interconnect [356], and new models, such as the BSP [684], and the $\log P$ model [136].

The bulk synchronous parallel (BSP) model is based on computation steps, and communication steps realized using $k$-relations (see Section 2.2.1). In BSP, the parallel program consists of supersteps, with synchronization of any subset of PEs done at the end of each "superstep." In contrast to the PRAM model, the BSP model allows processors to work asynchronously, models latency and limited bandwidth, and requires few machine parameters as long as a certain programming methodology is followed. The BSP model has recently become a communication standard consisting of ~30 BSP operations.

The latency overhead model ($\log P$) is also an asynchronous model, which takes its name from the following basic parameters that describe parallel time complexity.

- In the absence of stalls, $L$ is an upper bound on the *latency* experienced by any packet during communication.
- $o$ is the *overhead*, defined as the time interval that a PE is engaged in the transmission or reception of each message; during this interval, the PE cannot perform any other operation.
- $g$ is the *gap*, defined as the minimum time interval allowed between consecutive packet transmission or reception at a PE; notice that the reciprocal of $g$ is proportional to the available per-PE communication bandwidth.
- $P$ is the *number of PEs*, and equals the number of local memory modules.

For more information on parallel models of computation, including simulations among them, refer to [452].

### 1.3.3 Program Locality and Latency Hiding

Program locality in parallel computing takes several forms. *Spatial locality* occurs because memory accesses tend to refer to nearby memory words, while *temporal locality* occurs because a recently referenced memory word is likely to be referenced again. In addition, *processor locality* occurs because references to a memory word tend to come from the same PE. Properly exploiting all types of locality by the user, the system software, and the hardware, is essential to using parallel systems efficiently. More recently, latency hiding was addressed by MPP system designers through the use of nonblocking communication, efficient prefetching schemes, hardware implemented cache coherence, and larger, higher utilized caches (hidden memory) exploiting temporal and processor locality by limiting expensive (remote) main memory access, e.g., with a miss ratio of 10%, remote accesses decrease by a factor of 10.

Other software-based latency hiding techniques are multiple contexts (threads), simulating a number of virtual processors (processor overloading), data redundancy, and relaxing memory consistency requirements. However, Mark Hill argues against using aggressively relaxed consistency models because, with the advent of speculative execution, these models do not give a sufficient performance boost to justify exposing their complexity to the authors of low-level software [268]. Even without any instruction reorders at least three possibilities for compiler optimization exist: prefetching, multithreading, and caching.

*Prefetching* methods based on look-ahead, or long cache lines exploit spatial locality, thus reducing communication latency. Keeping coherent multiple copies of shared variables while prefetching takes place is very traffic intensive. With the *binding prefetch* policy, the fetch is assumed to have happened when the prefetch instruction is issued; thus, it is the responsibility of the fetching process to ensure that no other PE will update the prefetched value before it is actually used; this may require atomic swaps to manipulate a lock. With the *nonbinding prefetch* policy, the cache coherence protocol invalidates a prefetched value if it is updated prior to its use.

*Multithreading* provides another way to overlap computation and communication [296]. In a shared memory system, concurrent instruction streams executing on different PEs are called processes. With multithreading, a process at each PE is further broken up into a large number of relatively independent parts, called threads. When a thread issues a possibly nonblocking remote read operation, e.g., on a cache miss, then the PE does not have to sit idle or busy wait at all, but instead the PE switches context and executes a different thread. Thus, multithreading increases PE utilization, if a context switch costs less than a remote access.

One may use existing analytical models of multithreading and prefetching to predict upper bounds on processor utilization based on system and application parameters, such as remote cache miss latencies, and application run lengths. The models make the following assumptions:

- The impact of multithreading and prefetching on program synchronization performance and scheduling is ignored,
- For both models, remote cache miss latencies are constant. Thus, for multithreading, latencies are independent of context switch, and there are no additional cache misses during a context switch; this is a reasonable assumption, since context switch code is in cache (frequently used), or hardwired.

- Multithreaded processors switch context only on remote cache misses, and similarly, software prefetches are issued only for blocks that would otherwise cause a remote cache miss.

#### 1.3.3.1 Multithreading Model

The following multithreading model is due to Agarwal [13].

Let $n$ be the number of threads at a given PE, and $c$ the context switching overhead. Also, let $r(n)$ the average number of cycles (instructions) between remote accesses, and $l(n)$ the communication latency on a remote memory access with $n$ threads (including network delays, cache miss penalty, etc.).

Notice that $r(n) \leq r(1)$, due to cache interference and fine grain computation, and $l(1) \leq l(n)$, due to higher network and memory contention. Also, $r(n) = 1/m(n)$, where $m(n)$ represents the cache miss rate.

On a single-threaded PE, the maximum utilization is $r(1)/[r(1)+l(1)]$.

On a multithreaded PE, maximum PE utilization occurs if we have $(n-1)r(n) + nc > l(n)$, since there are enough threads to context switch after each remote access. Then, maximum utilization is $r(n)/[r(n) + c]$, which improves over the single threaded case if $r(1)c < r(n)l(1)$; this is just $c < l(1)$, if $r(1) = r(n)$. If $n$ is below the level for maximum utilization, then the multithreaded PE utilization is $nr(n)/[r(n) + l(n)]$, which is bounded by $n$, if $r(1)l(n) = r(n)l(1)$.

Agarwal has extended the above basic model to model the effect of network latency $l(n)$ and cache miss ratio $m(n)$ [13]. Bianchini and Lim have used Agarwal's basic multithreading model to explain empirically why a small number of contexts (two or three) is usually sufficient to achieve most of the performance improvement in the current generation of shared memory multiprocessors [69].

#### 1.3.3.2 Prefetching Model

Unlike multithreading, prefetching relies on the compiler or user to insert explicit prefetch instructions for data cache lines that they would otherwise miss. This gives prefetching a greater flexibility than multithreading in overlapping communication and computation, by adjusting the prefetch size (and iteration step size) appropriately. Bianchini and Lim developed a performance model for evaluating processor utilization due to prefetching. Their model is based on a common prefetch loop, in which execution alternates between prefetch and computation intervals. Each prefetch accesses cache blocks which are used one computation block ahead; while it is possible to prefetch more than one computation block, any such loop can be transformed into an equivalent loop that prefetches a single computation block ahead.

The model relates the optimal number of cache blocks ($n$) to prefetch with the following parameters: average amount of loop computation per basic prefetched block ($r$), communication latency on a cache miss ($l(n)$), and prefetching overhead due to possible block replacement prior to use ($c$) [69]. Notice that $l(0) \leq l(n)$, due to higher network and memory contention.

Similar to multithreading, there are two cases to consider: (a) enough blocks are prefetched to allow communication latency to overlap completely with loop computation, and (b) the computation block is too short to avoid processor idle time. In the former case, we must have $nr + 2nc > l(n)$, and the processor utilization is $r/[r+c]$. In the latter case, the processor utilization is $2nr/[nr + l(n)]$. Notice that processor utilization without prefetching is $r/[r + l(0)]$, which is smaller if, for example, $l(n) < 2nl(0)$, or $l(n) < nr$.

The prefetching and multithreaded models presented provide a useful tool for evaluating the potential benefits of latency hiding techniques. For example, we can determine the range of $r(n)$, such that prefetching results in improved processor utilization in order to justify architectural support for prefetching. Comparing the above multithreading and prefetching models, Bianchini and Lim point out that prefetching achieves better performance improvements than multithreading on today's parallel systems [69].

### 1.3.4 Customization of Interconnection Networks

One of the most critical engineering issues in the design of interconnection networks is packaging and layout of nodes and links. Depending on the network size, two-dimensional or three-dimensional VLSI layout, customization using intraboard design, interboard wiring, input-output, and maintainability must be considered.

For ease of maintenance and support of incremental scalability, network design follows a hierarchical approach.

Thus, at the individual board level, we have a planar layout which consists of many chips including processors, memory modules, network interface controllers, and other devices. VLSI design issues are usually addressed at this level.

At the interboard level, we have a three-dimensional layout. Most designs stack boards together, allowing for backplane (or side) wiring. In this case, critical issues involve minimizing wire length, and reducing the number of wires to improve maintainability. Such problems are considered both between different boards, and also between different cabinets. For large networks supporting a global clock, other important network design issues involve clock synchronization to avoid skewing, system diagnosis, and heat dissipation using properly designed cooling systems.

Relatively precise models are available only at the lowest packaging levels. When networks are hardwired into VLSI chips, the size of the chip and its capacity to process information by moving it through the chip are interesting complexity measures [419]. Thus, a well-defined VLSI complexity model that relates measures of hardware complexity with time efficiency provides a systematic approach to the analysis of parallel complexity. A layout of a network on a grid is a mapping of network nodes to the points of a lattice. Network connections are mapped to edge-disjoint paths on the lattice. Depending on the network, a number of paths are allowed to cross at a grid point, which represents the basic device, e.g., for multistage networks this is usually a $2 \times 2$ switch. Proofs of area layouts, even in two dimensions, are usually very interesting, since they are geometric illustrations, sometimes resembling chaotic pictures.

Today's VLSI technology is based on laying out circuits in two dimensions. The fact that VLSI chips are multilayer (with tens of layers) does not change the dimensionality; it merely increases the 2-D area of the chip by a constant factor. Let $A$ be the chip area, $s$ the problem size, and $T$ the time required to perform a computation on a VLSI chip. Thompson has shown that for certain computations, $A \times T^2 \geq \mathcal{O}(f(s))$ [663]. The bound can be considered as a *memory bound on chip area* $A$, an *I/O bound on volume* $AT$, or most commonly a *bisection communication bound* $\sqrt{AT}$. Bisection width refers to the number of wires that need to be cut in order to divide the network into two equal parts. Notice that bisection width is usually proportional to network performance, since it represents the maximum amount of information flowing between two halves of the chip. Since the area of a VLSI chip is proportional to the memory space the chip can hold, the above lower bound corresponds to the space complexity of the computation [419]. The VLSI complexity model can be easily extended to 3-D "cubic" chips; in this case, the bisection is $(AT)^{\frac{2}{3}}$. If $V$ is the 3-D chip's volume, then $V^2 \cdot T^3 \geq \mathcal{O}(f'(s))$, for an appropriate real function $f'(s)$.

Raghunath and Ranade [568, 569] considered customization of various interconnection networks, assuming hierarchical $N$-node network design based on network decomposition into $M$ modules with $N/M$ PEs each, called bundles. PEs in a bundle were considered to co-exist on the same chip, board, or cabinet. Raghunath and Ranade considered static and dynamic networks with a similar bisection width and computed several important technology-related metrics, such as (a) *module pin count*, i.e., the number of wires leaving or entering each module, and (b) *number of wire bundles*, i.e., number of wires interconnecting a pair of bundles together. From all networks considered, the complete graph has the minimum pin count, but also the maximum number of bundles for a given bisection. The next tradeoffs, involve 2- and 3-hop static networks; an $n$-hop network is a Cartesian product graph of $n$ complete graphs. The Cartesian product $G = (V, E)$ of two graphs $G_1 = (V_1, E_1)$ and $G_2 = (V_2, E_2)$ has $V = V_1 \times V_2$, and $E = \{((u_1,\ u_2), (v_1,\ v_2))\}$, where either $u_1 = v_1$ and $(u_2,\ v_2) \in E_2$, or $u_2 = v_2$ and $(u_1,\ v_1) \in E_1$.

### 1.3.5 Network Partitioning

In massively parallel computing, only a few applications are able to efficiently utilize all PEs. In order to support a diverse mix of applications with various sizes and characteristics in a dynamic fashion, an efficient PE allocation scheme is needed. Thus, most parallel systems support space sharing, with applications allocated distinct, independent subsets of PEs. An important issue in allocation is to provide a contention-free partition that maximizes PE utilization. PE fragmentation prevent idle PEs from being utilized and comes in two flavors: *internal* and *external fragmentation.* Internal fragmentation occurs when allocation assigns jobs to subsystems with certain size restrictions, thus causing any additionally allocated PEs to be wasted. External fragmentation occurs when allocation fails to assign idle PEs. This may happen due to:

- *Insufficient resources*, i.e., too few idle PEs,
- *Imperfect subsystem recognition* of the allocation method,
- *Dynamic job departures* which fragment the system into randomly scattered PE regions; although the total number of free nodes may be sufficient, there is no subsystem large enough to accommodate the incoming job.

Processor allocation has been studied especially for direct networks, such as meshes, hypercubes, and tori. Depending on the network architecture and topology, constraints may exist both on the shape, i.e., a hypercube or a mesh, and also on the size of the partition, e.g., the number of PEs must be an integer power of 2. However, most commercial systems, such as the Intel Paragon and Cray T3E, provide the capability to allocate variable size subsystems to incoming jobs.

### 1.3.6 Amdahl's Law

A consistent definition of scalability which considers the best performance possible, independent of cost vs. efficiency considerations, must take into account architectural issues, algorithmic issues, and physical constraints [510].

Parallel performance is usually related to communication patterns in applications. Amdahl's law tries to relate program speedup to the ratio of the scalar (sequential) part vs. the parallel part of the application.

Amdahl's classic paper argues that when the problem size is fixed, speedup is bounded by the reciprocal of the sequential fraction of the algorithm [26, 27]. Amdahl's argument assumes that all PEs are synchronous, and ignores cache coherence, load balancing, communication, and synchronization overheads. The derivation is as follows.

Let $s$ be the fraction of operations in a program that must be executed sequentially, and $p$ be the number of available PEs in the system. For some parallel algorithms $s$ can be evaluated, e.g., Flatt and Kennedy have measured the serial part in the LinPack parallel numerical library routines [206]. Thus, the maximum possible attainable speedup is $1/[s + (1-s)/p]$, with an upper limit of $1/s$, as $p$ increases. This places severe limits on the efficiency of large systems, for example, if $s = 0.1$ (90% parallelism) and $p = 16$ PEs, then Amdahl's Law implies a maximum speedup of 10, independent of how many PEs are used.

Gustafson et al. later argued that for many scientific problems, such as weather prediction, the data parallel part of a program is not static, but scales with the problem size, while the sequential time, e.g., arising from program loading, process creation, control parallelism (pipelining operations), and possibly I/O calls remains almost constant [245]. Thus, Gustafson et al. defined scaled speedup to account for parallel system users who actually increase the accuracy of the solution, or the problem size, to keep the elapsed time of the run more or less constant. In this case, suppose that the parallel execution time is $t_s + t_p$, where $t_s$ ($t_p$) is the time the program runs serially (respectively, in parallel). Then, since the problem takes $t_s + pt_p$ time on the 1-node system, the speedup is bounded by $1 + (p - 1)/(s + 1)$, which scales with a slope $1 - p$.

### 1.3.7 Superlinear Speedup

The term superlinear speedup is misleading, since it actually refers to superunitary efficiency (as actually suggested by Eugene Miya in the comp.parallel newsgroup). For superunitary efficiency, the total parallel work on the $N$ PEs must be less than the total work performed by the equivalent uniprocessor. Although parallel programs include overhead resulting from job scheduling, interprocessor communication and synchronization, and load imbalance, researchers have reported superunitary efficiency, and for randomized algorithms even unbounded speedup [190, 262, 526]. In the latter case, which includes parallel searching and branch-and-bound algorithms it may make sense for the sequential program to mimic the parallel version [381, 424, 468, 581].

Historically, some other arguments for superunitary efficiency are:

- The time spent servicing interrupts or switching contexts is reduced, since only one PE in the parallel system actually services interrupts [6], and

- Overhead introduced by the uniprocessor permits the $N$-node parallel system to be more than $N$ times faster; there are several examples: (a) faster remote access to cache compared to main memory access, (b) loop overhead [526], (c) extra causality relations between simulated events (total order) in sequential algorithms, and (d) inefficiencies in the sequential version caused by a sub-optimal system library (e.g., I/O), linear search of the free block list in dynamic memory allocation (malloc) [507], or inefficient uniprocessor hardware arising from improper cache alignment, too small a cache size, or too low a hit ratio [653].

### 1.3.8 Parallel Benchmarks

Benchmarks try to model a real job mix with a smaller set of representative programs. Benchmarks can be considered useful for comparing computer performance only when the benchmark problems are representative of the workload. While the arithmetic mean can be used for comparing execution times, the harmonic mean is appropriate when application performance must be expressed as flop rate in (flops/sec), e.g., in scientific computing [174].

Traditionally, supercomputer benchmarks have focused on Numerical Analysis routines. Existing benchmarks are the LinPack benchmark (linear algebra routines), the Numerical Aerodynamic Simulation (NAS) kernels, ScaLapack (message passing linear algebra routines), Splash (shared address space parallel programs), ParkBench (message passing parallel programs), TPC (transaction processing), and the Livermore Fortran kernels [702].

# Chapter 2

# Network Communications

## 2.1 Routing Strategies

With advances in VLSI technology it has become feasible to build multicomputers consisting of hundreds, or even thousands of processor nodes with local memory, which communicate with each other over a fixed interconnection network. For many parallel applications, the interconnection network determines overall performance. Thus, an essential condition for the efficient use of such machines are routines for exchanging data between the processors. In view of the many network topologies and the multitude of communication patterns, it is not surprising that a rich body of theoretical and practical studies has been developed around the theme of communication.

In this chapter we probe into several of the topics that are covered in more detail in later sections. We define general multicomputer and multiprocessor communications, for which we will provide known results and open problems later. We concentrate on *permutation routing* and related problems of *sorting, partial permutation routing, dynamic routing, routing to random destinations, $k$-relations, isotonic routing, and fault tolerant routing.* We discuss the importance of communication patterns in parallel applications, such as computing multidimensional Fast Fourier Transform (FFT), finite elements, matrix problems, and divide and conquer strategies.

### 2.1.1 What is a Packet?

A packet is the smallest part of a message which can be independently routed. Messages could be variable in length, consisting of several packets. Packets consist of various fields.

- The *header field* provides destination address(es), and sometimes includes source address. For variable size packets, it is convenient to represent the data length field first in the Header field. Path-select, or packet priority information may also be included.

- The *data field* provides an operation code which distinguishes: (a) request from reply packets, (b) read, write, or synchronization instructions, (c) blocking, or nonblocking instructions, and (d) parallel program instructions, from switch setup or system test instructions. It also provides a virtual memory address, and any actual data. Sometimes performance related information may also be included, such as a transaction identity/type, and epoch counters. Special flags are also needed for synchronizing accesses to local communication buffers (which may wait for network data), and for distinguishing sets of buffers, e.g., with sequences of non-blocking operations. In addition, if packets do not reach their destinations in their original order, a sequence number may be provided for appropriate packet reordering.

- The *checksum* (CRC) decodes header information (and sometimes data) for error detection, or correction.

- A *packet termination flag* may be used as an alternative to a packet length subfield for variable size packets.

Examples of communication instruction primitives, with their corresponding fields, are as follows.

- *Remote read or DMA*, with Header, memory address, and CRC.

- *Reply to remote read or DMA*, with Header, data, and CRC.

- *Remote write*, with Header, memory address, data, and CRC.

- *Reply from remote write*, with Header and CRC.

- *Synchronization*, with Header, address, data, and CRC. Synchronization includes Fetch&Add, Compare&Swap, and other read-modify-write operations.

- *Reply from synchronization*, with Header, data, and CRC.

- *Performance-related.*, e.g., remote enqueues.

Higher-level parallel software provides complete communication interfaces, as shown below.

- *I/O drivers.*
- *Virtual shared memory libraries.*
- *Message passing libraries.*
- *Data parallel languages:* e.g., HPF, Fortran 90, CMF, CMC*, and CRAFT.
- *Clustering services and operating systems*, e.g., DEC OSF/1.

## 2.1.2 Virtual Circuits - Flow Control

Virtual circuits (VC) provide dynamic sharing of physical wires. Thus, network bandwidth can be allocated in a fair manner between parallel processes. Another advantage of VCs is that they help in avoiding head of the line blocking effects. Since each VC is associated to a buffer, an appropriate flow control reservation protocol can guarantee that packets moving to a congested network region will not interfere with other network packets, even if they share a communication link.

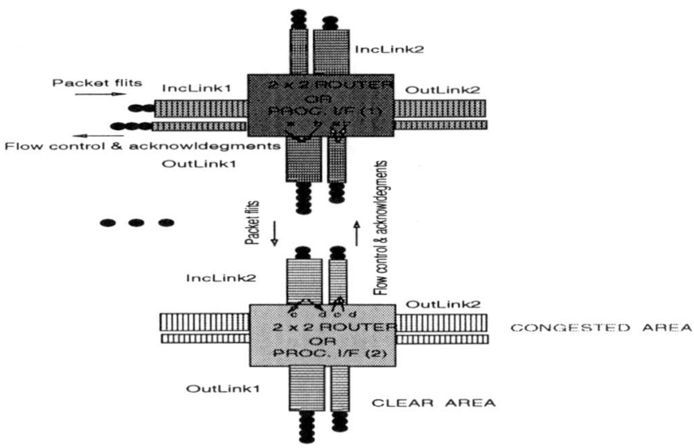


**FIGURE 2.1**
**Virtual channels for avoiding head of line blocking**

For example, in Figure 2.1, packets from incoming Link 1 (incLink1) headed to a congested area, will be stopped at the first router (or PE interface I/F (1)), while other packets headed to a clear area are able to move further.

While in virtual cut-through and store-forward model, flow control is per packet, with wormhole routing a packet is further subdivided into flits (flow control units), with each flit having a unique flow control. Each flit may also consist of several smaller transmission units, called physical units, or phits. A phit is usually a few bytes, while a flit can vary considerably, e.g., for the TMC CM-5, a packet consisted of 4 to 72 phits, while for the Cray T3E, a packet consists of 2 or 10 phits depending on the instruction type (single operand, or array) and the request/reply nature of the packet.

Two methods are common for flow control on *synchronous communication systems.* In Ding and Bhuyan's model, a packet may move only if the next buffer is nonfull, while in Jenq's cut-through model, a packet may move if the next buffer is full and a packet is simultaneously transmitted out of this buffer. As shown in Figure 2.2, with each flit consisting of $n \geq 1$ phits (physical units), Jenq's model requires control information to travel $n+1$ phits back. In Ding and Bhuyan's model, the same information travels only 2 phits back. Therefore, for long packets and network paths, flow control propagation time becomes significant, making Jenq's theoretically better model inefficient.

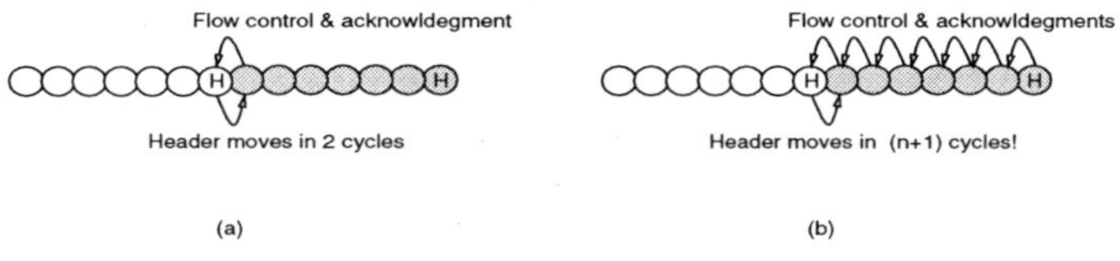


**FIGURE 2.2**
**Flow control: (a) Ding and Bhuyan's, (b) Jenq's cut-through**

Furthermore, practically in order to satisfy the cut-through condition, where consecutively transmitted packets continue to move without gaps between them, the packet size in flits should at least equal the sum of the (next) switch startup time plus the time in cycles ($t_{rt}$) between a packet departure from one switch and a ticket arrival to this switch triggered by the same packet; the time $t_{rt}$ is commonly referred to as packet-ticket round trip delay.

With *asynchronous transmissions*, enough buffering must be provided at each link to avoid bandwidth loss. Assuming unit delay, and taking into account the speed and length of transmission links, the buffer size per VC and per node should at least equal the packet size plus information floating in the network. This information is $2LB$, where $L$ represents the (uniform) signal propagation delay over the link (in sec), and $B$ the link speed (in Mb/sec).

### 2.1.3 Routing Strategies

In oblivious routing, packets are sent along a more or less cleverly chosen path. Furthermore, the routing path of any packet does not depend on the route of any other packet, i.e., packet routing decisions are carried out using only local information available at each network node. Regular topologies offer the advantage that all nodes have a global knowledge of the network, allowing for simple routing and scheduling decisions. Thus, in most parallel architectures, the router is simple enough to implement in hardware, and thus routing decisions are fast.

A very common oblivious deterministic routing strategy for many symmetric topologies, such as grids of various dimensions and hypercubes, is the so-called *dimension-order routing.* In this method, a packet is routed along dimension 0 until it has reached its correct position (along this dimension), then along dimension 1, and so on. Notice that the communication path is completely determined (*a priori*) by the source and destination addresses. Such nonadaptive, deterministic routing strategies are insensitive to network conditions, and thus extremely vulnerable to faults and hot spots in the network. Therefore, several adaptive strategies have been proposed.

In *hot-potato routing* (also called deflection routing) there are no buffers at intermediate nodes [44]. Packets continue to move, possibly in a wrong direction, until they ultimately reach their destinations. Deflection routing has been proposed for optical networks, since optical delay lines are expensive, and opto-electronic conversion for electronic buffering is slow [489].

With the *mad-postman* strategy, a packet is sent to the next node in the virtual network, even before compiling its address information [304, 305]. Packets moving in virtual networks are usually assigned additional buffers to ensure deadlock-free communication. However, extra address flits seem to cause extra traffic overhead with high load situations.

In the design of efficient routing algorithms, randomization plays a key role. *Probabilistic routing* uses a sequence of random bits for making routing decisions. The routing algorithm utilizes randomization to convert the input distribution, in which the deterministic algorithm suffers from unexpected worst-case instances, into a manageable and predictable input distribution, usually the uniform distribution. Thus, the randomization technique is useful when a deterministic algorithm runs on average much faster than in the worst case. This situation is common for instances of highly structured communication and computation problems, for which performance of deterministic techniques may suffer. Recently, Brassard and Bratley defined this type of randomization as *Sherwood* type [87]. Ben-David et al. [52] conjectured that every randomized algorithm has the same performance as an appropriately designed online adaptive algorithm.

### 2.1.4 Routing Functions

When a packet arrives at an intermediate network node, its header provides information on the path it must follow. In most cases, the packet header contains the destination address, which is generated at the sending node. For a network of $N$ nodes, this address is at most $\log_2 N$ bits. The link along which a packet is to be sent at an intermediate node is usually a function of this header information.

In asymmetric networks, a *routing table* with $N$ entries (equal to the number of possible headers) is allocated at each network node. Each table entry provides the output link that each packet (based on its header) should follow. For adaptive routing, a selection of outgoing links is given. Although this method is general, the number of entries of each lookup table ($\mathcal{O}(N)$) is a severe limitation.

Since classical routing tables are infeasible for large networks, *compact routing* techniques are of interest. In these schemes, the amount of routing information per network node is kept minimal ($\mathcal{O}(\log_2 N)$), by exploiting topological information. Using appropriate PE numbering, we incorporate topological information in the packet header, essentially grouping packets which must move along similar outgoing links into the same entity. For instance, in highly regular networks, such as hypercubes and meshes, it is possible to label PEs so that little information needs to be stored per network node to allow for shortest path routing.

Compact routing offers simplicity, efficiency, and scalability for the particular topology. Furthermore, by providing additional link/node status information a certain level of adaptivity is possible, e.g., fault-tolerant routing support. Most near-optimal algorithms consider regular structures, such as hypercubes or meshes, and few faults [102, 103, 104]. Examples of compact routing are mono-interval and multiple label interval routing, dimensional routing, prefix routing, and hierarchical routing.

*Interval routing* groups together packet headers, for all packets which use the same outgoing link. Packets are output to the link specified by an interval table, according to the interval their header belongs to. The intervals used in this routing scheme can be either circular [598, 686], or linear. A circular interval $(a, b)$, where $a, b \in [0, N-1]$ is defined as either $(a, b)$ if $a < b$, or $(a, N-1] + [0, b)$ if $a > b$. For *circular intervals* the packets with headers 0 and $N-1$ may be combined in a single interval, while for linear intervals this is impossible; notice that if one link is labeled 0 on all network nodes, the circular and linear schemes are identical.

Obviously, any network can support interval routing if an appropriate numbering of the nodes is applied. Multiple label interval routing uses more than one interval per outgoing link. Mono-interval routing is only optimal for a limited number of topologies [592]. Any network can be labeled in an optimal way, if $N-1$ mono-interval labels are provided [686]; however, the routing table is no longer compact. Proper numbering of PEs may produce the shortest path routing with a small number of intervals. Optimal and near-optimal labeling schemes in various graphs have been examined [81, 211]. New intervals can be introduced in the proximity of a faulty component to accomplish packet derouting.

*Multiple label interval routing* provides routing tables with a set of possible outgoing links. This structure is extremely general. The routing information per node is $\tau d \log_2 N$, where $\tau$ is the number of intervals. The $\log_2 N$ factor corresponds to coding the intervals for all source-destination pairs, while the $d$ factor represents the overhead of including a set of links, i.e., $d$ disappears from the formula for unique path interval routing.

The C104 switch is a low-latency $32 \times 32$ packet routing chip, which uses wormhole routing and supports linear multiple label interval routing [666]. The C104 is a building block for configuring parallel transputer arrays by connecting together T9000 transputers, in point-to-point or multistage topologies (e.g., Parsytec's SC320 nonblocking Clos network). This switch can also be used for internetworking to ATM networks, or FDDI rings. An example of mono-interval routing with 2 C104s and 6 T900s is shown in Figure 2.3 [162]. T9000.2 sends a packet to T9000.3, with the header equal to 3. When the packet arrives in the first C104 switch, the header is compared with all intervals. Since it is in the range

[3, 6) the packet is forwarded to the second C104 switch. At this switch since the packet header is in the range [3, 4) the packet is forwarded to its T9000.3 destination.

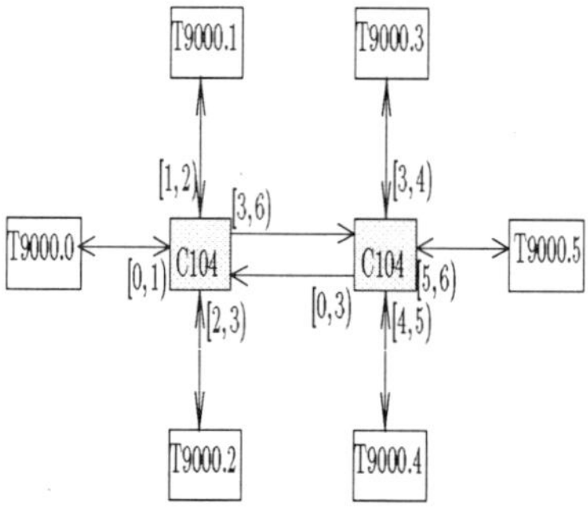


**FIGURE 2.3**
**Interval routing in networks**

In the following routing techniques, nodes are labeled so that packets move through network nodes with labels that are more and more similar to the packet header.

*Dimensional routing* is a special case of topology dependent routing. In dimensional routing, each packet corrects one dimension at a time. Since this imposes a strict order of dimension traversals, routing is also deadlock-free. Many massively parallel systems more or less follow this scheme. Examples include the CM-200 (hypercube), the Intel Paragon (2D mesh), and the Cray T3E (3D torus). The latest Cray T3E models use direction order routing which allows for routing adaptivity in case of hot spots or device faults [609].

*Destination tag routing* does not use interval tables. Instead, the complete path is included in each packet's header, usually as a sequence of output links for each intermediate node. If the network diameter is $D$ and the degree of each node $d$, the header requires $D \ \log_2 d$ bits. At each intermediate node, the first $\log_2 d$ bits are examined to determine the outgoing link. This definition directly relates to self-routing, since the routing problem is solved by the network itself, with requests computing their own trails on the fly (without any precomputation) as they move through the network. In certain cases, a function may first be applied to the header in order to compute the control tag, e.g., see ADM network routing in Section 3.4.3. Networks which are routed using such schemes are called FD-controllable networks [718]. The appropriate function is usually computed on the fly using table lookup.

In *prefix routing* as in destination tag routing, headers contain information about the path the packet has to follow based on the outgoing links. Prefix routing assigns fixed labels to all network nodes and links [39]. The routing table at node $u$ contains labels $l_1, l_2, \ldots l_d$, where $d$ is the degree of node $u$. A packet arriving at node $u$ follows the link whose label forms the maximum common prefix with the packet destination $v$. After the routing phase, the prefix which is no longer necessary is dropped from the packet header. Consider an example where nodes and labels are named over the string set $\{0, 1, ..., 9\}^*$. Suppose $v = 54321$, $l_1 = 5$, $l_2 = 54$, and $l_3 = 1$. The message will be routed along $l_2$. Dynamically changing networks can be easily accommodated into this scheme, with only constant adaptation cost. However, not all networks admit an optimum prefix routing scheme. Labelling of PEs and links can usually be computed using spanning trees. Labels for each outgoing link along a packet's path are concatenated together. Thus, the header is $\mathcal{O}(D\ d)$ bits long, which for many networks is $\mathcal{O}(\log N)$.

*Hierarchical routing* divides the network into smaller subnetworks and uses a different routing function on each subnetwork [154, 216]. Members of a group communicate with each other without crossing into other subgroups. This approach is quite flexible and can be implemented with minimal routing information, since each node needs only information about the subnetwork it belongs to and the supernetwork structure. With properly divided hierarchical networks, coding group information into the header requires only $\log_2 \log_2 N$ extra bits. If leading bits are dropped from the header, detection of arrival to a supernetwork is even simpler; one needs only examine the current size of the header.

Distributed algorithms are needed to label and set up a network in the most "optimal" way, given certain network node and routing table limitations. Hence, a lot of research effort focuses on uncovering the relation between structural network properties and routing methods, especially in the context of distributed systems [207, 350].

### 2.1.5 Deadlock, Livelock, and Starvation

To ensure correct functionality the parallel system communication facility must be always free of *deadlock*, *livelock*, and *starvation*.

A deadlock may be defined as a cyclic dependency of ungranted packet requests for buffer or channel resources [635]. Deadlock is detected as a cycle in the resource dependency graph; this graph shows resources (buffers or channels) as node components, and connects component $a$ to component $b$, if a packet may move from resource $a$ to resource $b$ (provided that resource $b$ is available). Deadlocks may occur when a common buffer pool is shared by incoming and outgoing traffic. A livelock refers to packets either indefinitely waiting at a network buffer (due to an unfair queuing policy), or circulating the network without making any progress towards their destination. Livelock usually does not occur with shortest path or randomized routing; it may be avoided with adaptive (usually nonshortest path) routing strategies by implementing age-based priority schemes. Both livelock and starvation, which refers to packets arriving at their destination but never having been consumed by the processor, reflect problems of fairness in network routing or processor scheduling policies.

Notice that not only routing should not introduce deadlock hazards, but also it is the parallel programmer's responsibility to avoid deadlocks caused by not properly posting receives for corresponding blocking send operations. Examples of deadlock (case 1) and starvation (case 2) are shown in Figure 2.4. In case 1, packets are routed in a clockwise manner with source-destination pairs (1,3), (3,1), (2,4), and (4,2); a central buffer is assumed. In case 2, packet 7 never reaches node 5 because other packets, e.g., packet 6, is always given priority; this phenomenon obviously causes program serialization.

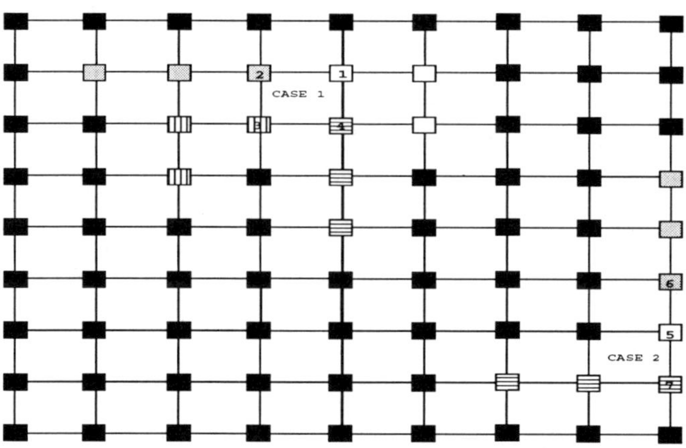


**FIGURE 2.4**
**Various routing models and examples**

Special care has been taken to address network deadlock. Separate buffers are necessary for type request and reply packets. Assuming only one type of packets, while for nonadaptive store-forward and virtual cut-through routing deadlocks can be avoided by using the concept of structured transmission buffer pools and restricting route assignment to avoid cycles [184, 243], special care must be taken for wormhole routing [184].

In the simplest case, since store-forward and virtual cut-through routing use flow control per packet, a structured pool may start packets at buffer $B_0$, and always move them from buffer $B_i$ to buffer $B_{i+1}$ possibly in a different switch. Deadlock may not arise for all communication patterns with cycles in the resource dependency graph, providing that a structured pool is a general deadlock avoidance technique for all communication patterns. *Reassembly deadlock* also arises for store-forward and virtual cut-through routing; when messages are split into packets, the routing technique may *not* guarantee the order of packet arrival. Then, assembly buffers may hold packets while other packets are received. The problem can be prevented by providing more buffers, or virtual channels.

The usual way to provide deadlock-free routing for wormhole routing, is to define a partial order on the buffer classes which result in cycle-free (virtual) channel dependency graphs [153]. Although this works fine for deterministic routing, in case of adaptive routing it is still possible that a cycle exists but the packets still do not deadlock. A theory has been developed for designing maximally adaptive, deadlock-free (but not necessarily livelock-free) routing algorithms in networks with network nodes implementing either central buffering, or input-output buffering [183]. Minimizing the number of buffers needed, while providing for maximal adaptivity, has been considered for $d$-dimensional tori [142].

One technique uses *initially adaptive*, followed by *deterministic dimension-order routing* after a certain number of dimension reversals [152]. Another possibility is to divide packets into different *virtual networks* depending on their destination. Virtual networks can be implemented in two ways: (a) by assigning separate physical channels at each node, one for each virtual network, and (b) by splitting physical channels into virtual channels, providing a separate buffer queue at each node for each virtual channel, and transmitting packets using time division multiplexing [151].

The virtual networks technique for a two-dimensional mesh is illustrated in Figure 2.5. The four cycle-free virtual networks $(A, B, C, D)$ shown can be combined in certain ways, e.g., either B and C, or A and

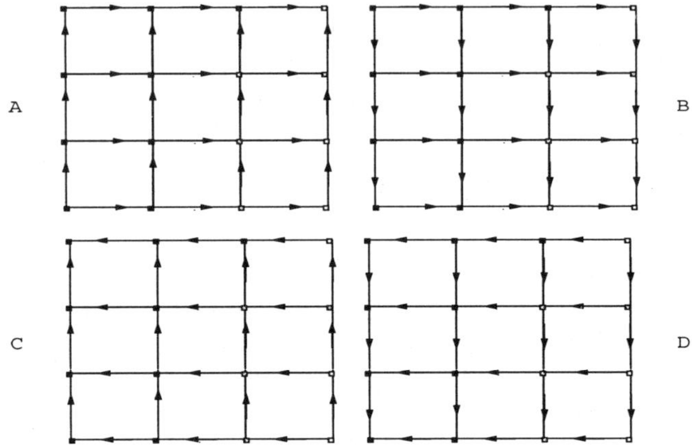


**FIGURE 2.5**
**Virtual networks for deadlock-free routing**

D, without introducing any cycles. Thus, only two buffers per node are needed for deadlock-free routing; the virtual network chosen for routing the packet (which allows for a certain level of adaptivity) is selected based on the source and destination of each packet. The technique can easily be extended to the $d$-dimensional mesh using $2d$ buffers per node. Furthermore, for $d$-dimensional tori, minimal, shortest path routing may be used within the appropriate mesh virtual network.

For wormhole routing several other techniques achieve deadlock-free routing by breaking up cycles in the channel dependency graph. The same techniques can be applied to multicasting (one to many communication). We partition the physical network into deadlock-free virtual subnetworks, and we multicast into each subnetwork using minimum spanning trees, as shown in Figure 2.6 for a two-dimensional mesh, the multicast process uses (X-Y) routing and is deadlock-free [432].

More efficient techniques allow for more routing adaptivity; there exists a necessary and sufficient condition for deadlock-free wormhole routing on any network [181]. The condition has been extended to wormhole multicasting [182]. For deadlock-free multicasting on meshes using virtual networks see [432, 433].

An interesting deadlock-free wormhole routing technique, called the *turn model*, which does not use virtual networks and allows for maximal adaptivity for a given network, is based on limiting certain packet turns

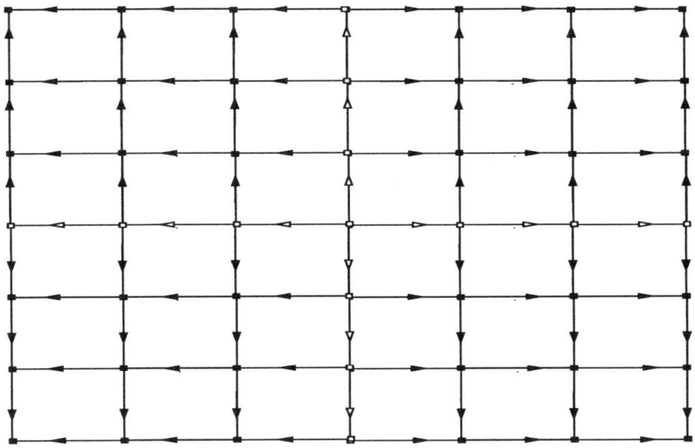

**FIGURE 2.6**
**Virtual networks for deadlock-free multicast**

which might cause cycles. An example of this technique is the west first, north last, and negative first routing technique for meshes, tori, and other sparse networks [224]; west first routes packets towards west first (using shortest paths) before moving them towards any other possible direction.

More recent techniques for achieving deadlock-free routing consider the sparsity of deadlocks in asynchronous networks. Thus, instead of avoiding deadlocks, precautions are taken to efficiently recover from them. In the *compressionless approach*, any established virtual channels which have not reached their destination within a given time delay are "torn off" [339]; this method is implemented using special test packets which follow flow control paths. A new and better approach called *Disha*[1] provides alternative paths on demand to deliver deadlocked packets [29, 545]. A practical example is the high-speed Myrinet network: if a packet is blocked for a certain period of time, the packet is dropped [80].

[1]Disha means "direction" in Hindi.

### 2.1.6 Fault Tolerance, Reliability, and Diagnosis Models

Several recent near-catastrophes show the immense risk in existing technological ventures:

- *Three-Mile Island:* Nuclear near-accident in U.S.
- *SDI:* Electronic false alarm (nuclear attack).
- *Union Carbide:* Gas accident at Bhopal, India.
- *Challenger shuttle explosion.*
- *Chernobyl nuclear accident.*
- *Exxon oil spill.*
- *Estonia ship disaster.*
- *Ariane-5 rocket crash.*
- *Mars Polar Lander.*

In the recently documented Ariane-5 rocket crash and several other cases, computer malfunction was identified as the main problem. Multicomputer systems are more susceptible to failure than conventional uniprocessor machines; transient, intermittent, and permanent hardware errors may occur at any time. Though the probability of failure of a single chip decreases due to better technologies, the statistical chance that one element in large parallel/distributed systems breaks down is not at all negligible. As the size of parallel systems increases, the probability of component failures (processing node, switches, or links) during the long computing times of parallel applications increases significantly.

As an example, consider memory access errors in a parallel system. Currently a DRAM memory chip has an ~1000-*fit* failure rate, under normal radiation levels (1 fit corresponds to 1 failure per billion hours). In a parallel machine with ~250 processors of 128 MByte local storage, there are some 100,000 such chips, resulting in 2-3 errors per day on the average! In more complicated systems, errors occur especially in corner situations, which arise from a mixture of cases. These errors are hard to predict using random or application-specific test vectors; thus, new test generation techniques have been developed [275].

For studying system fault tolerance and reliability, mathematical models based on graph theory and combinatorics have been proposed.

In the *p-faulty* model, component failures occur independently with a fixed probability ($p$) and may correspond to node failures, link failures, or both. These dynamic faults may correspond to a channel being unavailable, or to a full buffer.

In the *worst case* model, a set $F$ of faulty components (links and nodes) is chosen by an adversary. Such static faults could arise during fabrication runs. Notice that when a processing (or switch) node fails, all its adjacent links also fail. A fault-tolerant communication algorithm is called optimal if it finds a minimal feasible path for every packet, whenever such a path exists. A path is *feasible* if it contains no faulty nodes. A path is a *minimal* feasible path if it is the shortest feasible path. For the worst case, the *survival graph* $R$ has the same number of nodes as the network graph $G$, with any two nonfaulty nodes joined by an edge if there are no faults on the route between them. The diameter of this graph is a measure of routing overhead incurred by the faults.

Both hardware and software (algorithmic) fault tolerance approaches have been considered. With hardware fault tolerance extensive replication and adaptivity are used. Several techniques used for enhancing hardware fault tolerance are listed below:

- *Replicate ASICs for self-checking,*
- *Provide checksum methods (CRC, parity checkers) for end-to-end or switch-to-switch error detection,*
- *Provide address tables for source address validation,*
- *Use transaction acknowledgments and time-out counters for deadlock, livelock, and starvation detection,*
- *Employ link keep-alive protocol checkers for fault isolation,*
- *Use regroup algorithms for split-brain problems,*
- *Use epoch counters in packets to single out old packets from network, and*
- *Use built-in hardware fault injection for hardware and software testing.*

Mainly three algorithmic approaches to fault tolerance are used: the *dynamic*, the *static*, and the *information dispersal* approaches. All techniques try to maximize the probability of algorithm termination under faults.

The dynamic approach is based on *fault tolerant routing*. We are interested in routing a packet from its source to its destination, if there is a live path between these two nodes, i.e., the nodes belong to the same connected component. Hence, an interesting solution consists of routing in a fault tolerant fashion directly on the faulty network. We assume that the fault pattern remains fixed for the duration of routing. It is much harder to model algorithmic fault tolerance in the presence of transmission errors, or temporarily down links.

The static solution consists of reconfiguring the faulty network by embedding a fault-free network [257, 258, 317, 472, 659]. After the reconfiguration, classical communication algorithms can be used. A general (but idealized) semi-automated process for dealing with dynamic faults can be structured around reconfiguration. The process consists of four major steps [695]:

- *Error detection.* Two techniques are mainly used: (a) *structural detection* is based on test code directed towards detection of memory and communication network faults, e.g., parity, CRC, or self-tests, and (b) *behavior-based detection* compares test patterns collected at compile/assembly time with the exhibited behavior at run time.

- *Fault diagnosis and isolation.* Upon detection, the fault is analyzed and localized. Network devices and system routing tables are updated and recovery procedures are initiated.

- *Automatic reconfiguration.* Routing tables are adapted to detour permanently failed components. Fault tolerant routing is employed. Faulty processors are possibly replaced by spares. The application is repartitioned to a working set of processors, and program and data are downloaded on the newly defined fault-free partition within the injured network.

- *Consistent checkpointing during normal execution and rollback.* We avoid restart of application by doing backward recovery. This is implemented through rollback to fault-free, user-inserted checkpoints.

In general, we say that a system is *one step t-fault diagnosable*, if given the fault bound $t$ ($t > 0$) all faulty PEs can be correctly identified after a testing phase. Similarly, a system is called *pessimistically one step t-fault diagnosable*, if given the fault bound $t$ (($t > 0$), $t$ or fewer PEs that include all faulty PEs present and possibly some PEs of unknown status can be identified for replacement after a testing phase. Although pessimistic strategies involve a wasteful replacement of some PEs, the degree of diagnosability is generally higher.

Consider a graph $G$ with $N$ nodes, and node-connectivity $k(G)$. $k(G)$ represents the smallest number of nodes whose removal disconnects $G$; this is equivalent to having at least $k(G)$ disjoint paths between any pair of nodes in $G$. Graph $G$ is precisely one-step $t$-fault diagnosable if $N \geq 2t+1$, and $k(G) \geq t$ [246].

Fault diagnosis in general multicomputer graphs can be achieved in $\mathcal{O}(N^2)$ time with randomized algorithms, and $\mathcal{O}(N^{2.5})$ with deterministic algorithms [148, 149]. For a discussion of fault diagnosis algorithms, see [362, 519]. The design of easily diagnosable systems to within a set of at most $s$ faulty and fault-free units is discussed in [558, 636]. Precise diagnosis systems are discussed in [246].

### 2.1.6.1 Information Dispersal Algorithm (IDA)

In this section, efficient information dispersal and applications to fault tolerant routing for parallel computers are discussed. We pay attention to file transfer in a distributed network of computers along paths with possibly unreliable links.

Using error correction, one can assume error-free transmission, but there might be a probability of message loss (even if particular network links along paths fail with small probabilities). In case of transfer failures, one approach may be to transfer files along a chosen path, and retransmit along another path in case of confirmed failure. A more efficient way (time-wise) may be to simultaneously transmit several copies of the same file to the same destination along several paths. This approach increases network load accordingly.

The Information Dispersal Algorithm (IDA) [563] represents a reliable dispersing of file information into $n$ locations and subsequent file reconstruction from any $m$ pieces. This approach is based on error-correcting codes, as the message contained in a packet can be reconstructed at its destination from a small number of copies. Assume the file $F = b_1, \ldots, b_N$ is a string of length $N$. Then, each of $m$ pieces will be of size $N/m$, and the total number of characters is $nN/m$. If $n/m \sim 1$ the dispersal is space efficient. Moreover, both dispersal and reconstruction are computationally efficient as well.

Assume that we want to disperse the file $F$ so that $\leq k$ pieces will be lost during travel along paths in a communications network. Choose any upper bound $B$ (in $\mathbf{Z}$) for all $b_i$'s, and a prime number $p > B$. It is not necessary to choose the smallest prime number with that property (for example, if $b_i$'s are 8-bit bytes, $p = 257$ is a good choice). Now, for all $i$, we have $b_i \in \mathbf{Z}_p$ (a finite field). Actually, any finite field is applicable for the dispersion, reconstruction, and fault tolerance algorithms described in detail in the next paragraphs.

For 8-bit bytes, one can use the field $GF(2^8)$ with 256 elements. In order to define the corresponding arithmetics, one can use any degree-8 irreducible polynomial with coefficients in $\mathbf{Z}_2$.

In order to disperse $F$ using the IDA, we choose $m \in \mathbf{Z}$ such that $n = m + k$ and $n/m \leq i + \varepsilon$, for some $\varepsilon > 0$. Note that all arithmetic is done mod $p$. For each $i = 1, \ldots, n$, choose a vector $\mathbf{a}_i = (a_{i1}, \ldots, a_{im})$ in $\mathbf{Z}_p^m$, such that every (or randomly chosen) $m$ vectors in set $\{\mathbf{a}_1, \ldots, \mathbf{a}_n\}$ are (with high probability, respectively) linearly independent. We partition now the file $F$ into subsequences $F = (b_1, \ldots, b_m), (b_{m+1}, \ldots, b_{2m}), \ldots = S_1, S_2, \ldots$ of length $m$. Now, $F$ is dispersed into $n$ pieces (vectors) $\mathbf{F}_i = (c_{i1}, \ldots, c_{iN/m})$, where $c_{ik}$ is obtained using the following inner product:

$$c_{ik} = \mathbf{a}_i \cdot S_k = a_{i1} b_{(k-1)m+1} + \cdots + a_{im} b_{km} \,. \tag{2.1}$$

Obviously, the length of each $\mathbf{F}_i$ is $N/m$.

In order to reconstruct $F$ out of, say $\mathbf{F}_1, \ldots, \mathbf{F}_m$, we use the $m \times m$ matrix $\mathbf{A}$ whose $i^{th}$ row is vector $\mathbf{a}_i$, $i = 1, \ldots, n$. It follows that

$$\mathbf{A} \cdot \begin{bmatrix} b_1 \\ \vdots \\ b_m \end{bmatrix} = \begin{bmatrix} c_{11} \\ \vdots \\ c_{m1} \end{bmatrix},$$

and

$$\begin{bmatrix} b_1 \\ \vdots \\ b_m \end{bmatrix} = \mathbf{A}^{-1} \cdot \begin{bmatrix} c_{11} \\ \vdots \\ c_{m1} \end{bmatrix}.$$

If the $i^{th}$ row of $\mathbf{A}^{-1}$ is $(\alpha_{i1}, \ldots, \alpha_{im})$, then $b_j$ can be reconstructed using:

$$b_j = \sum_{j=1}^{m} \alpha_{ij} c_{jk} \,, \tag{2.2}$$

where $1 \leq k \leq N/m$, $1 \leq j \leq N$, $i = j \bmod m$, and $k = \lceil j/m \rceil$. Obviously, one needs to compute $\mathbf{A}$ only once, and use Eq. 2.2 for the reconstruction of the file $F$. It is known that size $m$ matrix inversion requires computational time of $\mathcal{O}(m^3)$, but $\mathbf{a}_i$'s can be chosen [563] so that this is decreased to $\mathcal{O}(m^2)$. However, for large files where $m^2 \leq N$, it takes much more time to calculate $b_j$'s (using Eq. 2.2), than to calculate $\mathbf{A}^{-1}$. Notice also that file dispersion using Eq. 2.1, and file reconstruction using Eq. 2.2 involve inner products only which are suitable for parallel computer communications.

To apply the IDA to fault-tolerant routing in parallel architectures, denote by $PC_n$ an $N$-node parallel computer with $N = 2^n$. Each node has a processor and a local memory module, and it can be visualized as an element $x$ of the Boolean $n$-cube $C_n = \{0,1\}^n$. In [682], a two-phase algorithm was proposed for routing between the memory modules. It was shown that, with (high) probability (equal to $1 - N^{-k}$), packets can be routed in time $c \log_2 N$, with $\leq d \log_2 N$ waiting packets in queues at individual nodes ($k$ is a constant which depends on $c$ and $d$).

In order to apply the dispersal of information to $PC_n$ one must have sufficiently large packets $Px$ at each node $x$. We want to route $Px$ from $x$ to $\pi(x)$ when buffers are of fixed size, and in a fault-tolerant fashion (even if many network links fail). We partition a packet $Px$ from the node $x \in C_n$ into $n$ pieces $Px_1, \ldots, Px_n$, so that every $m = \lfloor 5n/6 \rfloor$ pieces are sufficient for the reconstruction of the entire $Px$. For each $i$, we route $Px_i$ independently. We also assign a (vector) *ticket* $\mathbf{Tx_i} = (t_1, \ldots, t_{2(n+1)})$ to each packet piece $Px_i$, where each $t_i$ is an integer, $1 \leq t_i \leq n$.

For $x \in C_n$ and $1 \leq i \leq n$, denote by $x^{\langle i \rangle}$ the node $y \in C_n$ with all (binary) coordinates equal to those of $x$, except the $i^{th}$ one, i.e.:

$$x^{\langle i \rangle} = \begin{cases} x_j \ , j \neq i \\ \overline{x}_j \ , j = i \end{cases}$$

In the routing pattern, all pieces $Px_i$ (of packet $Px$) are routed simultaneously to their destinations $\pi(x)$. During this, each cube node $y$ contains a number of pieces (together with the corresponding tickets) $(P^{(1)}, \mathbf{T}^{(1)}), (P^{(2)}, \mathbf{T}^{(2)}), \ldots$, while time $t$ runs from 1 to $2(n+1)$. For each $t$, the processor at node $y$ scans the $t$-th coordinates $\mathbf{T}_t^{(1)}, \mathbf{T}_t^{(2)}, \ldots$ of $\mathbf{T}^{(\cdot)}$, and for every $j$, sends all the packets $P^{(r)}$ with $\mathbf{T}_t^{(r)} = j$ to the node $y^{\langle j \rangle}$ for $1 \leq j \leq n$ [658]. In case $\mathbf{T}_t^{(r)} = 0$, $P^{(r)}$ does not leave the node $y \in C_n$ (at time $t$).

*IDA Routing Algorithm* routes all packets $Px$ from $x \in C_n$ to $\pi(x)$.

cobegin for $x \in C_n$
Partition $Px$ into $n = \log_2 N$ pieces $Px_1, \ldots, Px_n$;
Choose randomly $n$ pairwise different nodes $R_1(x), \ldots,$
$\ldots, R_n(x)$;
Choose pairwise vertex disjoint paths (of length $\leq n+1$)
$P_1(x), \ldots, P_n(x)$ from $x$ to $R_1(x), \ldots, R_n(x)$;
Choose vertex disjoint paths (of length $\leq n+1$) $Q_i(x)$
from $R_i(x)$ to $\pi(x)$;
To each $Px_i$, attach a ticket $\mathbf{Tx_i}$ for routing along $P_i(x)$
followed by $Q_i(x)$;
Use previously described way of sending all pairs ($Px_i$,
$\mathbf{Tx_i}$) to $\pi(x)$.
coend

In this construction, it is assumed that the size of every packet is $|Px| = L$. Therefore, $|Px_i| = L/m$, $m = \lfloor 5n/6 \rfloor$, and $\sum_i |Px_i| = 1.2 \cdot L$. We also assume that each node's buffer $BFy$ can hold up to six packets $Px$.

***THEOREM 2.1***

*For any given permutation $\pi$, under the IDA routing, all packets will reach their destination with probability $\geq 1 - (1/N^4)$ [563].*

**PROOF** For a fixed $t$ and $y \in C_n$, denote by $X(y, x, t)$ the random variable describing the number of $Px_i$'s arriving at node $y$ at time $t$. Assuming that we "pad" tickets $\mathbf{Tx_i}$ which are shorter than paths $P_i(x)$ (or $Q_i(x)$) by zeroes, all the pieces $Px_i$ must travel simultaneously along their respective ways, either along $P_i(x)$ or $Q_i(x)$. Since paths $P_{(\cdot)}(x)$ and $Q_{(\cdot)}(x)$ are vertex disjoint (except possibly at the origin $x$), the random variable $X$ is binary valued (for each $x$, zero or one $Px_i$'s arrive at $y$ at time $t$). Denote by $p(y, x, t) = \Pr[X(y, x, t) = 1]$. At time $t$, the buffer $BFy$ will overflow when:

$$\sum_{x \in C_n} X(y, x, t) \geq 5n \; . \tag{2.3}$$

Without loss of generality, one can assume that no $Px_i$'s are lost in buffer overflows. It is also easy to prove the statistical independence of $X(y, x, t)$ for fixed $t$ and $y \in C_n$, and for arbitrary $x \in C_n$. From the symmetry of the IDA algorithm (for arbitrary $t$ and $y$) and since at the beginning time $t = 1$ we have $n$ pieces at each node of $C_n$, one has:

$$\sum_{x \in C_n} p(y, x, t) = n \; . \tag{2.4}$$

It is known [564] that for $N$ independent Bernoulli trials $X_1, \ldots, X_N$ with:

$$E\left[\sum_i X_i\right] = n \; , \tag{2.5}$$

and a given $\delta > 0$, one has:

$$\Pr\left[\sum_i X_i \geq (1+\delta)n\right] \leq \left[\frac{e^{\delta}}{(1+\delta)^{1+\delta}}\right]^n \; . \tag{2.6}$$

Since the random variable $X(y, x, t)$ satisfies the above theorem (Eq. 2.4 yields Eq. 2.5, and $\delta = 4$), we conclude that $(e^4/5^5)^n \sim 2^{-5n} \cdot 1.8^{-n}$ is the upper bound on the probability of having buffer $BFy$ overflow (at time $t$) during the IDA routing on $C_n$. It follows that $2^{-5n} \cdot 1.8^{-n} \cdot N \cdot 2(n+1) = N^{-4} \cdot 2(n+1)/1.8^n$ is the upper bound on the probability of overflow of some buffer at some time $t$. If $n \geq 4$, we get the upper bound of $\geq 1 - (1/N^4)$, what completes the proof. ∎

***THEOREM 2.2***

*Assume that we route packets on an $N$-node parallel computer $PC_n$ with $\leq N/n$ randomly failed links ($N = 2^n$), and that each packet $Px$ is partitioned into $n$ pieces in such way that the packet can be reconstructed from any $m = \lfloor n/2 \rfloor$ pieces. We apply the IDA routing with large buffer size $|BFy| = p \cdot |Px|$, making the buffer-overflow negligible. Under those assumptions, $1 - 2N \cdot (4 \cdot e/n)^{0.25n}$ is the lower bound on the probability that all $Px$ will reach their respective destinations [563].*

**PROOF** In order to route $Px$ from $x \in C_n$ to $\pi(x)$, the IDA algorithm constructs paths $P_i(x)$ from $x$ to $R_i(x)$, and paths $Q_i(x)$ from $R_i(x)$ to $\pi(x)$, for $1 \leq i \leq n$. $Px_i$, the $i$-th piece of $Px$, travels along $(P_i(x), Q_i(x))$ from $x$ to $\pi(x)$. Obviously, $Px_i$ will be lost if any of links on paths $P_i(x)$ or $Q_i(x)$ are faulty. Concerning the reconstruction of $Px$ (from $Px_i$'s), if $\geq k = n - m$ pieces are lost, $Px$ will not be reconstructable. In this case, either $(k+1)/2$ of $P_i(x)$'s, or $(k+1)/2$ of $Q_i(x)$'s are blocked by faulty links $(1 \leq i \leq n)$.

Assume now that all links fail independently with equal failure probability of $(N/n)/(nN) = 1/n^2$. We consider the event of $(k+1)/2 \geq 0.25 \cdot n$ of paths $P_i(x)$ are disconnected, $1 \leq i \leq n$. It can be proved [563] that the failure probability of a path $P_i(x)$ with $\leq n+1$ links is bounded from above by $(n+1)/n^2 \sim 1/n$. It follows that the expected number of disconnected paths is $n \cdot 1/n = 1$. From inequality 2.6, for $\delta = 0.25n - 1 \sim 0.25n$, and replacing the exponent $n$ by 1, one gets:

$$\Pr\left[\geq 0.25n \text{ paths } P_i(x) \text{ fail}\right] \leq \left(\frac{e}{0.25n}\right)^{0.25n} \leq \left(\frac{4e}{n}\right)^{0.25n} .$$

Similarly, the same upper bound of $(4 \cdot e/n)^{0.25n}$ for the paths $Q_i(x)$'s can be obtained. The analysis shows that the probability of nonreconstructability of $Px$ at $\pi(x)$ is twice the probability in inequality 2.6. Since we have $N = 2^n$ packets, the probability that a packet will fail is $\leq 2N \cdot (4 \cdot e/n)^{0.25n}$. This proves the theorem. ∎

Similar to the IDA, Huang and Abraham have introduced algorithmic fault tolerance based on data encoding [287]. They have used checksum techniques to detect and correct errors when matrix operations such as addition, multiplication, scalar product, LU-decomposition, and transposition are performed on parallel systems. The effect of injecting artificial static bit/word errors and dynamic transient errors on real parallel systems, focusing on the correctness of parallel algorithms such as matrix multiplication, Gaussian elimination, and FFT has also been examined using elaborate data encoding techniques [40, 186, 428, 492, 493].

#### 2.1.6.2 Reliability Models

Reliability models try to evaluate the probability of an adverse event. Several models have been used.

- *Binomial model.* We assume independent failure probabilities. The system fails when at least $k$ failures occur. This model is equivalent to a birthday surprise problem.

- *Fault trees using and/or gates.* Fault trees are useful in evaluating static faults. General fault trees can be converted to binary trees, and the probability laws for union and intersection of events may be used. The intersection law in its simple product form can be used only if events are relatively independent; this is not true, for example, if "loss of power" replaces "sabotage" in Figure 2.7.

- *Reachability matrices.* These techniques can be used to evaluate both static, and dynamic faults. For the dynamic case, models both without repairs, and with repairs (with a number of automatic repair mans) are considered.

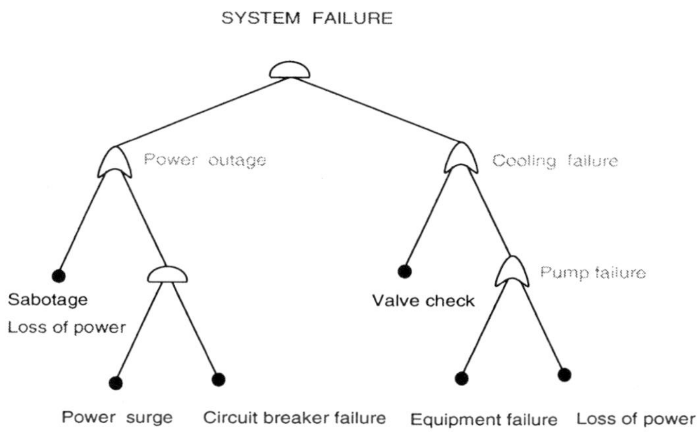


**FIGURE 2.7**
**An example with fault trees**

## 2.2 Communication Patterns

Communication patterns have been examined first in social experiments. If the communication pattern, i.e., the sender, the receiver, and the type of communication exchange is known in advance, an optimal routing schedule can be precomputed and the routing is called *offline*, while if the permutation is known on the fly the routing is *online*. In the former case the communication pattern is called *static*, while in the latter it is called *dynamic*. Offline routing is useful when data dependencies are known before run time, as in many scientific applications [221].

Online routing may be further classified into *oblivious* and *adaptive* strategies. A routing algorithm is said to be *deterministic oblivious* if the route of any packet depends only on its origin and destination, while it is *randomized oblivious* if the route of any packet is independently chosen according to a probability distribution (which is a function of its origin and destination) [17]. An *adaptive* strategy makes routing decisions for a packet depending on its current node and destination of packets encountered on its way. In a faulty system, adaptive routing may also check the status of adjacent links and nodes. Randomized and adaptive routing has been implemented in hardware, initially on the Chaos router [345], and more recently on commercial systems, such as the CM-5 [416] and Cray T3E [607] data networks, and the C104 $32 \times 32$ routing chip [666].

An alternative approach to probabilistic routing which minimizes contention of messages in the interconnection network involves randomly hashing the shared memory, i.e., storing data at random modules in the shared memory. Although this approach destroys data locality, it makes typical routing problems appear as random. Thus, it dramatically reduces network and memory contention, since requests are randomly distributed throughout the network links and memory modules. The approach of randomly hashing shared memory, has been used in the Tera multithreaded supercomputer, the IBM RP3 which implemented a combining network, and the German hardware PRAM prototype which included segmented scan (multiprefix) and synchronization operations [3, 513]. Although, lacking theoretical support, linear hash functions like $ax \bmod m$, for properly selected $a$, do very well in practice [187, 335].

### 2.2.1 General Routing

Many communication problems are special instances of the following $(N, p, k_1, k_2)$-routing problem. $N$ packets, each with its own source and destination, must be routed such that at most $k_1$ packets are initially at any node, and at most $k_2$ packets are finally at any node. The $N$ packets reside on $p$ PEs. A *k-relation*, also called $k$-$k$ routing, corresponds to $(kN, N, k, k)$-routing.

Special algorithms are also interesting if many of the packets that are sent or received by a node are the same. Of particular importance are the following intensive communication operations: (a) a *single-node broadcast* involves the transfer of a message from a particular node to all other network nodes; (b) a *single-node scatter* is similar to single-node broadcast except that $N$ different messages are broadcasted; (c) a *multinode broadcast* (also called *gossiping*) involves the simultaneous single-node broadcast from all network nodes (there are $N$ different messages); (d) a *total exchange* is similar to multinode broadcast, except that all the packets sent are different; (e) a *single-node accumulate* (also called *gather*) is the dual operation to single-node scatter; and (f) a *multinode accumulate* is the dual operation to multinode broadcasting.

Multinode broadcast can be implemented by performing a single-node gather followed by a single-node scatter. On the other hand, single-node gather, which has the same complexity as single-node scatter, is a subproblem of multinode broadcast. Therefore, all three problems have the same complexity up to constant factors. The same is true for the multinode accumulate. For problems (a)-(f) above, pipelining techniques and edge-disjoint spanning trees (or Hamiltonian paths) effectively reduce the delay time [212, 308, 364, 640].

### 2.2.2 Permutation Routing

Another special case of $(N, p, k_1, k_2)$-routing is *permutation routing.* In this problem, there are initially $N$ packets, one at each of the $N$ processor nodes. The packet at node $x$, $0 \leq x \leq N-1$ is destined to node $\pi(x)$, where the mapping $x \mapsto \pi(x)$ is a *permutation.* The aim is to *route* the packets in *parallel,* given that at most one packet can traverse each link at each time step.

Frequently used permutations belong to the *BPC*, Omega, inverse Omega, and Ascend/Descend classes [389, 418, 495, 647]. Suppose that an arbitrary packet initially located at node $x \equiv x_1\, x_2\, \ldots\, x_n$ is destined to the node $\pi(x) = x' \equiv x'_1\, x'_2\, \ldots\, x'_n$. For the *Bit Permute and Complement (BPC)* class of permutations $\pi$ is uniformly defined; for any $x$, $0 \leq x \leq N-1$, $\pi(x)$ is a permutation of $w \equiv w_1\, w_2\, \ldots\, w_n$, where either $w_i = x_i$, or $w_i = \bar{x}_i = 1 - x_i$ (complement of $x_i$). Examples of *BPC* permutations are

1. Bit reversal:

$$\pi_1(x) = x_n\, x_{n-1}\, \ldots\, x_1$$

2. Matrix transposition:

$$\pi_2(x) = \begin{cases} x_{l+1}x_{l+2}\ldots x_{2l}x_1x_2\ldots x_l, & \text{if } n = 2l \\ \\ x_{l+1}x_{l+2}\ldots x_{2l+1}x_1x_2\ldots x_l, & \text{if } n = 2l+1 \end{cases}$$

3. Shuffle:

$$\pi_3(x) = x_2\ x_3\ \ldots\ x_n\ x_1$$

4. Vector reversal:

$$\pi_4(x) = \bar{x}_1\ \bar{x}_2\ \ldots\ \bar{x}_n$$

5. Bit shuffle:

$$\pi_5(x) = \begin{cases} x_1x_3\ldots x_{n-1}x_2x_4\ldots x_n, & \text{if } n = 2l \\ \\ x_1x_3\ldots x_nx_2x_4\ldots x_{n-1}, & \text{if } n = 2l+1 \end{cases}$$

6. Unshuffle:

$$\pi_6(x) = x_n\ x_1\ x_2\ldots\ x_{n-2}\ x_{n-1}$$

7. Shuffle row major:

$$\pi_7(x) = \begin{cases} x_1x_{l+1}x_2x_{l+2}\ldots x_lx_{2l}, & \text{if } n = 2l \\ \\ x_1x_{l+2}x_2x_{l+3}\ldots x_lx_{2l+1}x_{l+1}, & \text{if } n = 2l+1 \end{cases}$$

8. Butterfly:

$$\pi_8(x) = x_n\ x_2\ x_3\ \ldots\ x_{n-1}\ x_1$$

9. Exchange:

$$\pi_9(x) = x_1\ x_2\ \ldots\ x_{i-1}\ \bar{x}_i\ x_{i+1}\ x_{i+2}\ \ldots\ x_n$$

The *BPC* permutations also include the *s sub-class* and *s super-class* of permutations (an *s* sub-bit reversal corresponds to a bit reversal permutation applied to the *s* least significant digits). *BPC* permutations are common in scientific applications [307]. The task of computing the transpose of a matrix arises often in matrix algorithms like in multidimensional FFT, computing congruence matrix transformations $UAU^T$, or executing a step of the $QR$ algorithm for finding eigenvalues of a matrix. The matrix transposition permutation is also used for conversion of a matrix stored with one row per processor (called consecutive row) to the storage of one column per processor (called consecutive column), and for changing array allocation from consecutive to cyclic storage representation (and vice versa). Other *BPC* permutations are useful for conversion between: (a) consecutive row matrix representation, (b) consecutive column storage, (c) cyclic row, (d) cyclic column, (e) combinations of cyclic/consecutive row (or column) representation, and different matrix embeddings, such as binary and binary-reflected Gray code on hypercubes.

*BPC* permutations have been extended to *affine* permutations, under which $i'$, the index of the node to which the $i^{th}$ node routes its packet, is given by $\overline{i'} = A \cdot \overline{i} + \overline{b}$. Here $\overline{i}$ and $\overline{i'}$ denote the binary expansion of $i$ and $i'$, respectively, $A$ is an invertible 0-1 matrix, and $\overline{b}$ is a 0-1 vector. Most communication patterns in numerical analysis application deal with affine transformations. The affine permutations contain the subclass of *bit-oriented permutations*. For these permutations, $A$ is a permutation matrix with exactly one 1 in every row and column. For example, the following important permutations are affine (and also bit oriented):

- Shuffling and unshuffling,
- Adding bitwise a fixed number to the processor indices, and
- Changing from row-major to shuffled row-major indexing.

Omega (or inverse Omega) permutations are realizable by a $\log N$-stage[2] Omega network (respectively, inverse Omega) without conflict at any stage. More precisely, consider $N = 2^n$ nodes represented by binary numbers from 0 to $N-1$. For any two nodes $x$ and $y$, let $L(x, y)$ (and $M(x, y)$) represent the number of similar least significant bits, up to the first different bit (respectively, similar most significant

[2]All logarithms in this book are to the base 2, unless otherwise indicated.

bits). Notice that $0 \leq L(x,y), M(x,y) \leq \log N$. A permutation $\pi$ is an *Omega permutation* if and only if for any nodes $x$ and $y$, we have $L(x,y)+M[\pi(x),\pi(y)] < \log N$. We call $\pi$ an *inverse Omega permutation* if for any $x$ and $y$, we have $M(x,y) + L[\pi(x),\pi(y)] < \log N$. Omega and inverse Omega permutations are useful in parallel applications such as parallel load and store on matrices, simultaneous conflict-free access to all rows, columns, diagonals, and blocks of arrays, and divide and conquer strategies [389]. Some examples of Omega and inverse Omega permutations, which do not belong to the *BPC* class, are

10. Cyclic shift of amplitude $k$, $1 \leq k \leq N$,

$$\pi_{10}(x) = (jx + k) \bmod N \ ,$$

where $j$ is odd.

11. Cyclic shift within segments,

$$\pi_{11}(x) = \delta_2(x + k) \bmod 2^{n-j} \ ,$$

where $\delta_2$ is the (decimal) number equivalent to the $j$ most significant bits in the binary representation of $x$, and $1 \leq k \leq N$.

12. Unscrambling $j$-ordered vectors,

$$\pi_{12}(x) = (jx) \bmod 2^k + (x_n, x_{n-1}, \ldots, x_{k+1})\, 2^k \ ,$$

$j$ is odd, $1 \leq k \leq N$.

With *normal Ascend* (and *Descend*) *permutations*, node $i$ ($0 \leq i < N = 2^n$) communicates with nodes ($i$ $XOR$ $2^b$) for all $b = \{0,\ 1,\ \ldots\ , n-1\}$ (respectively, $b = \{n-1,\ n-2,\ \ldots\ , 0\}$) [559]. In contrast, in the superclass of $2^b$ *Ascend* (and $2^b$ *Descend*) *computations*, node $i$ communicates with nodes $(i + 2^b) \bmod N$ and $(i - 2^b) \bmod N$, for all $b = \{0,\ 1,\ \ldots\ , n-1\}$ (respectively, $b = \{n-1,\ n-2,\ \ldots\ , 0\}$) [495]. Both normal Ascend/Descend permutations, and $2^b$ Ascend/Descend computations arise in efficient parallel algorithms based on the divide and conquer primitive for many problems, including polynomial evaluation, inner product, linear recurrence, triangular system evaluation, Batcher's odd-even mergesort and bitonic sorting [286, 471, 486, 487, 495]. These patterns also occur in software implementations of parallel scan (prefix) operations, and logarithmic complexity dissemination barriers [471, 486]. The basic multiscan (prefix) operators are usually associative.

### 2.2.3 Isotonic Routing

Several other schemes are derived from permutation routing. *Partial permutation routing* corresponds to routing with at most one packet originating from any node and at most one packet destined to any node.

*Isotonic routing* refers to any of the following routing schemes. A *semi-contraction* is a special case of partial permutation routing; the absolute distance $|x-y|$ between the initial locations of any two packets $x$, $y$ is always greater than or equal to the absolute distance $|\pi(x)-\pi(y)|$ between their final destinations. Distances refer to integer differences between node representations, not Hamming distances. A *semi-expansion* is the dual operation to a *semi-contraction. Packing* corresponds to partial permutation routing of $M \leq N$ packets onto a single interval (e.g., the first $M$ locations), while preserving their relative order. *Unpacking* (or spreading) is the dual operation to packing. Thus, packings and unpackings are special cases of semi-contractions and semi-expansions. A *semi-broadcast* refers to routing $M \leq N$ packets, when the final destinations of packets from different nodes are ordered, i.e., packets at nodes $x$ and $y$ ($x < y$) are headed to a set of destinations $S_x$ and $S_y$, such that any member of $S_x$ is smaller than any member of $S_y$. A *semi-gather* is the dual operation to a semi-broadcast. In Figure 2.8, we illustrate these basic communication patterns.

### 2.2.4 Other Communication Patterns

As explained above, communication patterns usually correspond to one-to-one, one-to-many, many-to-one, and many-to-many routing situations. There are several other patterns which arise in many common computation problems. One of the most useful patterns is the parallel scan (prefix) operation [78, 79]. A common example of a scan is the parallel computation of partial sums of $N$ numbers which are initially distributed to PEs, e.g., via a scatter operation. As shown in Figure 2.9, a parallel scan operation (shown as a partial sum instantiation) can be performed using several approaches, such as (a) recursive doubling, (b) divide and conquer, (c) odd-even reduction, and (d) recursive broadcasting. Three algorithms require only one variable per PE, while divide and conquer requires two. The change of values during the various communication/computation phases of each of these patterns is also shown in the figure. The best pattern to realize a scan operation clearly depends on the topology, and the relative cost of arithmetic, send/receive vs. remote copy and broadcast operations. The aim is to keep all PEs

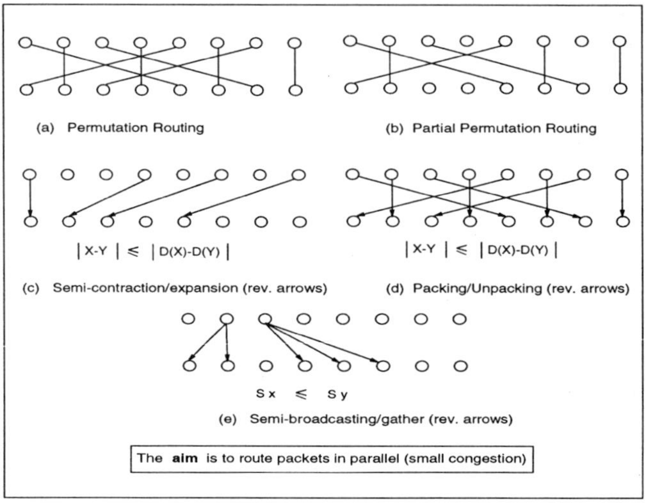


**FIGURE 2.8**
**Permutation-based patterns (1-d vector, $N = 8$)**

busy, while minimizing the latency costs.

In Figure 2.10, we show similar symmetric communication patterns arising in practice when solving tridiagonal matrix systems and recurrence equations [382].

Dynamic patterns are also common. *Routing to random destinations* corresponds to routing $N$ packets from distinct nodes to independently chosen random destinations. In *dynamic routing* packets are generated at every node with a fixed rate and head to independently chosen random destinations. Whether this assumption is justified depends on the particular application, although saturation rate and latency are performance metrics used often in comparing network topologies.

### 2.2.5 Embedding Relations

Graph embedding (called simulation or mapping) is an important network transformation technique. In fact, issues like space-time embedding and time dilation arise in the definition of Einstein's general theory of relativity. Used together with graph partitioning, graph embedding is a binding element between parallel algorithms and architectural topology.

In parallel systems, graphs are used to model architectural issues, such

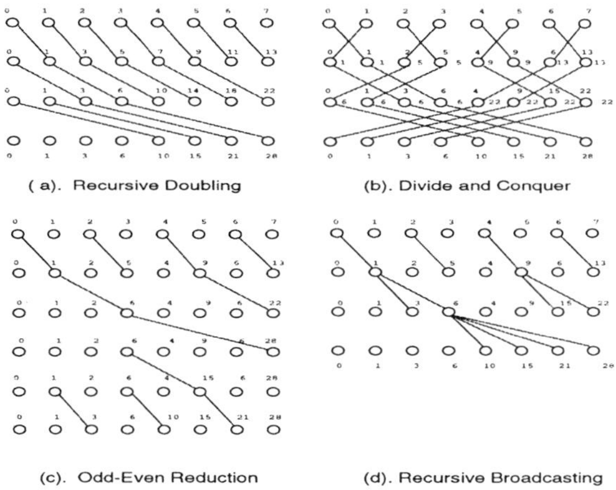


**FIGURE 2.9**
**Scan-related communication patterns**

as the topology of the interconnection network. Embeddings provide a generic framework for developing parallel algorithms for new topologies by porting, with little additional programming overhead, existing strategies to other topologies. Furthermore, task graphs (called concurrency graphs) express the necessary communication and synchronization patterns for realizing a particular computation. An embedding of a given task graph onto a particular topology could lead to substantial savings in communication time. Actually, this is the main approach used for the design and evaluation of parallel algorithms, such as static process allocation. Finally, embeddings are used for reconfiguring faulty networks, i.e., providing fault-free physical subgraphs in "injured" physical system graphs. In this approach, the operating system hides the system graph and allows users to define virtual topologies. Since fault-tolerant routing is not required, if network redundancy is provided it is possible to maintain network performance (bandwidth and latency) in the presence of a limited number of faults.

Mathematically, an embedding of a problem graph $G = (V_G, E_G)$ onto another target graph $H = (V_H, E_H)$ is an injective function $f : V_G \to V_H$. Most theoretical work on embedding concerns static scheduling [484]. For a given *node expansion* (also called load factor, or compres-

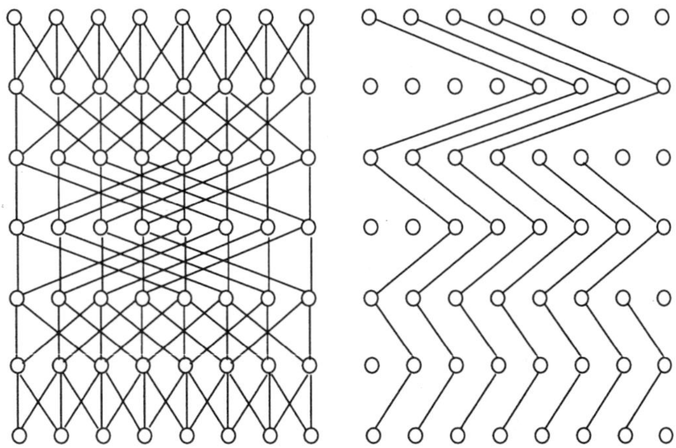

(e). Balanced Cyclic Reduction (f). Parallel Cyclic Reduction

**FIGURE 2.10**
**Other complex symmetric communication patterns**

sion ratio) $c_f = \frac{|V_H|}{|V_G|} \geq 1$, the quality of graph embedding depends on minimizing both *edge congestion*[3] which is a measure of edge contention during global intensive communication, and *dilation* which is a metric of latency overhead during point-to-point communication in the target graph. Let $v_1, v_2 \in V_G$. Then, the above measures are defined as follows:

- *Edge congestion is the maximum number of edges of $G$ mapped on a single edge of $H$.*
- *Dilation $d_f = max_{v_1 \neq v_2} \frac{d(f(v_1),f(v_2))}{d(v_1,v_2)}$ is the length of the longest path that any edge of G is mapped.*

Two embedding examples are illustrated in Figure 2.11. The dilation, expansion, and congestion metrics are also listed.

Most recent work on embedding concentrates on designing embedding with small dilations. When both the application graph and the system graph are arbitrary, embedding is NP-complete [132]. To overcome this,

[3]Node congestion can be defined similarly. Edge congestion is simply called congestion.

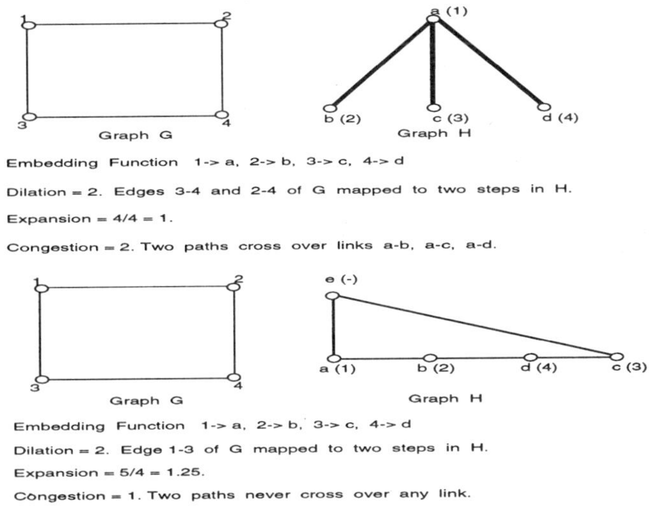


**FIGURE 2.11**
**Examples of embedding graph $G$ onto graph $H$**

one has to require at least one of the two graphs is regular. For example, one can show that any mesh can be embedded onto another mesh with dilation equal to the ceiling of the node expansion ratio, which is optimal.

## 2.2.6 Reduction of Communication Problems

Some reductions of the routing problems are possible. Permutation routing can be reduced to sorting, when sorting is on the unique final destination of the packets. Thus, if we can sort $N$ data elements in a given amount of time, we can perform any permutation in the same time.

Similarly, partial permutation routing can be reduced to sorting and unpacking as follows. First, nodes without messages send a dummy packet to destination $N+1$. After sorting the packets according to their destination address, $M$ genuine packets move to the nodes indexed 0 through $M-1$. The dummy packets can now be discarded. Since the remaining packets are ordered according to their final destination, their routing corresponds to performing an unpacking operation.

Clearly, routing-by-sorting imposes some conditions on the problem and the hardware. For example, a sorting problem should be solved statically, and this almost excludes that such an algorithm can be extended to dynamic inputs. In reaction to this, Borodin [83] gives a detailed analysis of the so-called *pure packet routing*, a notion which he tries to define so that impractical tricks are excluded. Similar limitations are considered in [352, 450].

Integer sorting is equivalent to partial permutation routing if the integers to be sorted are consecutive numbers, i.e., $\{1,\ 2,\ \ldots\ , N^{\varepsilon}\}$, $0 < \varepsilon < 1$ [353].

## 2.3 Routing in General Graphs

### 2.3.1 Performance of Routing Algorithms

In oblivious routing, we consider a model consisting of an undirected network $G = (V, E)$ of maximum degree $\Delta$. Each node in $V$ sends and receives up to one packet from each of its neighbors. The choice of routing paths between any two nodes depends only on the source and destination and does not depend on the rest of the communication pattern.

Lower bounds on the time delay for oblivious permutation routing have attracted much interest. Borodin and Hopcroft considered the possibility that some permutations cause node congestion, leading to large queuing delays [84]. Assuming a graph with $N$ nodes, and degree $\Delta$, equipped with $n$-port communication they proved a lower bound of $\Omega(N^{1/2}/\Delta^{3/2})$ for the class of deterministic oblivious permutation routing algorithms. More recently, Kaklamanis et al. and Borodin et al. improved this lower bound by considering link congestion.

**THEOREM 2.3**

*Deterministic permutation routing on a graph with $N$ nodes and degree $\Delta$ requires:*

1. $\Theta(N^{1/2}/\Delta^{1/2})$ *steps for oblivious algorithms with SIMD model [84, 85]; or*

2. $\mathcal{O}(\log N)$ *steps for adaptive algorithms with SIMD model [85].*

3. *$\Theta(N^{1/2}/\Delta)$ steps for oblivious algorithms with MIMD model [85, 321]; or*

4. *$\Theta(\log_\Delta N)$ steps for adaptive algorithms with MIMD model [85].*

**PROOF** We prove only part 3 of the theorem, namely that deterministic permutation routing on a graph with $N$ nodes and degree $\Delta$ requires $\Theta(N^{1/2}/\Delta)$ steps for oblivious algorithms with MIMD model [321].

Assume that network $G = (V, E)$ has $N$ nodes. This proof relies on the edge congestion in $G$. Denote by $G_t$ the subgraph of $G$ (the *destination graph for* $t$) of all paths (specified by the routing algorithm) from an origin $s$ to a fixed destination node $t$. Because each $t$ generates its $G_t$, the total number of destination graphs is $N$. For given $t$, consider the subset $C_k \subseteq G_t$ of all $\geq k$-congested edges (i.e., edges belonging to $\geq k$ $(s-t)$-paths). We claim that $N/[2 + 2(k-1)(\Delta - 1)]$ is a lower bound for the number of such edges.

Denote by $V(C_k)$ all nodes incident to an edge in $C_k$. Obviously, if $k \leq N/\Delta$, then $t \in V(C_k)$, and $|V(C_k)| \leq 2|C_k|$. In order to prove the lower bound, we assume that $k \leq N/\Delta$. The bound follows then from the inequality $|V \setminus V(C_k)| \leq (k-1)(\Delta - 1)|V(C_k)|$. In order to prove the inequality, choose any node $v \in V \setminus V(C_k)$. Then, there is a path $v\, v_1 \, \ldots \, v_n \, t$. Since $t \in V(C_k)$, there exists a node $v_i$, the very first node on that path belonging to $V(C_k)$. Because $v_{i-1} \notin V(C_k)$, the edge $v_{i-1}v_i \notin C_k$. It follows that $v_{i-1}v_i$ is at most $(k-1)$-congested. Obviously, one of $\Delta$ edges incident to $v_i$ must be $k$-congested. As a consequence, for each $v_i \in V(C_k)$, one has $\leq (k-1)(\Delta-1)$ of nodes $v$ in $V \setminus V(C_k)$ having $v_i$ as a first such node. Therefore, $|V \setminus V(C_k)| \leq (k-1)(\Delta-1)|V(C_k)|$. Now, $N = |V(C_k)| + |V \setminus V(C_k)| \leq |V(C_k)| + (k-1)(\Delta - 1)|V(C_k)| \leq [2 + 2(k-1)(\Delta-1)]|C_k|$, and the lower bound for the number of all $\geq k$-congested edges follows. This lower bound implies that the number of $k$-congested edges in each of the $N$ destination graphs is $\geq N/(2k\Delta)$.

For an edge in $e \in E$, we define its weight as the number of destination graphs $G_t$ in which $e$ is $k$-congested. It follows that the sum of weights of all $N\Delta$ edges in $E$ is $\geq N^2/(2k\Delta)$, or that the average weight of $e$ is $\geq N^2/(2k\Delta^2)$. Therefore, there must be at least one edge with the weight $\geq N^2/(2k\Delta^2)$. If one chooses $k$ to be equal to the lower bound for edge weights, $k = N^2/(2k\Delta^2)$, it follows that $k = \sqrt{N/2}/\Delta$. This means that there exists at least one $(\sqrt{N/2}/\Delta)$-congested edge $e$ in

$\geq \sqrt{N/2}/\Delta$ destination graphs $G_t$. By keeping $\sqrt{N/2}/\Delta$ $(s-t)$-paths fixed we obtain a permutation in which $\sqrt{N/2}/\Delta$ packets will use the edge $e$. ▮

Given a set of packet routes on a given network, we define the *dilation* $(W)$ as the maximum distance traveled by any packet, and the *congestion* $(U)$ as the largest number of packets that must traverse a single link. Leighton et al., using a nonconstructive proof, which can be viewed as an offline algorithm, proved:

***THEOREM 2.4***

*There exists a schedule for routing a set of packets whose paths have congestion $U$ and dilation $W$ on an MIMD network in $\mathcal{O}(U+W)$ steps using constant size queues [409].*

Recently, a constructive proof and improvements have been provided [410, 562]. Leighton et al. provided an algorithm for routing packets on a leveled network of constant degree $\Delta$ and depth $L$. All links in such a network are directed, from a level $i$ node to a level $i+1$ node, where $0 \leq i \leq L-1$. Considering permutations, from the $N$ inputs at level 0 to the $N$ outputs at level $L$, they proved the following theorem.

***THEOREM 2.5***

*There is an online scheduling algorithm which routes $N$ packets whose paths have congestion $U$ on any bounded-degree leveled MIMD network in $\mathcal{O}(U+L+\log N)$ steps using constant size queues, with high probability [408].*

The algorithm is a generalization to Ranade's permutation routing strategy on the hypercube [577], which was originally intended for implementation of an efficient simulation of shared memory on a distributed memory machine.

It is based on ghost packets and random rankings assigned to packets, and uses only $\mathcal{O}(\log^2 N)$ random bits. For messages $b$-bits long this algorithm can be pipelined, achieving $L + cb(U + L + \log N)$ time delay, with high probability ($c$ is a constant). It is assumed that each queue can hold at least $b$ packets. This result was later used for permutation routing on the hypercube, mesh, $n$-dimensional mesh, and shuffle/exchange graph in asymptotically optimal time. However, the multiplicative constant of the dominant term was ignored. Furthermore, the above result provides the best-known approximation ratio for the general job-shop scheduling problem.

Recently, Palis et al. considered nonconstant degree leveled networks [520]. By selecting a shortest path for all (source, destination) pairs on the original network, a new leveled network is constructed with $L+1$ levels, $N$ nodes per level, and two types of links between: (a) node $u$ at level $i$ and node $u$ at level $i+1$, and (b) node $u$ at level $i$ and node $v$ at level $i+1$, if there is a shortest path whose $i^{th}$ link connects $u$ and $v$ in the original network. A two-phase routing algorithm first forwards the packets to independently chosen random destinations, and then sends them to their final destinations along a unique shortest path determined by (b)-type links first, and then, if the path is shorter than $L$, by (a)-type links. The two-phase routing strategy is optimal, provided that the leveled network has degree $\Delta \geq 2$, and it is *nonrepeating*, i.e., any two paths that share a link and then diverge never share a link again.

***THEOREM 2.6***

*Any permutation of $N$ packets can be routed on a leveled MIMD network, with $LN$ nodes and $L$+1 levels, in $\mathcal{O}(L)$ steps, with high probability [520].*

Although applying the above algorithm to the mesh, $n$-dimensional mesh, hypercube, shuffle/exchange, or star graph yields an asymptotically optimal time delay, the two-phase routing algorithm does not guarantee constant queue size.

Valiant pioneered the work on probabilistic routing. Let a symmetric scheme refer to traffic being uniformly distributed to all network links. We have:

***THEOREM 2.7***

*In any oblivious, nonrepeating, and symmetric routing scheme, an arbitrary $k$-relation can be routed on an MIMD $N$ node, degree $\Delta$ network, with maximal route length $\mu$, and expected route length $\eta$ in just $T$ steps (provided that $T \leq (k\,\mu\,\eta)/\Delta$), with high probability ($\geq 1 - k\,N\,\left((e\,\eta\,\mu\,k)/(T\,\Delta)\right)^T$, where $e = 2.718281\ldots$) [682].*

**PROOF** If we label edges in the network by $\{e_1, e_2, \ldots, e_{N\Delta}\}$, a fixed packet route will consist of edges $R = e_1 e_2 \ldots e_r$. For $P_X$, the probability that packet $X$ traverses at least one edge of path $R$, we have

$$P_X \leq \sum_{i=1}^{r} P_{Xi} \tag{2.7}$$

where $P_{Xi}$ denotes the probability that packet $X$ traverses the edge $e_i$. Total amount of traffic through $e_i$ is equal to Traffic$(e_i) = \sum_X P_{Xi}$. Because of the symmetry of our routing scheme, it follows that $\sum_X P_{Xi} = \sum_X P_{Xj}$ for all $i, j$ $(1 \leq i, j \leq N\Delta)$. From (2.7) we have:

$$\sum_X P_X \leq r \sum_X P_{X1} \tag{2.8}$$

and the total expected traffic is

$$\begin{aligned} k\eta N &= \sum_{i=1}^{N\Delta} \text{Traffic}(e_i) \\ &= \sum_{i=1}^{N\Delta} \sum_X P_{Xi} \\ &= N\Delta \sum_X P_{X1} \end{aligned} \tag{2.9}$$

Relations (2.8) and (2.9) yield:

$$\sum_X P_X \leq \frac{rk\eta}{\Delta} \tag{2.10}$$

For a chosen packet $Y$ traversing the route $R$, the expected number of remaining $kN - 1$ packets whose paths will intersect any edge of $R$ is obviously $\leq rk\eta/\Delta$. Because the routing is oblivious, we have $kN$ independent Poisson trials with same upper bound on the probability of success. By Hoeffding's Theorem, if $T > kr\eta/\Delta + 1$, then $B(T, kN, r\eta/N\Delta)$ is the upper bound for the probability of having at least $T$ successes. It is easy to prove now that

$$B\left(T, kN, \frac{r\eta}{N\Delta}\right) \leq \left(\frac{erk\eta}{T\Delta}\right)^T. \tag{2.11}$$

It follows that for a fixed packet $Y$, RHS of (2.11) is an upper bound for the probability that the routes of $T$ other packets have at least one common edge with the route of $Y$.

Recall that a *queue line* in a network is a directed path $\{v_1, v_2, \ldots, v_{n+1}\}$ together with a queue $q_i$ at each node $v_i$ $(1 \leq i \leq n)$, and a set $\{(i_j, t_j, r_j) | 1 \leq j \leq m\}$ defining $m$ packets $X_1, \ldots, X_m$ such that packet $X_j$ (with destination $v_{i_j + r_j}$) is added to the queue $q_{i_j}$ at time $t_j$. If a queue line contains $m$ packets, then no packet is delayed by more than $m - 1$. As a consequence, the probability of packet $Y$ being delayed for $T$ is bounded by the RHS of (2.11). We considered $kN$ packets; henceforth the probability that at least one of them is delayed for at least $T$ is $kN\,(e\eta\mu k/T\Delta)^T$, which proves the lower bound for the probability $p$. ∎

Valiant also showed that any oblivious routing algorithm on a graph with an asymptotically optimal diameter (or mean route length), either requires $\Omega(N^\varepsilon)$ steps (for some constant $\varepsilon > 0$), or makes packets travel a path of length at least $2\,\log_\Delta N$ [683]. This, in turn, implies the asymptotic optimality of the probabilistic two-phase routing strategy for the shuffle/exchange, the $n$-dimensional $n$-way shuffle network, and other sparse graphs.

Many graphs, including the mesh and the hypercube, can be written as a Cartesian product graph $G \times H$. By factoring permutations into elementary transpositions (called matchings), offline routing on the $G \times H$ graph can be performed without any packet congestion, beyond that caused by permutation routing on $G$ and $H$.

***THEOREM 2.8***

*Any permutation can be offline routed on the MIMD $G \times H$ network in $T(G) + T(H) + \min\{T(G), T(H)\}$ steps, where $T(G)$ $(T(H))$ is the time required for offline permutation routing on graph $G$ (respectively, on graph $H$) [24, 47].*

**PROOF** Let $|G| = n$, $|H| = m$. Denote by $G_1, G_2, \ldots, G_m$ the copies (*columns*) of $G$ in $G \times H$, and denote by $H_1, H_2, \ldots, H_n$ the copies (*rows*) of $H$. Let $N = \{1, \ldots, n\} \times \{1, \ldots, m\}$, and denote by $S_N$ the permutation group of all pairs $\{(i,j) | 1 \leq i \leq n\ ,\ 1 \leq j \leq m\}$. Permutation cycles are denoted by brackets of the form $[(i,j)(k,l)]$.

We consider a number of important subgroups of $S_N$; the subgroup $K$ of all permutations which for all $i$ keep each column $G_i$ setwise fixed, and the subgroup $R_i$ (for all $1 \leq i \leq n$) which for all $j$ keep each row $H_j$ setwise fixed. Obviously, $R_n$ contains permutations which keep every row setwise fixed. Let $R_0 = S_N$ and choose an arbitrary permutation $\rho \in R_0$. We want to express $\rho$ as $\rho = \delta_n \sigma_n \delta'_n$, where $\delta_n, \delta'_n \in K$ and $\sigma_n \in R_n$. We factor $\rho$ in $n$ phases through a subgroup sequence $S_N = R_0 > R_1 > \ldots > R_{n-1} > R_n$, where $<$ denotes the subgroup operation. In $0^{th}$ phase, $\rho = \delta_0 \sigma_0 \delta'_0$, where $\delta_0, \delta'_0$ are the identity permutations. In the $i^{th}$ phase, $\rho = \delta_{i-1}\sigma_{i-1}\delta'_{i-1}$ can be written as $\rho = \delta_i \sigma_i \delta'_i$ with $\delta_i, \delta'_i \in K$ and $\sigma_i \in R_i$.

We construct now a bipartite directed multigraph $X = (C \cup C', A)$ on sets $C = \{c_1, c_2, \ldots, c_m\}$ and $C' = \{c'_1, c'_2, \ldots, c'_m\}$ (representing the columns of $G \times H$). If permutation $\rho$ maps a point in column $i$ to a point in column $j$, there exists an arc from node $c_i$ to node $c'_j$ in $X$. Now, $X$ is a regular bipartite multigraph of degree $n$, and it can be partitioned into a set of disjoint perfect matchings $M_1, M_2, \ldots, M_n$.

In the $i^{th}$ phase, we consider the matching $M_i$. For that matching, denote by $(x_1, 1), (x_2, 2), \ldots, (x_m, m)$ the points in $G \times H$ corresponding to the vertices in $C$, and let $(y_1, 1), (y_2, 2), \ldots, (y_m, m)$ denote the points in $G \times H$ corresponding to the vertices in $C'$. The factorization $\rho = \delta_i \sigma_i \delta'_i$ is achieved (for all $1 \le j \le m$) by mapping of each point $(x_j, j)$ to $(i, j)$, concatenated by mapping of $(i, j)$ to $(y_j, j)$, and finally, by mapping of $(y_j, j)$ to $(x_j, j)$. It follows that $\delta_i = \delta_{i-1}\ \lambda$, where $\lambda$ is the product of the $m$ disjoint 3-cycles $[(i,j)(y_j,j)(x_j,j)]$. Note that one can get 2-cycles in case when two or all three points involved coincide. In a similar fashion, $\delta'_i = \lambda\ \delta'_{i-1}$, and in the $i^{th}$ phase,

$$\begin{aligned}\rho &= \delta_{i-1}\sigma_{i-1}\delta'_{i-1} \\ &= (\delta_{i-1}\lambda)\ (\lambda^{-1}\sigma_{i-1}\lambda^{-1})\ (\lambda\delta'_{i-1}) \\ &= \delta_i \sigma_i \delta'_i\ .\end{aligned}$$

It is easy to prove that $\lambda^{-1}\sigma_{i-1}\lambda^{-1} = \sigma_i \in R_i$, and that for each $i$, $\delta_{i-1}\lambda$, $\lambda\delta'_{i-1} \in K$. We constructed the routing algorithm for the $G \times H$ in $T(G) + T(H) + \min\{T(G), T(H)\}$ steps.

From the construction of the routing algorithm for $G \times H$, it follows that the time complexity is dominated by the complexity of perfect matching decomposition of the graph $X$. It is well known that there exist a number of sequential polynomial [522], or randomized polylogarithmic [490] algorithms for the later graph-theoretical problem. This completes the proof. ∎

Theorem 2.8 also holds if each $i^{th}$ copy of $G$ in $G \times H$, $1 \le i \le N_G$ is permuted, where $N_G$ is the number of nodes in graph $G$. A similar theorem applies when, each copy of $G$ in $G \times H$ contains a fixed number $f$ of faulty nodes [47]. Recently, Youssef proved:

***THEOREM 2.9***
*Any Omega or inverse Omega permutation can be offline routed on the MIMD $G \times H$ network in $T(G) + T(H)$ steps [717].*

For randomized sorting, first Reif and Valiant obtained optimal randomized sorting on a nonexpander graph, called the cube-connected cycle [585]. Leighton and Plaxton provided efficient randomized sorting algorithms on the butterfly (the multiplicative constant is only 7.45).

***THEOREM 2.10***
*$N \log N$ packets can be sorted on the MIMD $N \log N$ node butterfly in $\mathcal{O}(\log N)$ steps using constant size queues, with high probability [414].*

For messages $b$-bits long, pipelining this algorithm achieves $b+\mathcal{O}(\log N)$ bit steps, with high probability [414]. Leighton et al., using Theorem 2.10 and efficient embeddings of butterfly networks, proved that probabilistic sorting on the $n$-dimensional binary hypercube, and $d$-dimensional mesh ($n \geq 2$) can be performed in asymptotically optimal time [408].

### 2.3.2 Fault Tolerant Routing

When faults are placed maliciously, we are interested in certain graph properties [278, 281]. In particular, good routing ($\rho$) on network $G$ corresponds to bounding the diameter ($d_R$) of the *surviving route graph* $R$ [173]. The directed graph $R$ has the same number of vertices as graph $G$, and any two nonfaulty nodes $x$ and $y$ in $R$ are joined by a link if there are no faults on the route $\rho(x, y)$ between them. The surviving graph diameter is a measure of worst case performance degradation caused by faults, for a given routing algorithm.

Recall that a graph $G$ is *k-node connected* if there exist $k$ node-disjoint paths from any node $x$ to any node $y$ in $G$. As it turns out, several bounds on $d_R$ depend on $k$.

***THEOREM 2.11***
*If $|F| \leq k$ (node or link faults) and $G$ is $(k+1)$-node connected, then*

*there exists a routing algorithm $\rho$ such that the diameter of the surviving route graph $R$ is $d_R \leq \max\{2k, 4\}$ [173]. In addition, if $|F| \leq k/2$, this bound becomes $d_R \leq 4$ [534].*

***THEOREM 2.12***

*If $\rho$ is a shortest path routing on network $G$, and the set $F$ consists of $f$ faulty links and $k$ faulty nodes with degrees $\Delta_1, \Delta_2, \ldots, \Delta_k$, then the diameter of the surviving graph is $d_R \leq 2f + 1 + \sum_{i=1}^{k}(\Delta_i - 1)$ [193].*

**Problem 1** *Find an upper bound on $d_R$ if routing is of almost minimal length?*

**Problem 2** *Find an upper bound on $d_R$ if randomized routing is used?*

A shortest path routing $\rho$ is $(d, f)$*-tolerant*, if for every choice of $f$ (node or link) faults in $G$ the surviving route graph has diameter $d_R \leq d$. A graph $G$ is $(d, f)$*-tolerant*, if there exists a shortest path routing $\rho$ on $G$ which is $(d, f)$-tolerant.

***THEOREM 2.13***

*If $G$ is $(d_G, f_G)$-tolerant and $H$ is $(d_H, f_H)$-tolerant, then graph $G \times H$ is $(\max\{d_G, d_H, 2\},\ f_G + f_H + 1)$-tolerant [88].*

Peleg and Simmons studied graph properties which lead to good bounds on the diameter $d_R$ [534]. One can also pad graphs to build larger fault tolerant graphs [88].

### 2.3.3 PRAM Emulation on Distributed Memory

Since PRAM algorithms are relatively easy to design, and most existing parallel computers have a distributed memory, except for parallel machines with a small number of nodes, a natural need arises to efficiently simulate PRAMs on distributed memory systems (DM). While many to one routing can be used for simulating the CRCW and CREW PRAM models on parallel interconnection networks, permutation routing is employed for EREW PRAM simulations. A rich theory has been developed, and here we provide pointers to the most relevant literature.

A PRAM simulation on DM is performed in rounds. In every round each node performs a step for all the PRAM processors it is simulating (possibly just one), and then a routing is performed to write results and to read new data. If the access pattern is assumed to be randomly distributed, and if there is a lot of parallel slackness (implying that each node of the DM simulates a large number of PRAM nodes), one can hope that things work out fine. Because of Chernoff bounds, one can estimate that with high probability the random access pattern is actually highly balanced. The maximum number $k_{\max}$ of packets that any node sends to any other node can easily be estimated.[4] Therefore, one has a simple routing pattern in which each node sends at most $k_{\max}$ packets to each other node.

However, if one can not assume a random access pattern, or if there is insufficient parallel slackness, PRAM simulation implies more than just a routing problem. The real problem is the distribution of data over the memories of the DM, so that the necessary routing can be performed fast for all or most of the accesses.

One approach, which eliminates the assumption of random inputs, is to distribute data according to some hash function which is chosen uniformly from a universal class of hash functions [577]. Even sharper upper bounds can be obtained if in addition to some clever distribution technique, data are copied [146, 166, 227, 323, 467, 476]. In these papers, it is assumed that the network is completely connected, e.g., by a broadcasting optical crossbar switch. Other authors have dealt with bounded degree networks [263, 542], or even with simulations on meshes and mesh-like networks [264, 543, 544, 420]. More recently the focus appears to be no longer on achieving even faster simulations (in [146], a step of a PRAM with $N$ processors is simulated with a delay of $\mathcal{O}(\log\log\log N \cdot \log^* N)$ only), but on more diverse aspects such as fault tolerant simulations [56], and simulations on reconfigurable networks [147].

### 2.3.4 Optimality and Lower Bounds

Many papers have considered the "optimality" of algorithms which may be a more or less strong notion. Sometimes it means that the time consumption of the algorithm exceeds some lower bound by a constant factor only [62, 65, 133]; it may also mean that the leading constants of a lower bound and the algorithm match [318, 320, 625]; and sometimes it means that the time consumptions are identical [412, 573, 630].

[4] If in a given round of a PRAM simulation on DM with $N$ nodes, $k$ is the expected number of packets, then with high probability, $k_{\max} \leq k + 2 \cdot (k \cdot \log(T \cdot N^2))^{1/2}$, where $T$ is the total number of rounds that are performed.

The latter two types of optimality are mainly applied to meshes (because it is mostly trivial to achieve the first optimality on meshes). The most commonly applied lower bounds for store-forward communication algorithms simply exploit limitations of the topology, such as:

- *Distance bound*: A packet travels at most over distance 1 during a single routing step. Thus, a routing algorithm takes at least as many steps as the maximum distance a packet has to go.
- *Throughput bound*: At most one packet can go over a network link during one routing step. Thus, a routing algorithm takes at least as many steps as the maximum number of packets that have to pass through a single link.
- *Bisection bound*: A subset $\mathcal{C}$ consisting of $c$ links can transfer at most $c$ packets during one routing step. When the links of $\mathcal{C}$ "bisect" the PEs of the mesh into two (not necessarily equal) parts, with $m$ packets going from one part to the other for a certain routing problem, a routing algorithm takes at least $m/c$ routing steps. We say that $c$ forms a **bisection** of the mesh.
- *Connection availability bound*: Any link in the mesh can be used for the transmission of at most one packet during one routing step. There are $2dn^d$ links in the mesh (even less in a mesh without wraparound links). Thus, routing of a certain distribution of packets on the mesh requires at least $D/(2dn^d)$ routing steps, where $D$ is the sum over all packets of the distance they have to go.

In most cases where it is possible to prove a lower bound at all, the argument relies on one of the above four bounds. Only rarely, for example in some routing problems on rings and sorting problems on meshes, a higher bound can be proven by an argument based on the initial lack of global knowledge [366, 626].

## 2.4 Performance of Packet Routers

Parallel and distributed systems consist of processors communicating through an interconnection network. The network is usually configured with communication switches (also called routers), responsible for reliable and efficient message transmission. Routers are usually compared to the ideal crossbar switch; this switch never delays a packet waiting for transmission over an idle link. Additional switch functionality such as packet multicasting or combining is sometimes provided.

We analyze $n \times n$ packet switches which may simultaneously send (and receive) one packet to (from) each of their $n$ output (input) ports. We consider (a) one FIFO buffer at each input port, (b) FIFO buffers at each input port, partitioned per each output port, (c) one FIFO buffer at each output port, and (d) a central data buffer. We assume fixed size packets, preventive flow control (no packet loss or retransmission), and similar switch startup time ($T_s$ clock cycles), spent in buffering and prerouting before packet transmission. A packet is transmitted only if there is free buffer space at the subsequent switch. If we were to adopt Jenq's model, packets headed to a full buffer could also be accepted if another packet is simultaneously transmitted out of this buffer. However, as discussed before, flow control propagation time becomes significant, especially for large, sparse networks, making this model inefficient [167, 170].

The $n \times n$ input data buffered switch (IDB) has $n$ FIFO buffers. The packet to be transmitted through an output port is selected from all input buffers in round robin fashion. Any packet directed to a full buffer is stalled.

The $n \times n$ split input data buffered switch (SIDB) has $n^2$ FIFO buffers, one for each input-output port combination. The SIDB switch can preroute a packet to determine if it is directed to a full buffer. If all packets corresponding to a given output port are directed to a full buffer, traffic is temporarily stalled. Otherwise, the packet to be transmitted is selected using round robin.

The $n \times n$ output data buffered switch (ODB) implements FIFO buffering at the output ports. All packets arriving simultaneously can be wired to the appropriate output buffer(s) within $T_s$ clock cycles. The ODB switch also uses prerouting to determine whether packets are directed to a full buffer. All such packets are temporarily stalled.

The $n \times n$ central data buffered switch (CDB) provides one dedicated buffer for each source-destination pair in the network. Flow control is per virtual circuit (VC), where VC header information identifies uniquely the packet's source and destination. Round robin is implemented at each output port, for selecting the packet to be transmitted. Simultaneous data buffer writes/reads upon VC arrival/departure are possible.

In continuous routing, a constant packet arrival rate is applied at each input port. Generated packets head to independently chosen random output ports. This rate is reduced (and may drop to zero) whenever buffers become full, thus taking into account preventive flow control (back pressure) of the switches. We focus on the following performance criteria:

- Switch throughput $S$ is the average number of packets routed per clock cycle through an output port of the switch. For replicated switches or switches with dilated ports, $S$ is the probability that a given output port is active.

- Switch latency $L$ is the average packet delay in the switch, measured from packet header arrival to departure time of the last packet bit.

- Network throughput $S_{Net}$ is the average number of packets routed per clock cycle through an output link of a last stage switch. Network throughput is usually normalized by the maximum possible throughput, corresponding to the network bisection throughput.

- Network latency $L_{Net}$ is the average delay, measured from packet header arrival to departure of the last packet bit from the network. It is computed by adding the corresponding switch latencies for all network stages.[5]

### 2.4.1 Continuous Routing Models

Next, we present performance models for all four $n \times n$ switches under continuous routing assumptions. At first, we review the Markov analysis for the IDB switch. Then, we extend existing analyses for the SIDB and ODB switches, and develop a Markov model for the CDB switch.

Continuous routing forms a reasonable traffic model for parallel computers and communication networks. Each processor generates packets at a fixed rate, with each created packet assigned to an independently chosen random destination. Packets are routed in parallel to their final destinations. We adapt the same discrete-time model for switches, taking into account finite buffering and preventive flow control.

We assume that the buffer size is $k$ packets and buffers are initially empty. The packet generating rate at each input port is held constant ($0 < p \leq 1$), and each generated packet heads to an independently chosen random output port. Since preventive flow control is used, the

[5]Sometimes network latency may include processor arbitration delay.

packet arrival rate is reduced appropriately (it may drop to zero) whenever corresponding buffers become full.

Markov chains provide a probabilistic method for studying variations which occur in an ordered sequence. The model consists of a finite number of states and a collection of transition probabilities between states. By applying Markov chain theory to continuous routing, we can derive the probability of buffer size $i$, $0 \leq i \leq k$, and compute the throughput and average latency for each switch. In the following analysis, the switch startup time is $T_s = 1$ clock cycle, and the packet size is one unit.

### 2.4.2 Input Buffered Switch

The $n \times n$ IDB switch has been analyzed [167, 256, 271]. We include this standard analysis here for completeness; an extension to actual switches is discussed later.

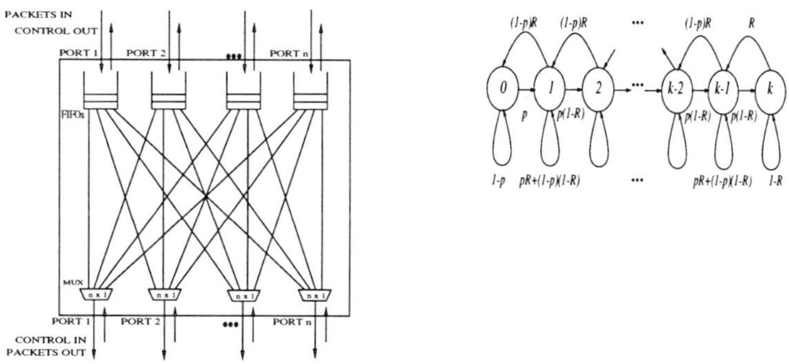


**FIGURE 2.12**
**(a) An $n \times n$ IDB switch; (b) Corresponding Markov chain**

The packet arrival rate to an input buffer is $p$. This rate drops to zero when the buffer is full. Let $R$ be the probability that a packet at the front of a buffer can move forward, and $\phi_i$ represent the probability that the queue length equals $i$. Figure 2.12 (b) shows the Markov chain which models the queue length of any input buffer in Figure 2.12 (a). We have, $(1 < i < k - 1)$:

$$\phi_0 = (1 - p)\,\phi_0 + (1 - p)\,R\,\phi_1$$
$$\phi_1 = p\,\phi_0 + (p\,R + (1 - p)\,(1 - R))\,\phi_1 + (1 - p)\,R\,\phi_2$$
$$\phi_i = p\,(1 - R)\,\phi_{i-1} + (p\,R + (1 - p)\,(1 - R))\,\phi_i + (1 - p)\,R\,\phi_{i+1}$$

$$\phi_{k-1} = p\,(1-R)\,\phi_{k-2} + (P\,R + (1-p)\,(1-R))\,\phi_{k-1} + R\,\phi_k$$

$$\sum_{i=0}^{k} \phi_i = 1\,. \tag{2.12}$$

We now compute the throughput at a given output port. The probability that an input buffer is nonempty is $1-\phi_0$. Since traffic from each input port is equally distributed to all output ports, the probability that a packet from this buffer is directed towards a given output port is $(1-\phi_0)/n$. Hence, the probability that no packet (from $n$ input buffers) selects the given output port is $(1-(1-\phi_0)/n)^n$, and the throughput is

$$S = 1 - \left(1 - \frac{1-\phi_0}{n}\right)^n. \tag{2.13}$$

The probability that a packet at the front of a buffer moves forward is equal to the packets transmitted out of the switch per clock cycle ($nS$) divided by the number of nonempty buffers $n\,(1-\phi_0)$. Hence,

$$R = \frac{S}{1-\phi_0}\,. \tag{2.14}$$

Let $C$ be the event that a packet can move forward. A packet can move forward only if (a) it leads its buffer (event $D$), and (b) this buffer is selected for transmitting a packet. Hence, $Prob\,[C] = R\,Prob\,[D]$.

The probability that the packet leads its buffer depends on the average queue length. Since that buffer contains at least one packet (given packet), we obtain:

$$Prob\,[D] = \frac{\sum_{j=1}^{k} \phi_j/j}{1-\phi_0}\,.$$

The latency is $L = \sum_{i=1}^{\infty} (1 - Prob\,[C])^{i-1} \cdot Prob\,[C] \cdot i = 1/Prob\,[C]$, or:

$$L = \frac{1-\phi_0}{R\sum_{j=1}^{k} \phi_j/j}\,. \tag{2.15}$$

### 2.4.3 Split Input Buffered Switch

Our analysis of $n \times n$ SIDB is a simple extension of the analysis of the $2 \times 2$ statically allocated fully connected switch previously considered

[167, 169]. Since the packet arrival rate at each input port is $p$, the packet arrival rate at each of the $n$ partitioned buffers is $P = p/n$. This rate $P$ drops to zero when the buffer is full. Let $R$ be the probability that a packet at the front of a buffer can move forward.

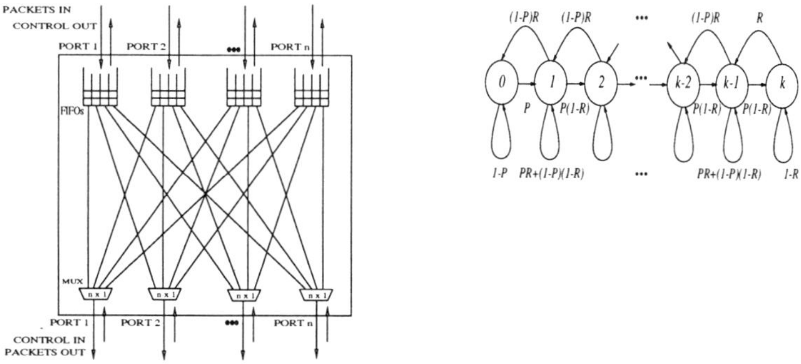


**FIGURE 2.13**
**(a) An $n \times n$ SIDB switch; (b) Corresponding Markov chain**

Figure 2.13 (b) represents the Markov chain which models the queue length of any buffer in Figure 2.13 (a). Let $\phi_i$ represent the probability that the queue length equals $i$. Assuming $k \geq 3$ we have, $(1 < i < k-1)$:

$$
\begin{aligned}
\phi_0 &= (1-P)\,\phi_0 + (1-P)\,R\,\phi_1 \\
\phi_1 &= P\,\phi_0 + (P\,R + (1-P)\,(1-R))\,\phi_1 + (1-P)\,R\,\phi_2 \\
\phi_i &= P\,(1-R)\,\phi_{i-1} + (P\,R + (1-P)\,(1-R))\,\phi_i + (1-P)\,R\,\phi_{i+1} \\
\phi_{k-1} &= P\,(1-R)\,\phi_{k-2} + (P\,R + (1-P)\,(1-R))\,\phi_{k-1} + R\,\phi_k \\
\sum_{i=0}^{k} \phi_i &= 1\,. 
\end{aligned} \tag{2.16}
$$

The throughput equals:

$$S = 1 - \left(1 - \frac{1-\phi_0}{n}\right)^n . \tag{2.17}$$

The probability that a packet at the front of a buffer moves forward ($R$) is

$$R = \frac{S}{(1-\phi_0)\,n} . \tag{2.18}$$

Let $C$ be the event that a given packet can move forward, and $D$ the event that this packet leads its buffer. Notice that $Prob\ [C] = R\, Prob\ [D]$. The probability that the packet leads its buffer depends on the average queue length. Since this buffer contains at least one packet (given packet), we obtain:

$$Prob\ [D] = \frac{\sum_{j=1}^{k} \phi_j / j}{1 - \phi_0}.$$

The latency is $L = \sum_{i=1}^{\infty} (1 - Prob\ [C])^{i-1}\ Prob\ [C]\ i = 1/Prob\ [C]$, or:

$$L = \frac{1 - \phi_0}{R \sum_{j=1}^{k} \phi_j / j}. \tag{2.19}$$

### 2.4.4 Output Buffered Switch

Since the packet arrival rate at each input port is $p$, the probability $c_j$ ($j = 0, 1, \ldots, n$) that a particular output buffer receives $j$ new packets follows the binomial distribution $c_j = B(n, j, p/n)$, with $j$ successes in $n$ trials, and $p/n$ success probability of a single trial.

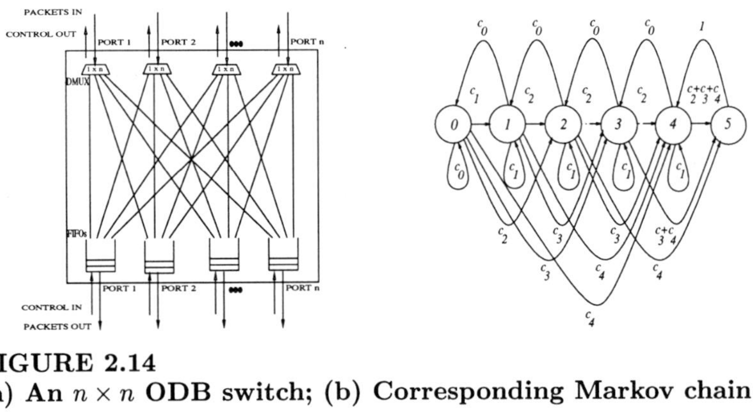


**FIGURE 2.14**
**(a) An $n \times n$ ODB switch; (b) Corresponding Markov chain**

Figure 2.14 (b) represents the Markov chain describing the queue length of any buffer of an $n \times n$ ODB switch with maximum buffer size $k = 5$ shown in Figure 2.14 (a). Notice the transitions controlled by binomial coefficients, due to simultaneous arrivals of packets headed to the same output port. Let $\phi_i$ denote the probability that the queue length is $i$, $0 \leq i \leq k$. Also, we extend, $\phi_l = 0$, for $l < 0$, and $c_l = 0$, for $l > n$. We have, $(1 \leq i < k - 1)$:

$$\phi_0 = c_0\,\phi_0 + c_0\,\phi_1$$

$$\phi_i = c_i\phi_0 + c_1\phi_i + c_0\phi_{i+1} + \sum_{j=1, j\neq i}^{n-1} c_{j+1}\,\phi_{i-j}$$

$$\phi_{k-1} = c_{k-1}\phi_0 + c_1\phi_{k-1} + \phi_k + \sum_{j=1, j\neq i}^{n-1} c_{j+1}\,\phi_{k-1-j}$$

$$\sum_{i=0}^{k} \phi_i = 1\,. \tag{2.20}$$

The throughput represents the probability that an output buffer is busy. Hence,

$$S = 1 - \phi_0\,. \tag{2.21}$$

Let $C$ be the event that a given packet leads a buffer, and therefore it can move forward. The probability of event $C$ depends on the average queue length. Since this buffer must contain at least one packet (given packet), we obtain:

$$Prob\ [C] = \frac{\sum_{j=1}^{k} \phi_j/j}{1 - \phi_0}\,.$$

The latency is $L = \sum_{i=1}^{\infty} (1 - Prob\ [C])^{i-1}\ Prob\ [C]\ i = 1/Prob\ [C]$. Hence,

$$L = \frac{1 - \phi_0}{\sum_{j=1}^{k} \phi_j/j}\,. \tag{2.22}$$

### 2.4.5 Central Buffered Switch

In our flow model, incoming VCs are evenly distributed to output ports, with $\lambda$ VCs per port. We may view each $VC$ as a buffer of size one. Only if this "buffer" is empty, a packet with this VC header can be accepted. Since the packet arrival rate at each input port is $p$, the corresponding arrival rate for a given $VC$ is $P = p/\lambda$ if the VC "buffer" is empty, otherwise it is zero. Let $R$ denote the probability that a packet with given VC can move forward.

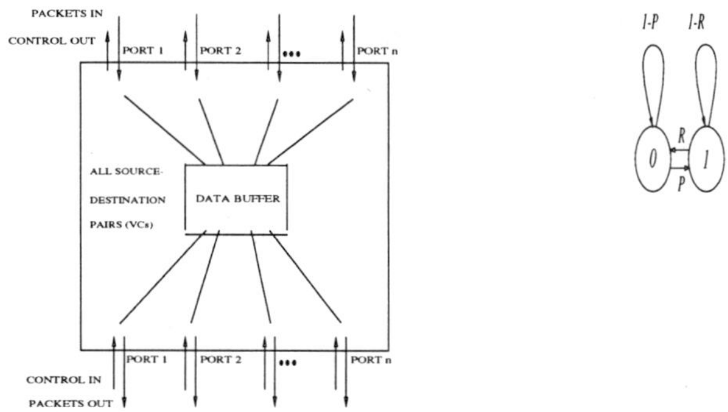


**FIGURE 2.15**
**(a) An $n \times n$ CDB switch; (b) Markov chain for VC "buffer"**

Figure 2.15 (b) represents the Markov chain describing the queue length of any VC "buffer" of the switch shown in Figure 2.15 (a). Let $\phi_i$ represent the probability that the queue length equals $i$. We have:

$$\phi_0 = (1 - P)\, \phi_0 + R\, \phi_1$$
$$\sum_{i=0}^{1} \phi_i = 1\,. \tag{2.23}$$

The throughput $S$ corresponds to at least one of $\lambda$ "buffers" being nonempty:

$$S = 1 - {\phi_0}^{\lambda}\,. \tag{2.24}$$

$R$ is related to the event that the correct VC "buffer" is selected. We obtain:

$$R = \frac{S}{(1 - \phi_0)\, \lambda}\,. \tag{2.25}$$

The latency is $L = \sum_{i=1}^{\infty} (1 - R)^{i-1}\, R\, i$, or:

$$L = R^{-1}\,. \tag{2.26}$$

### 2.4.6 Performance Comparisons for Generic Switches

For a single IDB switch the definition of $R$, as the probability that a packet *at the front of a buffer* moves forward, hides dependencies introduced by head of the line (HOL) blocking. However, the analytical approximation for IDB is usually within a small range (1-2%) of simulation results. Notice that such HOL phenomena do not occur with the other three switches.

We now assume that all switches have similar buffer space requirements, startup time, and process packets of equal (fixed) length. We compute throughput and latency for various packet arrival rates by solving systems of nonlinear equations using an iterative (trial and error) approach; if $\phi_0$ is chosen in $[0, 1]$ with at least 4 digits of accuracy, then the error on $S$ and $L$ is always less than 0.5%. Because of the monotonic relation of the involved functions with respect to $\phi_0$, it is much faster to binary search the solution space for the best $\phi_0$ value.

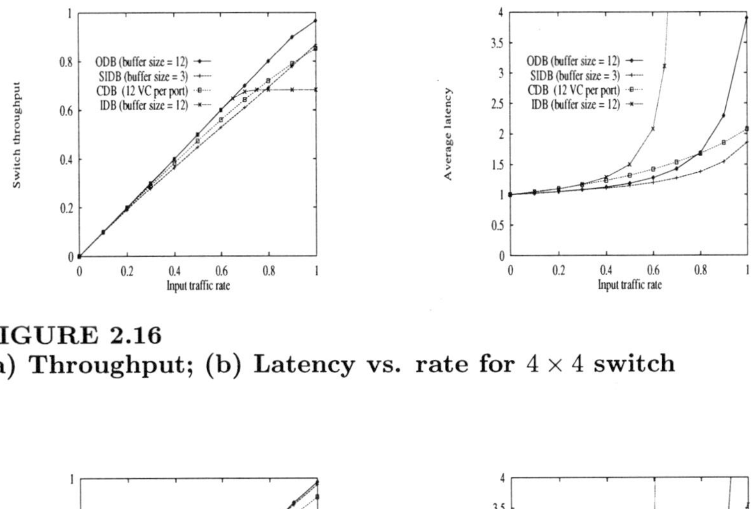


**FIGURE 2.16**
**(a) Throughput; (b) Latency vs. rate for $4 \times 4$ switch**

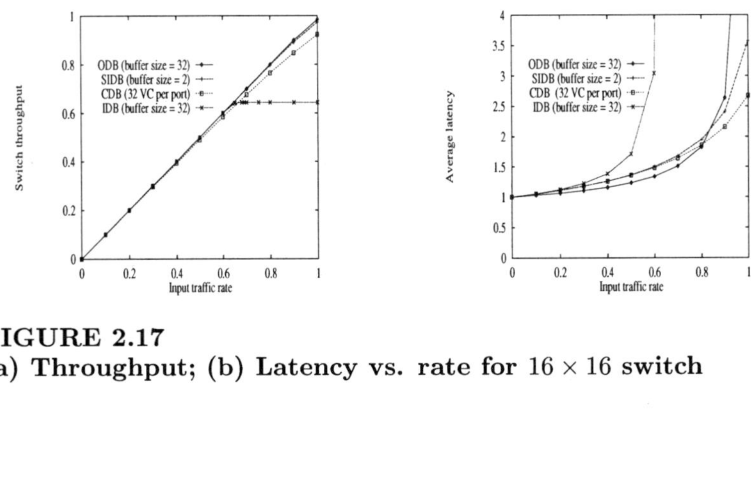


**FIGURE 2.17**
**(a) Throughput; (b) Latency vs. rate for $16 \times 16$ switch**

We compare the throughput and latency for various packet generating rates, and switch sizes. In Figure 2.16 (a) and (b), we consider $4 \times 4$ switches which can store 48 packets. Hence, for the SIDB switch, each of the 16 input buffers has size $k = 3$, while for the IDB and ODB switches each of the four buffers has size $k = 12$, and for the CDB switch there are $\lambda = 12$ VCs per each output port. Similarly, in Figure 2.17 (a) and (b), we consider $16 \times 16$ switches which can store 512 packets. Hence, for the SIDB switch, each of the 256 input buffers has size $k = 2$, while for the IDB and ODB switches each of the 16 buffers has size $k = 32$, and for the CDB switch there are $\lambda = 32$ VCs per each output port.

From the figures, we observe that the $4 \times 4$ (and $16 \times 16$) IDB switch saturate at input traffic rates above 0.68 (respectively, 0.64). The saturation rate for IDB approaches ~0.586 (for large switches); this is the maximum offered load for which the average latency remains bounded. To avoid falling into this oversaturated, unstable operation, admission control techniques such as back-pressure flow control, and traffic management policies, e.g., priority queuing, should be properly implemented.

For large input rates, the ODB maintains the highest throughput, while the SIDB and CDB switches have comparable throughput; SIDB seems to outperform CDB for large switches.

For large arrival rates, both SIDB and CDB offer smaller latencies than the ODB switch. However, in all four cases, the average latency does not consider packets stalled due to a full buffer. In reality, all such packets would be retransmitted, slightly increasing the average latency [703]. For a packet generating rate approaching one, the average latency increases $(1 - S)/S\%$, e.g., for the $4 \times 4$ switches the increase is 3, 11, 15, and 46% for the ODB, CDB, SIDB, and IDB switch, respectively.

Central buffering is an expensive option for high-performance interconnections; for symmetric systems with shortest path routing, buffer space is grossly unutilized; e.g., 0.21% of all VCs are used in a 7-dimensional hypercube, and 6.25% are used in a $16 \times 16$ mesh. Split input or output buffer switches, could achieve similar or better performance at a fraction of buffer cost.

All switches use statically allocated buffers, thus each buffer serves a single input stream. For better buffer utilization more complicated switches should be considered, e.g., the dynamically allocated fully connected (DAFC) switch [169]. The buffer in this switch may serve different input streams. Figure 2.18 illustrates the complex Markov transitions for a $2 \times 2$ DAFC switch (transition probabilities are omitted). Although, this switch has the highest throughput for random traffic, its

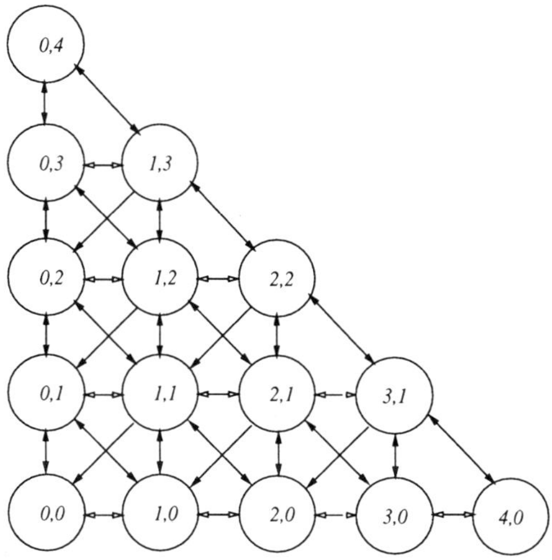


**FIGURE 2.18**
**Markov chain for a $2 \times 2$ DAFC switch (buffer size 4)**

design requires complex hardware data structures (pointers), especially for $n > 2$. Also, the DAFC design fails to sustain acceptable bandwidth in the presence of nonuniform traffic spots (NUTS), i.e., link contention for special communication patterns, since the DAFC switch places no limit on the number of slots a single input stream may take.

Also notice, that by providing separate buffers for different priority packets it is possible to allow a network to select high-priority packets to bypass others, and thus be delivered first. In a simple nonpreemptive implementation, additional buffers (for each priority level) are needed to store high-priority messages when several high-priority packets arrive destined to the same output port. Such switches are very interesting, especially within new telecommunication systems implementing quality of service (QoS).

For very high-speed link protocols, non-FIFO crossbar switches are attractive. A crossbar matches incoming packets to output links, avoiding bottlenecks associated with shared bus lines and centralized shared memory switches. The saturation rate of a non-FIFO crossbar which serves packets in a totally random basis asymptotically approaches ~63%. A simple iterative approach, which is based on distributed random scheduling, improves switch performance close to that of an ODB switch with just three or four iterations [28].

### 2.4.7 Large Routers and Networks

So far, we have assumed unit packet size and switch startup time. For real routers, the packet size is expressed in bits, the throughput in megabits per second, and the latency in seconds. Thus, real bandwidth and latency measures must be appropriately adjusted.

Assuming no congestion, the minimum packet latency through a real switch is $L_0 = T_s + K_1\, b$ seconds, where $T_s$ is the switch startup time (the interval between header arrival and header departure from the switch), $K_1$ is the number of bits after the header (including any termination bits), and $b$ is the transmission delay for one bit along the outgoing link. The switch bandwidth takes into account only the number of data bits ($K_2$); we are ignoring the overhead from header, checksum or parity, and termination bits (these bits are used in computing latency). Thus, the actual bandwidth ($S'$) in bits per second, and average latency ($L'$) in seconds, can be obtained using the previously computed values of $S$ and $L$ as follows:

$$S' = \frac{S\, K_2}{L_0} \tag{2.27}$$

$$L' = L\, L_0 \tag{2.28}$$

Switch models are not sufficient for selecting a particular buffer strategy. Instead, it is necessary to consider complete networks. For example, a unique path multistage topology can be derived from the binary butterfly (or the shuffle-exchange) by grouping buddies of $2 \times 2$ switches. Figure 2.19 (a) shows a decomposition of the 64-node binary butterfly into a three-stage network configured with 16, $4 \times 4$ switches per stage. Grouping of switches is based on the network "buddy" property that the four output links from two-stage $j$ switches constitute the four inputs for two-stage $j + 1$ switches; these four switches can be replaced by a more "powerful" $4 \times 4$ switch. An advantage of unique path networks, besides their incremental network design, is their deadlock-free routing property; essentially, one needs only allocate separate buffers (or VCs) for request and reply packets.

The unique path property of both network configurations, along with the symmetric traffic assumptions of the continuous routing model, imply that buffers assigned to different network paths are independent. However, buffers on a given network path are interdependent, since a packet can move forward only if the next buffer is not full.

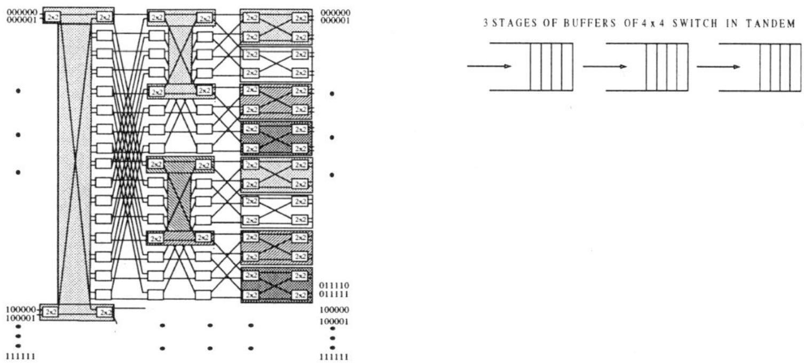


**FIGURE 2.19**
**(a) A 64-input unique path multistage network; (b) Corresponding queuing model**

Hence, network throughput (traffic rate at the output link of a last stage switch) and network latency (sum of individual-stage latencies) can be obtained by considering a queuing system with buffers connected in tandem. The number of buffers equals the number of network stages, e.g., three buffers for the network of Figure 2.19 (a), as shown in Figure 2.19 (b). Analytical models for networks configured with IDB and CDB switches correspond to systems of network equations; there is one such system for each network stage. Furthermore, for all stages but the last, network equations are appropriately adjusted: (a) the probability of a packet moving forward ($R$) is reduced by multiplying the right-hand side of the two (IDB and CDB) equations computing $R$, with the probability of a nonfull buffer at the next stage $(1 - \phi_k)$, and (b) throughput rates at one stage are mapped to input traffic rates at the next stage.

Analytical modeling is a powerful technique for studying the performance of networks and computing systems. However, any real system is too complex to be modeled accurately, especially when considering software latency. Thus, trace-driven or execution-driven simulation based on real applications is often considered. Furthermore, analytical results tend to show slightly better performance than simulation results; in reality, a queued packet contends for access to a given output port, while in analytical models, this packet contends for access to any port with the same probability.

Continuous routing can also be used to evaluate buffered switches with retransmissions, by comparing cell loss probabilities $(\phi_k)$. An overhead in this case, is that, in-order packet delivery from the same source (transaction completion) is not guaranteed. It is also very interesting to extend this model to hot-spot, bursty traffic [96], and continuous multicasting traffic [202].

### 2.4.8 Technological Constraints - Cost vs. Performance

Performance estimates from analytical modeling or simulations must take into account technological constraints and cost. Frequently, switch layout area mainly depends on the buffer space allocated to hold VCs; small variations for different switch size cancel out because of proportional differences in the number and size of control and routing circuits. In this case, constant pin-out constraints imply that the throughput of a $4 \times 4$ switch is twice as much as that of an $8 \times 8$ switch. An extension to unique path, multistage networks of switches (e.g., a $64 \times 64$ network configured as a three-stage butterfly network with 16, $4 \times 4$ unicast switches per stage, or a two-stage network with 8, $8 \times 8$ unicast switches per stage) will yield similar results; saturation throughput is essentially determined by the first network stage [238].

Pin-out constraints also imply that the "contention-free" latency of the $4 \times 4$ switch is approximately twice smaller than the latency of a $8 \times 8$ switch; the approximation is due to time dependencies for handling header information (e.g., strobe signals and routing table read). Thus, smaller switches also achieve smaller latency. One must be careful when compiling these results in terms of multistage networks. As noted earlier, latency computations do not consider packets which are not accepted because a buffer in the switch is full.

The situation seems to change when taking into account cost vs. performance considerations. A unique path multistage network configured with $2 \times 2$ switches is four times as costly in terms of switches, and asymptotically four times as costly in terms of links as the same network configured with $4 \times 4$ switches. Thus, considering the cost-performance ratio, it appears that the balance shifts towards larger, more cost-efficient VC-based switches.

# Chapter 3

# Multistage Interconnection Networks

## 3.1 Definitions

Single-stage networks connect $n$ input to $m$ output terminals using an $n \times m$ communication switch. In multiprocessors, input terminals connect to PEs and output terminals connect to memories, while in multicomputers, both input and output terminals connect to PEs, thus forming a wrapped multistage interconnection network. An arbitration

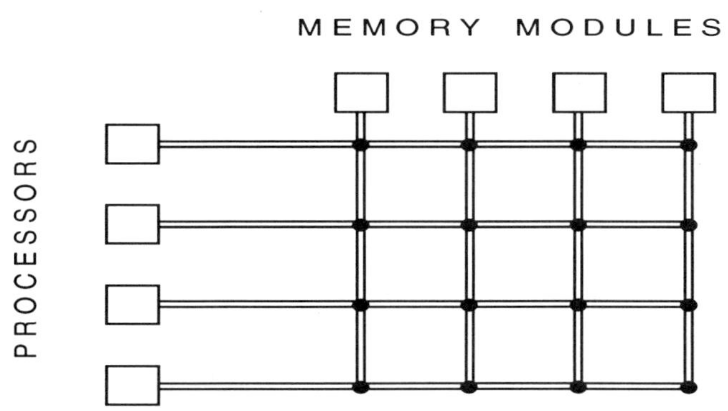


**FIGURE 3.1**
**A $4 \times 4$ crossbar with 16 crosspoints**

mechanism ensures that each output connects to only one input while, in general, one input may connect to several outputs, if multicast is possible. Single-stage networks are quite expensive in terms of the number of input/output (I/O) connections (called crosspoints). Thus, for example, the $n \times m$ communication switch would have $nm$ crosspoints, as shown in Figure 3.1 for a $4 \times 4$ crossbar switch.

In order to facilitate construction of parallel computers with many PEs, *multistage interconnection networks* (MINs) were introduced. MINs provide a cost-effective way towards network scalability, modular extensibility, and partitionability by providing I/O terminal connectivity through a disjoint union of sets of switch stages where each switch is usually a small crossbar (see Figure 3.2). MINs provide full I/O terminal connectivity: $N$ input and output terminals can be connected together using an $\mathcal{O}(\log N)$ stage interconnection network. The first applications in 1950s comprised telephone switching networks equipped with circuit switching. Circuit switching eliminates switching delays after the physical connection has been established, thus providing guaranteed service needed for time-critical applications, such as voice transfer. Later, after the 1970s, more applications were introduced by realizing large multiprocessors and multicomputers and introducing packetized radio networks. Depending on the underlying network protocol, unsuccessfully routed messages may be buffered, misrouted, or dropped, and using an acknowledgment protocol detected and retransmitted.

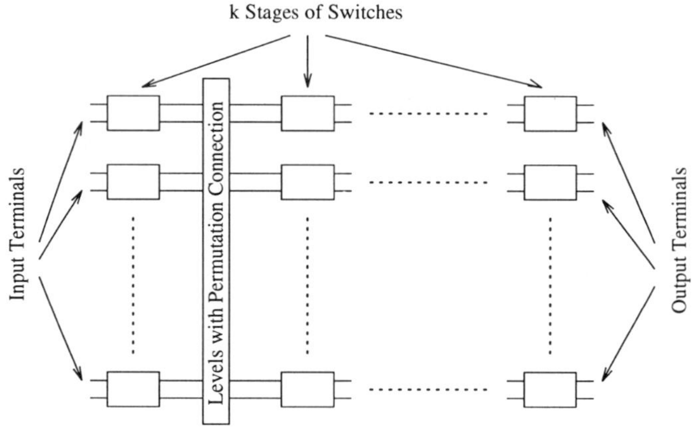


**FIGURE 3.2**
**A $k$-stage multistage interconnection network**

We characterize MINs by the interstage pattern. Let $V = V_0 \cup V_1 \cup \ldots \cup V_{k-1}$ be the disjoint union of $k$ sets of switches, each set being a *stage* of switches. For $0 \leq i \leq k$, $E_i$ is a multiset of directed links; each link goes from a vertex in $V_i$ to a vertex in $V_{i+1}$, and $E = E_0 \cup E_1 \cup \ldots \cup E_{k-2}$. $(V, E)$ is a directed acyclic network called a *k-stage MIN*. If we have to interconnect $N$ input ports with $M$ output ports, we can use a *generalized MIN with N inputs and M outputs*. If each stage $V_i$ consists of $n_i$ switches of size $\alpha_{i,j} \times \beta_{i,j}$ $(1 \leq j \leq n_i)$, the stage $V_i$ will have $l_i = \sum_{j=1}^{n_i} \alpha_{i,j}$ inputs and $m_i = \sum_{j=1}^{n_i} \beta_{i,j}$ outputs. Obviously, $l_0 = N$ and $m_{k-1} = M$.

## 3.2 Representation and Functionalities of Multistage Networks

Although a network can be described mathematically by enumerating the *entire set of connection assignments* realized by the network, e.g., a full crossbar realizes all input permutations and graphical network representations help in deriving important network properties, such as network isomorphism.

As shown in Figure 3.3 MINs can be represented graphically in different ways. Besides the usual way of representing switches with boxes and adjacent incoming/outgoing links with lines (shown on top), two other graphical representations are common. In the *link graph model (LGM)*, nodes represent switches and I/O terminals, while edges represent adjacent network links. Since the order of connections of links to switches is not taken into account, the LGM model can not be used to represent networks built with incomplete crossbars, i.e., crossbars for which there is at least one input-output pair which can not be connected. In the *crosspoint graph model (CGM)*, nodes represent network links, while edges represent crosspoints. An example is shown at the bottom of Figure 3.3.

While most MINs are built using small crossbars, other basic multistage network building blocks exist, such as *concentrators*, *hyperconcentrators*, and *superconcentrators*.

An $n \times m$ *concentrator* provides $m$ disjoint paths from any $m < n$ input wires to the $m$ output wires. Partial concentrators have also been defined [681]. Concentrators can be used to multiplex a large number of callers to a small number of high bandwidth channels; AT&T's SLC-40, SLC-96, and SLC-2000 switches are examples of concentrators.

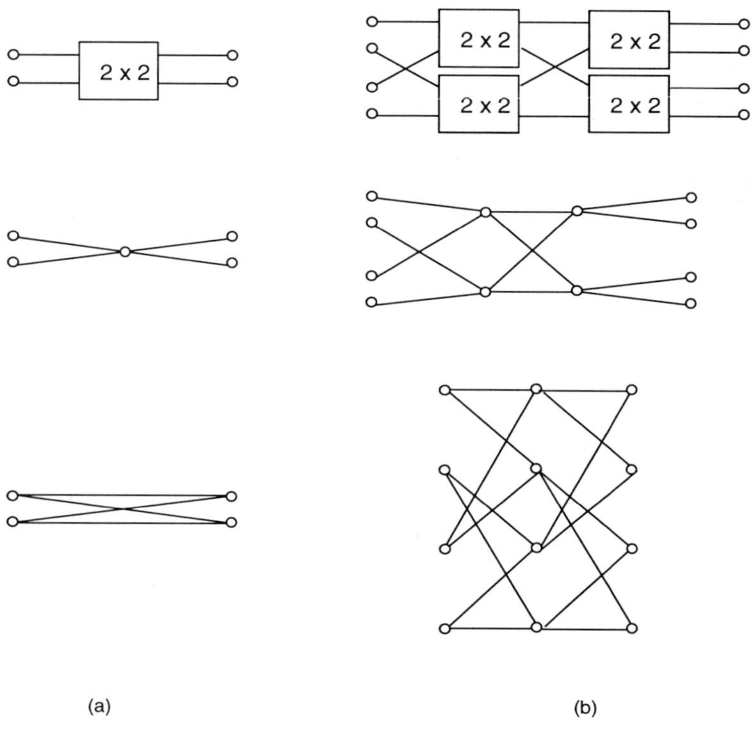


**FIGURE 3.3**
**Graphical representations of (a) $2 \times 2$ switch, and (b) MIN with $N = 4$ nodes**

Concentrators are also used to build specialized MINs, such as superconcentrators, permutation and broadcasting networks, for communication, bit manipulations, shifts, image compression and expansion, and bit interleaving.

The switch or concentrator size, defined as the number of switch crosspoints, is an adopted metric of hardware complexity. The probabilistic existence of $n \times m$ concentrators of size $29n + \mathcal{O}(1)$ is shown in [244, 546, 547]. Deterministic constructions exist, e.g., the $504n+\mathcal{O}(1)$ size construction by Gabber and Galil [219]. Constants in the expander graph constructions have been reduced [23, 548]. Masson showed that there are $n \times m$ concentrators ($m \geq 4$) with an optimal number of crosspoints $(n-m+1)m$ [461]; the design of an optimal $6 \times 4$ concentrator is shown in Figure 3.4. Similar results have been derived for partial concentrators by deleting elements from full size concentrators [462]. The expected capacity of sparse crossbar concentrators corresponds to the capacity achieved by almost all sets of inputs [551, 601].

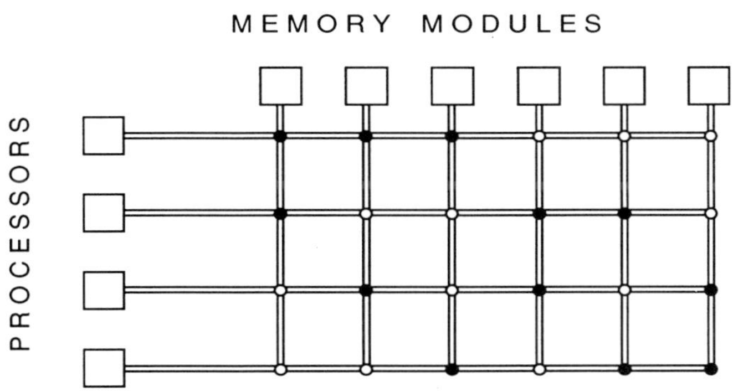


**FIGURE 3.4**
**A $6 \times 4$ concentrator with an optimal number of crosspoints**

Using balanced sparse crossbars, bipartite concentrators can be constructed with fixed fanout, $\mathcal{O}(\log N)$ concentration time, and a minimum number of crosspoints $(n - m + 1)\ m$, for any number of inputs and outputs [516, 244]. Bounds on I/O buffered concentrators and superconcentrators have been considered in [242]. Bounds on concentrators consisting of multiplexing instead of arithmetic and switching circuits are discussed in [109].

An $n \times n$ *hyperconcentrator* provides a set of disjoint paths from any set of $k$ input wires ($1 \leq k \leq n$) to the first $k$ output wires. Notice that a hyperconcentrator which uses only the top $k$ input wires behaves as an $n \times n$ concentrator. A hyperconcentrator can be implemented using a sorting network, or a parallel prefix network together with a butterfly network [134].

Finally, an $n \times n$ *superconcentrator* provides a set of disjoint paths from any set of $k$ input wires ($1 \leq k \leq n$) to any arbitrarily chosen set of $k$ output wires. An $n \times n$ superconcentrator may be implemented by connecting two full duplex $n \times n$ hyperconcentrator switches in tandem. The first hyperconcentrator brings the $k$ given inputs to the top $k$ positions, and the second (inverse) hyperconcentrator distributes these inputs to their final destinations [681]. Lower bounds on superconcentrators are provided in [172]; the $\mathcal{O}(N)$ upper bound provided is also the best possible, although constructions are impractical since they are based on nonuniform expander graphs.

## 3.3 Design and Classification of MINs

Assuming a circuit switching routing mode, a functional classification of MINs based on their relative power in realizing connection assignments is attempted in Figure 3.5. Notice that little is known in regard to the relative power of the generalized connector class vs. the nonblocking class.

The families of unique path and multipath *blocking networks, rearrangeable permutation networks, wide-sense* and *strict-sense nonblocking networks, sorting networks*, and *generalized connectors* are shown with corresponding examples. Specific MIN examples wil be provided in this chapter. The classes of *banyan, Delta*, and *bidelta* networks shown, which are commonly used for categorizing blocking networks, are defined below.

Three main functional classes of MINs comprise blocking, nonblocking, and rearrangeable networks. In *blocking* MINs, the available number of switches does not allow all combinations of I/O connection assignments to be realized at the same time. We distinguish between *unidirectional* blocking MINs, when all channels and switches only allow for simplex communication, and *bidirectional* blocking MINs, when duplex

**FIGURE 3.5**
**A classification of circuit-switched MINs**

communication is allowed [506]. Examples of unidirectional MINs are NEC Cenju-3 and IBM RP3, while the IBM SP-1/2, TMC CM-5, and Meiko CS-2 are examples of bidirectional MINs.

In *nonblocking and rearrangeable* MINs all I/O combinations can be realized at the same time. In *nonblocking* MINs a request to establish a path from a source node to a destination node, which may arrive at any time and asynchronously with other requests, can be satisfied without perturbing existing connection paths in the network. Nonblocking networks are *strict-sense nonblocking* when an I/O path is always possible irrespective of already established connections, and *wide-sense nonblocking* where a routing protocol is needed for realizing new I/O paths in order to avoid interference with established paths. Notice that general sorting networks, such as multibutterflies and multi-Beneš can also be used as strict-sense nonblocking networks. Nonblocking networks have often been used for configuring both parallel interconnection networks, and also high-speed computer networks based on small switches, e.g., Starburst, Bellcore's Sunshine, and AT&T's Starlite. A *rearrangeable* MIN can connect any input to any free output, but with rearrangement of existing I/O paths. Such MINs also require multipaths between every I/O pair, but only a smaller number than nonblocking networks. In case of asynchronous operation, rearrangeable MINs show the same characteristics as blocking MINs.

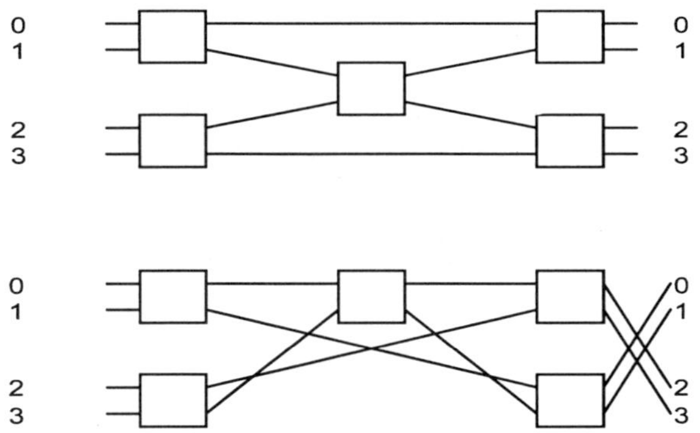


**FIGURE 3.6**
**Wide-sense topological equivalence of two networks**

The similarity of many blocking networks raises the need to characterize relationships between MINs which may be essentially identical, even though they may be drawn differently. The *topological equivalence* of

MINs follows from the graph theoretical property of isomorphism. We say that two MINs are topologically equivalent if their underlying graphs are isomorphic. We distinguish two cases of topological equivalence: in *wide-sense topological equivalence* one MIN is obtained from the other by relabeling (redrawing) switches and renumbering I/O terminals (see Figure 3.6), while in *strict-sense topological equivalence*, also called terminal equivalence, the numbering of I/O terminals is preserved. Formally, two networks are wide-sense (strict-sense) topologically equivalent if their LGM representations are identical under renaming of all nodes (respectively , internal, nonterminal nodes). Notice that concatenation of two topologically equivalent networks may not yield equivalence of the resulting network. Procedures exist to test whether two multistage permutation networks are topologically equivalent. The algorithm given by Oruc et al. takes $\mathcal{O}(kN^2)$ time [517], while the algorithm by Sridhar takes $\mathcal{O}(kN \log kN)$ [641], where $N$ is the size of the networks and $k$ the number of permutations they realize.

Two MIN networks are caled *functionally equivalent* if they realize the same set of connection assignments at the inputs and outputs, i.e., permutations for a permutation network. This definition does not consider the internal switching structure of the MIN. Obviously, MIN functional equivalence is a generalization of topological equivalence, and thus, topological equivalence implies functional equivalence [524, 710].

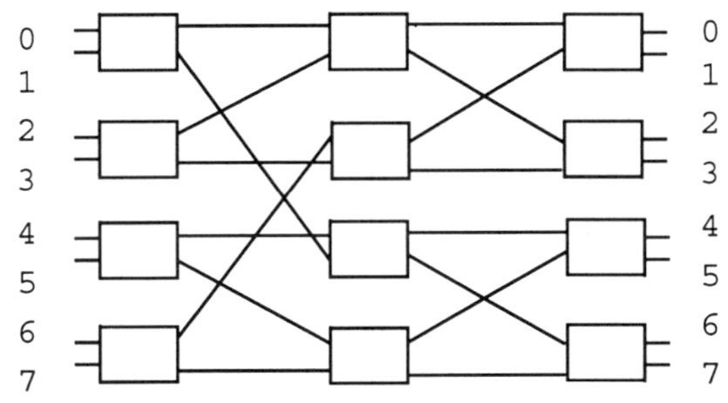


**FIGURE 3.7**
**A nonbanyan MIN**

The family of *banyan networks* is characterized by a unique path for each input-output pair, while all outputs are reachable from any input [225]. Any full-access MIN with such a property is called a banyan network. Banyan networks usually consist of $k \times k$ switching elements arranged in stages in such way that each input-output path traverses each stage only once. For a MIN with $N$ inputs and $N$ outputs, $\log N$ stages, with $(N/2)$ $2 \times 2$ switches at each stage are required to obtain the full-access property. An example of a banyan network is the MIN shown in Figure 3.3 (b). For example, the route from input terminal 0 to output terminal 2, follows two switches with the appropriate settings (cross, straight). However, the network in Figure 3.7 is not a banyan network, since it does not provide for *full connectivity*, i.e., there is no connection possible between input terminal 2 and output terminal 4. Also recall that a MIN which provides multiple paths between any input-output connection pair is not a banyan network, since banyan networks provide only a unique route from any input node to any output node. There are other important issues in the design of MINs, such as the *buddy property*, which will be presented soon.

Multistage banyan networks constructed from a number of fixed size $n \times m$ switches are called *regular*. Otherwise, they are *irregular*. *Rectangular* banyan MINs are regular banyans with $n = m$. Rectangular banyans built recursively are called *SW-banyan* networks. An SW-banyan is a *Delta network* if it has a certain level of distributivity and a unique routing path descriptor, determined by destination node only. More precisely, a Delta network consists of either,

- A single switch (e.g., an $n \times n$ crossbar), or
- One stage of $n \times n$ switches, followed by one stage of $n$-disjoint Delta networks: $\Delta_1, \Delta_2, \ldots, \Delta_n$; notice that each Delta network is recursively defined, if there exists a smaller Delta network with the same number of inputs. Furthermore, the following *buddy property* defines the necessary connection requirements: input nodes in the network are connected to the first stage switches, and output $j$ of each switch in the first stage is connected to an input node of the $\Delta_j$ Delta network [527]. The buddy property can be tested by identifying loops in bipartite subgraphs formed at each stage [14]. This distributivity property makes it possible to control any switch in the $i$th stage using the $i$th digit of the destination address (in the base-$n$ number representation system). Thus, switch settings for realizing a particular connection request are uniquely (and easily) determined from the switch stage and the destination tag associated with each message (refer to *destination tag routing* in Chapter 2).

In general, nonsquare Delta networks consist of $n$ stages of $k \times l$ switches, with $k^n$ inputs and $l^n$ outputs, with the $i$th stage having

$(k^{n-i}l^{n-i})$ $k \times l$ switches. The MIN shown in Figure 3.3 (b) is a Delta network. A nondelta 8-input banyan network is shown in Figure 3.8. Notice that the buddy property does not hold for its last stage; thus, this network lacks distributivity, and a unique path decriptor. Not all Delta networks, even for $N = 8$, are topologically equivalent to each other. Sridhar and Raghavendra proved that there is an exponential number of nonequivalent Delta networks [642].

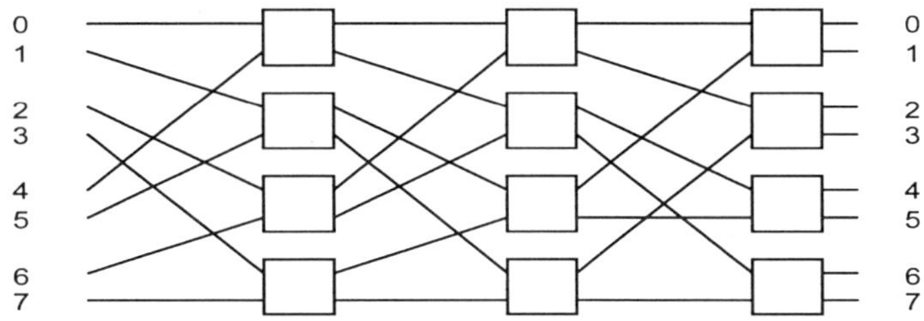


**FIGURE 3.8**
**A nondelta network for $N = 8$**

A Delta MIN is also a *bidelta network* if its inverse is also a Delta network, i.e., the Delta property is preserved when reversing inputs and outputs. Thus, the link labeling is done in such a way that both the network and its reverse network are Delta networks. A bidelta network's dual path descriptor is the sequence obtained by joining its path descriptor with the path descriptor of the network's inverse. The MIN in Figure 3.3 (b) is a bidelta network.

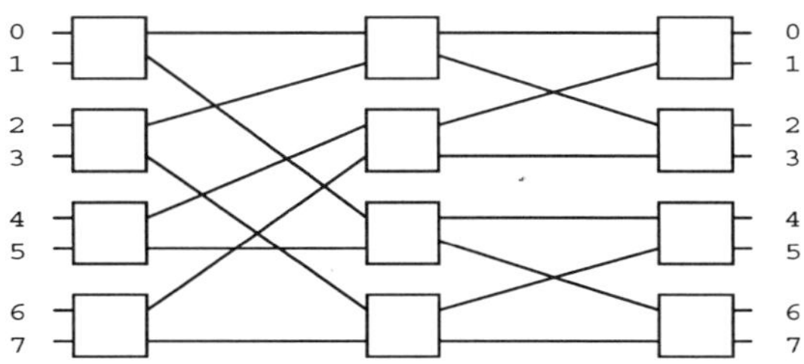


**FIGURE 3.9**
**A nonbidelta, Delta network**

An example of a Delta network which is not bidelta is shown in Figure 3.9. It can be proved that all $N = n^k$ input, $k$-stage bidelta networks configured with $n \times n$ switches are isomorphic to each other, and thus both topologically and functionally equivalent. The proof considers mapping a general bidelta network to the butterfly network (refer to the Section 3.4).

## 3.4 Blocking Networks

*Blocking networks* do not admit all permutations. That is, no edge-disjoint paths can be drawn for an arbitrary permutation. Several MINs defined in this section, such as the *Omega*, *butterfly*, *baseline*, *indirect cube*, *reverse-exchange*, and *indirect binary cube* network (and their inverses) are all bidelta networks.

### 3.4.1 Shuffle-Exchange and Omega Network

If a MIN uses the same topology in all stages, it can be built as a single-stage, recirculating network with memory available at each switch. Thus, the 1-stage recirculating shuffle-exchange MIN shown in Figure 3.10, can pass data from any input to any output port in at most $n = \log_2 N$ steps. While cheaper than a full *Omega* network, the recirculating network can *not* be circuit-switched, i.e., it cannot directly pipeline the transfer of data, unless it is augmented with proper buffering.

Shuffle-exchange MINs are based on the perfect shuffle interconnection pattern [649]. Many important applications of parallel algorithms can be linked to a shuffle communication pattern. For $N = k^n$ input ports, let $x = x_1 x_2 \ldots x_n$ be an arbitrary port number, $0 \leq x \leq N-1$ where $0 \leq x_i \leq k-1$, $1 \leq i \leq n$. The *perfect k-shuffle permutation* $\pi^k$ is defined as $\pi^k(x) = x_2 x_3 \ldots x_{n-1} x_n x_1$. It cyclically shifts the digits in $x$ to the left for one position. Thus, a perfect 2-shuffle can be interpreted as perfect shuffling an $N$-card deck.

It can be proved that any permutation can be realized on the shuffle-exchange network in $2 \log N - 1$ passes. One can reduce the number of passes, by increasing the number of network stages. As a consequence, if we increase the number of stages to $\log_2 N$, we obtain a unique input-output path for all input-output combinations. This MIN is called the *Omega network* [389]. An example of this popular MIN is shown in Figure 3.11.

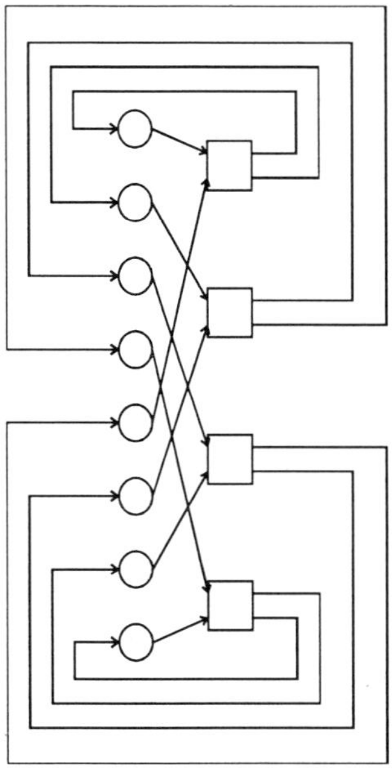

**FIGURE 3.10**
**One-stage shuffle-exchange for $N = 8$ PEs (shown as circles)**

Omega networks can also be constructed recursively. A $2 \times 2$ switch represents a 2 input Omega network, while an $N$ input Omega MIN is obtained from two $N/2$ input Omega networks, together with an extra switching stage added in front which distributes the $N$ input ports, as shown in Figure 3.12. The MIN obtained using this recursive decomposition is essentially an $N$ input Omega network.

Omega networks, proposed by Lawrie [389], have the nice feature of a replicated interconnection pattern in all stages. Thus, they are representatives of highly symmetric, blocking MINs. Another convenient property is the self-routing capability (destination tag routing): an $i$th bit 1 in the binary destination tag sets the $i^{th}$ stage switch in straight mode, while an $i$th bit 0 sets the $i^{th}$ stage switch in cross mode.

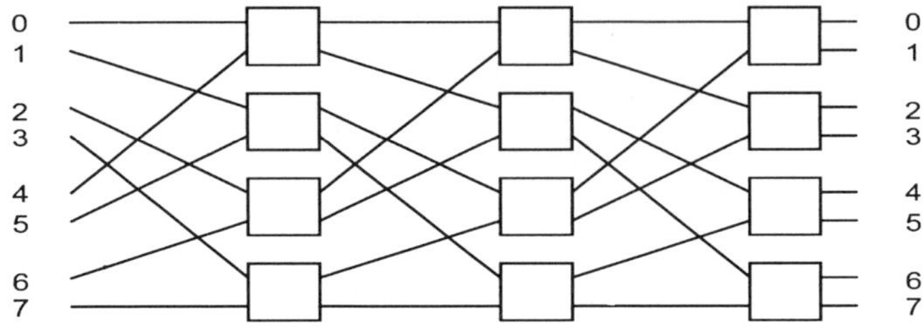


**FIGURE 3.11**
**The Omega network for $N = 8$**

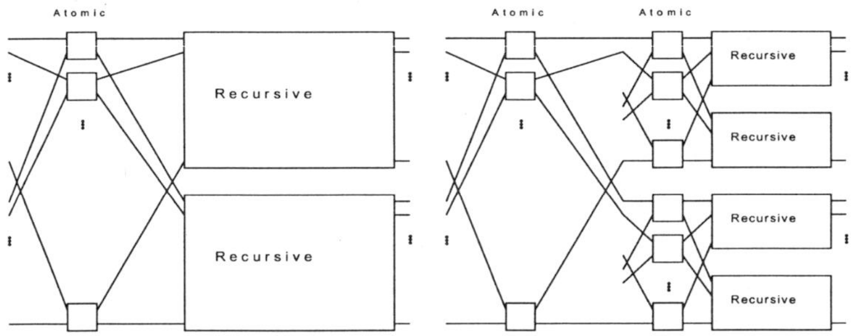


**FIGURE 3.12**
**Unfolding the recursive construction of Omega network**

The $N$-node circuit-switched Omega network realizes (in one pass) affine permutations of the form $(ax + b) \rightarrow (cx + d)$, $x = 0, 1, \ldots, N - 1$, if $gcd(a, N) = gcd(c, N) = 1$, i.e., the two pairs: $a$, $N$ and $c$, $N$ are relatively prime to each other. Furthermore, by using broadcasting switches, the network can realize all affine communication patterns of the form $(ax+b) \rightarrow (cx+d)$, $x = 0, 1, \ldots, N-1$, if $gcd(a, N) \leq gcd(c, N)$, By using graph coloring-based heuristics, Bernhard proved that the number of rounds for routing $N$ messages on the $N$ node Omega network in a conflict-free manner is only $\Theta(logN)$ from optimal [60].

In general, *network partitioning* corresponds to realizing edge-disjoint routes to support multiple user access to disjoint subsets of the resources. Omega networks can be easily partitioned to implement multitasking. For example, an Omega network can be partitioned into four independent parts by setting all switches in the first two stages to pass. Since each partition is independent, different user communication requests do not share the same physical links, and therefore all users obtain predictable performance. However, notice that banyan networks can not realize all possible partitions. If this is the case, more general network structures are needed, such as generalized connectors (see Section 3.6).

There are numerous other MIN topologies similar to Omega network. For example, the generalized Omega network consists of stages of square switches, with shuffles between stages implemented so that outputs from an individual switch in one stage connect to consecutive switches in the following stage. The routing is based on mixed radix numbers, as each stage may have its own switch size.

### 3.4.2 Butterfly Network

The *ith k-ary butterfly permutation* of a port number $x = x_1x_2 \ldots x_n$ is defined by $\rho_i^k(x_1x_2 \ldots x_{i-1}x_ix_{i+1} \ldots x_n) = x_1 \ldots x_{i-1}x_nx_{i+1} \ldots x_{n-1}x_i$, where $1 \leq i \leq n$. This permutation swaps the $n^{th}$ and the $i$th bits of the port address as shown in Figure 3.13.

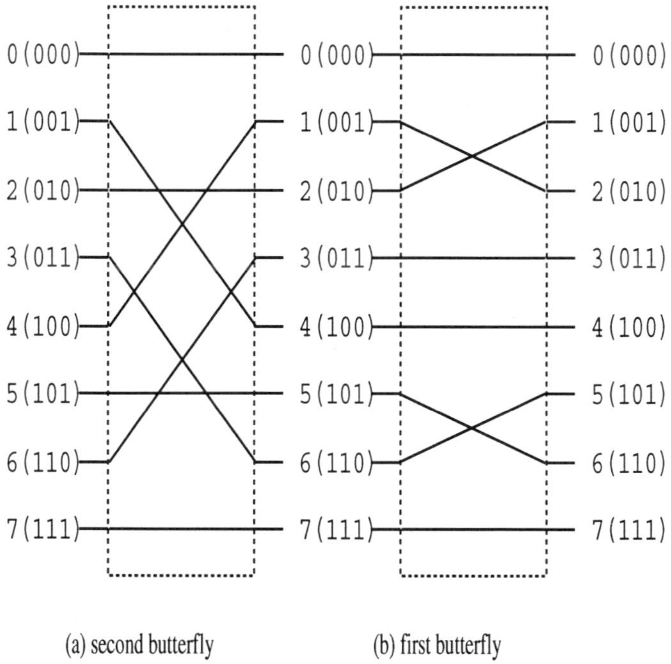


**FIGURE 3.13**
**The first and second butterfly permutations for $N = 8$**

Notice that the butterfly permutation corresponds to the computation pattern of a one-dimensional FFT.

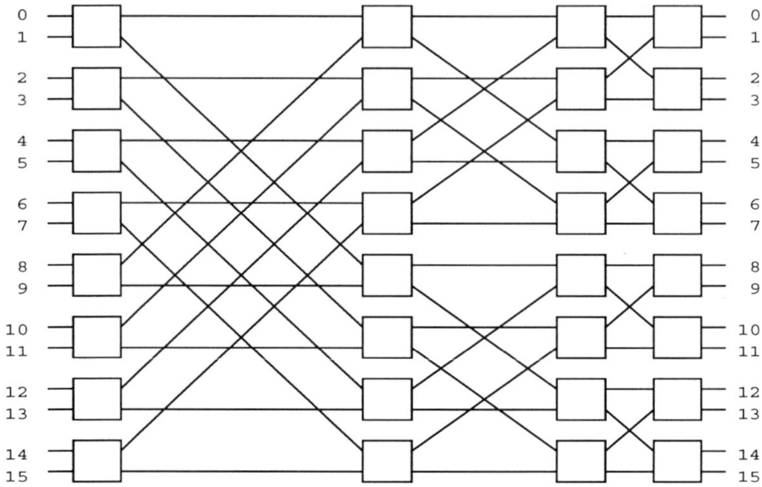


**FIGURE 3.14**
**The butterfly network for $N = 16$**

A butterfly network with 16 nodes is shown in Figure 3.14. Butterfly networks are essentially Omega networks drawn differently. For similar size networks, the mapping from a stage $l$ Omega switch to a stage $l$ butterfly switch consists of performing $l$ right cyclic shifts on the binary representation of each switch. Hence, these two networks are wide-sense topologically equivalent. With more or less sophisticated techniques, it can be shown that all bidelta MINs, including the Omega, butterfly, baseline, and indirect binary cube are topologically equivalent.

### 3.4.3 Inverse Omega, Baseline, Indirect Cube, etc.

The *inverse perfect shuffle* permutation is defined using the expression $\pi^{k^{-1}}(x) = x_n x_1 x_2 \ldots x_{n-2} x_{n-1}$. By concatenating such permutations, we can obtain the *inverse Omega* network (also called *flip network*) which is useful in fast Fourier transforms and matrix operations. Each stage in this network consists of exchange switches followed by the inverse perfect shuffle (see Figure 3.15). Notice that an inverse Omega network equipped with broadcast switches realizes all affine communication patterns of the form $(ax + b) \to (cx + d)$, $x = 0, 1, \ldots, N-1$, if $gcd(a, N) \geq gcd(c, N)$.

By combining butterfly and Omega network properties one obtains the *baseline* MIN [709] (see Figure 3.16). The *ith k-ary baseline permutation*

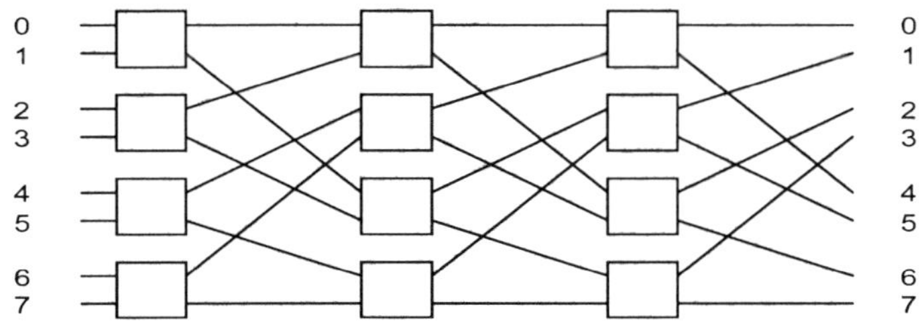


**FIGURE 3.15**
**The inverse Omega network for $N = 8$**

of $x$ is defined as $\delta_i^k(x_1 \dots x_i \dots x_n) = x_1 \dots x_{i-1}x_n x_i x_{i+1} \dots x_{n-1}$. This operation performs a cyclic shift of the $i+1$ least significant digits to the right by one position. In a baseline MIN, the link pattern $E_i$ is the $(n-i)$th baseline permutation $\delta_{n-i}^k$ for $1 \leq i \leq n$, and the $E_0$ is the perfect $k$-shuffle. The *inverse baseline network* is obtained from a baseline network by reversing the order of stages.

In the *indirect binary cube* MIN, the interconnection pattern $E_0$ is defined to be the perfect shuffle $\pi^k$, and $E_i$ for $1 \leq i \leq n_k$ is the $(n-1)$th butterfly permutation $\rho_{n-i}^k$. The *indirect binary n-cube* MIN is based on the link patterns of an $n$-dimensional binary hypercube [532]; this MIN complements the $i$th bit of the input port address $x$ into $\epsilon_i(x_1 x_2 \dots x_{i-1} x_i x_{i+1} \dots x_n) = x_1 x_2 \dots x_{i-1} \overline{x_i} x_{i+1} \dots x_n$. The link stages in an indirect binary cube correspond to links of a hypercube; the last stage of links is connected according to $\pi^{k^{-1}}$. This stage is necessary in order to pass the identity permutation when all switches are set in a straight switching pattern. An inverse cube, obtained by reversing the order of stages, is shown in Figure 3.17.

*Data manipulator* MINs represent another important family of blocking interconnection networks. Each stage has $N$ switching elements, each of which has three outgoing links. The link pattern is the so-called plus-minus $2^i$, because each switch is connected to a switch at the same position and to switches at distances $\pm 2^i \pmod{N}$ in the next stage. Therefore, such MINs can be defined as PM2I networks. The functionality of the terminal switches ranges from (de)multiplexer to full $3 \times 3$ crossbar switches, while always a limited number of alternate paths, usually two, is provided.

The first of the data manipulator network had the role of bit manipulations within data words, such as permuting, replicating, spac-

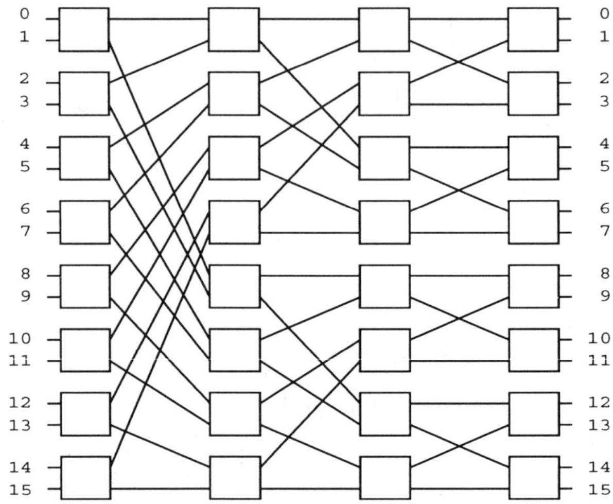


**FIGURE 3.16**
**The baseline MIN for $N = 16$**

ing, masking, and complementing [197]. This network was a four-stage manipulator with the ability of performing cyclic shifts of any size on any $N$-bit word using settings of the three stages according to the shift amount. The switches were not controlled individually. Instead, each stage of switches was divided into two equal groups of $N/2$ switches, with each group being controlled by the same set of signals.

The $N$-node augmented data manipulator (ADM) is the generalization of this shifting network [634], and consists of $m = \log_2 N$ stages. As shown in Figure 3.18, stage $i$ and stage $i+1$ switches which either agree in all bits, or disagree only in the $i^{th}$ bit are connected together. Notice that $+2^{m-i}$ and $-2^{m-i}$ connections at the first stage ($i = 1$) are symmetric (not replicated), and thus not shown.

ADM uses ternary switches, with each switch set in a *straight*, *down*, or *up* pattern. Multicating to a selected subset of links may also be allowed. In contrast to the initial data manipulator network, each switch in an ADM can be controlled individually, thus the ADM can receive packets from only one of its inputs at a time. A control tag routing algorithm can be used on the ADM network. It is based on binary functions, and is similar to destination tag routing [466]. In general, a packet's destination node $D$ is written in terms of the source node $S$ as follows: $D = S + (-1)^{f_{2n-1}} f_{n-1} 2^{n-1} + (-1)^{f_{2n-2}} f_{n-2} 2^{n-2} + ... +$

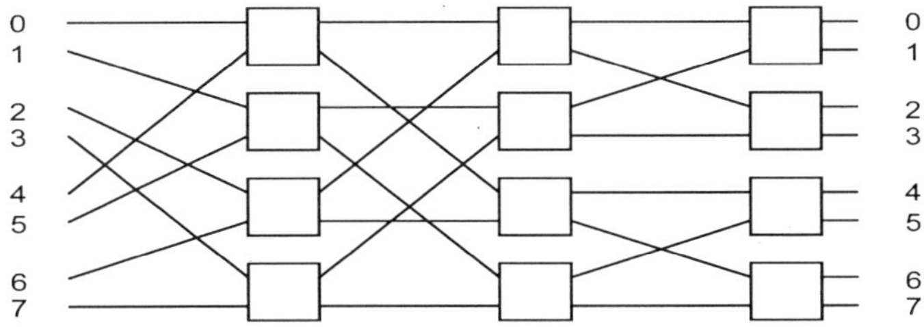


**FIGURE 3.17**
**Shuffled butterfly for $N = 8$ (or inverse indirect cube)**

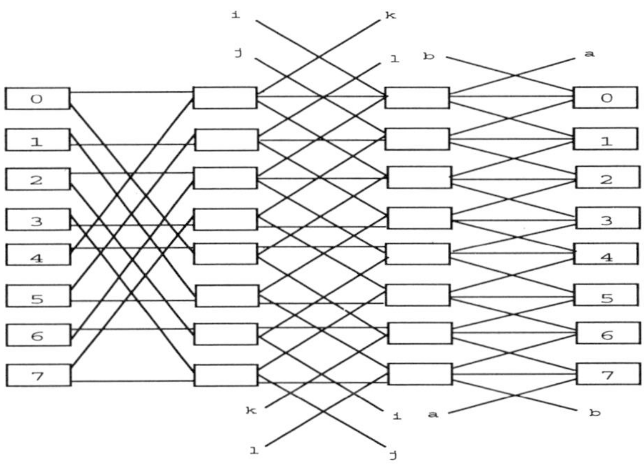


**FIGURE 3.18**
**The augmented data manipulator for $N = 8$**

$(-1)^{f_n} f_0 2^0 \pmod{N}$, where $F = (f_{2n-1}, \ldots, f_0)$ is a binary function, with each of the first $n$ terms representing a $(0,1)$ sign, and each of the last $n$ terms representing a magnitude. Thus, for example, if the source is $S = 13$ and $D = 6$, then $F_1 = 00000111$ and $F_2 = 00011001$ provide two disjoint routing paths between $S$ and $D$. At stage $i$ the switch examines the packet's $f_i$ and $f_{n+i}$ bits. If $f_i = 0$, then the packet is routed to the next stage by following a straight path. Otherwise, if $f_i = 1$ then, (a) if $f_{n+i} = 0$, the packet is routed to the next stage by following a $+2^i$ link, or (b) if $f_{n+i} = 1$, the packet is routed to the next stage by following a $-2^i$ link [466]. Notice that in ADM the reply can be routed similarly by just complementing the sign bits of the $F$ function.

By reversing the ADM, we obtain the *inverse ADM* (IADM) [582]. The Gamma network is an IADM network configured with $3 \times 3$ crossbar switches [525]. This allows all switch inputs to be connected to a unique switch output, concurrently. As a result, one obtains a much larger number of network states, which yield better performance and reliability. Notice that, as shown in the previous routing example, when the distance $D$ in a (source, destination) pair ($S$, $D$) is odd, there always exist two disjoint paths from $S$ to $D$. However, when their difference is a multiple of some $2^i$, all possible paths share some straight links. This property limits fault tolerance of ADM networks.

Several techniques allow for performance improvements over the simple, unique, or limited path-blocking networks. For example, buddy groups of switches can be replaced with larger switches which offer better performance. Multiple paths can also be created, either by providing additional switches, or even chaining switches at each stage. Also notice that by assigning PEs to each network switch, rather than just I/O terminals, one really obtains a point-to-point interconnection network which looks more like a MIN.

Minimal full access (MFA) networks are MINs with $N = 2^n$ input and output terminals which require a minimum number of switching elements and still provide full access capability. Examples of MFA networks include the butterfly and the Omega network. Any MFA network configured with $b \times b$ switches must have at least $\log_b N$ stages, since each stage increases the number of accessible outputs from a particular input by a factor of $b$. It is known that any MFA is also a banyan network.

Certain classes of topologically equivalent MINs are remarkably small. This raises the question on the number of nonequivalent MFA networks. In case of equal interstage link patterns, the MFA networks are called *uniform* MFA networks (UMFA). It can be shown that, for $N = 8$ the Omega network is wide-sense topologically equivalent to the UMFA [642]. For $N = 16$, there are three nonequivalent UMFA networks.

Patel also provided a nonuniform, nonequivalent MFA network with 8 inputs (see Figure 3.19). Notice that the buddy property for this network does not hold (examine the first level). In the case of such nonuniform MFA networks, the number of nonequivalent networks is very large. Sridhar and Raghavendra enumerated exponentially many nonisomorphic MFA networks with $2 \times 2$ switches and provided procedures for testing nonequivalence [642].

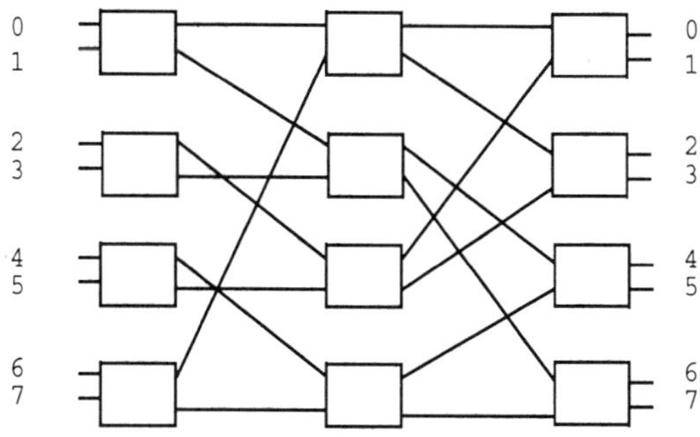


**FIGURE 3.19**
**A nonuniform, nonequivalent MFA network**

### 3.4.4 MIN Performance - Static and Dynamic Patterns

Certain classes of permutations may be self-routed, i.e., routed without conflicts, using an online destination tag routing algorithm [418]. Such permutation classes include, the BPC, Omega, and inverse Omega families [500, 566], as well as supersets of BPC, such as the family of linear transformation and complementation permutations (LTC) [612].

One may also consider the design of networks which pass given permutations. MIN design is usually based on the buddy property which pairs switches connected to each other in consecutive network stages [14] and is closely related to network partitioning [633].

### 3.4.5 Permutation Capability - Static Traffic

The *permutation capability* of a circuit-switched MIN is defined as the number of permutation requests from all inputs to all output that can be routed in edge-disjoint manner. The *combinatorial power* refers to the fraction of permutations that can be routed in edge-disjoint manner. These metrics are often used as an indication of the MIN data routing capability.

Notice that an $n \times n$ crossbar switch can implement any of the $n!$ permutations in constant time. Thus, it's combinatorial power is unity. Also, the permutation capability of $N = f^L$-node, regular SW-banyans configured with $L$ stages of $f \times f$ switches, is $f^{Lf^{L-1}}$ [107, 589]. The proof uses the fact that, due to the unique path MIN property, each crossbar setting corresponds to a distinct permutation. Thus, using $f = 2$ and $L = \log N$ for the Omega network, and blocking MINs topologically equivalent to it, we see that only $N^{N/2}$ can be realized using edge-disjoint paths from the $N!$ possible permutations. For example, if $N = 8$, only $8^4/8! = 10.16\%$ of the permutations can be realized on the 8-node Omega network in a conflict-free manner.

The permutation capability of nonrectangular SW-banyans is considered in [589]. Also, the permutation capability of extra-stage rectangular SW-banyan networks, such as the extra-stage Omega network, relates to an enumeration problem of partial subgraphs of hypercubes [222]. Despite the symmetry of the hypercube topology, no closed form analytical expression exists, except the trivial $N^{N/2}$ lower bound which corresponds to the Omega network, and the $2N^{N/2}$ upper bound which corresponds to the maximum number of settings of switches for the extra-stage network. Similarly, for $N \leq 2048$, only relatively good bounds for the permutation capability of the $N$-node ADM network are known [9], although the exact number of distinct permutations within each network stage has been obtained in closed form. This computation relates to an enumeration of $N$-bit binary numbers with no circularly adjacent 1's. An empirical evaluation has shown that the ADM MIN has better permutation capability than the Omega network [9].

Furthermore, considering multipath blocking packet-switched MINs configured with extra stages, replicated $k \times k$ switches, or $k \times k$ switches with dilated ports, an important issue is to compute the permutation capability. Szymanski and Hamacher empirically developed a relation between the blocking probability of the multistage network and its permutation capability [657, 658]. Assuming statistically independent permutations or random destination requests for large networks, the combinatorial power $C_N$ of an $N$-node multistage network can be related to the blocking probability $p_b$ as $C_N \sim 1 - P_b^n$ [658]. Notice that the latency may be computed as $(1-p_b) + 2p_b(1-p_b) + 3p_b(1-p_b)^2 + \ldots$ which evaluates to $1/(1-p_b)$. The probability $p_b$ can be approximated either experimentally, or analytically as in the case of $d$-dilated packet-switched banyans with buffers of size one and randomized routing, when $p_b \sim \frac{\log_b N}{d^d}$ [656]. Notice that under the assumption that links block independently, the bounding methods developed in [255] apply equally well to blocking or failure. Thus, these techniques, initially targeted to estimate reliability, may prove useful in obtaining estimates of permutation capability without requiring simulations to determine the blocking probability.

Notice that multistage networks, such as the Omega network, can provide conflict-free circuit-switched simultaneous access from $N$ processors to $M$ memory modules ($M > N$) for accessing row, column, diagonal, reverse diagonals, or cross section elements of an $N \times N$ allocated matrix [389]. Memory skewing schemes provide necessary gaps in allocating matrix elements into memory modules. Network switches which are not involved in the memory skewing allocation scheme may be neglected for all routing purposes.

### 3.4.6 Bandwidth and Latency - Dynamic Traffic

Wu and Lee considered the performance analysis of *circuit-switched MINs* using Markov chain analysis [708]. However, the combinatorial explosion of Markov states in modeling edge-disjoint paths makes this technique too complex to use. Furthermore, the study of circuit switched MINs has gradually diminished since various packet routing techniques for MINs have been proposed and used.

Thus, most past analytical performance studies concentrate on *buffered packet-switched MINs.* Notice that when the buffer size is only one, the MIN is usually called an unbuffered packet-switched MIN in the literature, instead of a single-buffered packet-switched MIN.

#### 3.4.6.1 Single-Buffered Packet-Switched MINs

Generally, the analytical models described below assume: (a) statistically independent input requests, (b) a constant message generation rate, and (c) (usually) a synchronous MIN operation, i.e., all switches use a synchronized global clock.

A number of authors considered the description of MIN performance in terms of bandwidth requirements, usually related to interleaved memory [260, 583]. In order to develop the bandwidth performance measure for a generic MIN with $p$ PEs and $m$ memory modules, note that the probability for a PE $P_x$ accessing the memory module $M_y$ is $1/m$ (for all $x, y$-combinations) [704]. It follows that the probability that no PE makes a request for access for the memory module is $(1 - 1/m)^p$, and the probability that the memory module has at least one request is $(1 - (1 - 1/m)^p)$. Now, the bandwidth of a single stage crossbar switch (equal to the number of memory modules with at least one PE request) will be

$$\mathrm{BW} = m\,(1 - (1 - 1/m)^p)\,. \tag{3.1}$$

From Eq. 3.1 it is easy to prove asymptotic linearity in either $p$ or

$m$, for a constant ratio $p/m$ [37]. One generalization of Eq. 3.1 can be obtained by assuming that PEs do not always generate the memory module requests during bus cycles. If $r$ denotes the probability that a PE makes a request, then $r/m$ is the probability that a PE tries to access the memory module $M_y$. In this case:

$$\mathrm{BW} = m\,(1 - r/m)^p \tag{3.2}$$

is the general form for the bandwidth of a single crossbar switch in a MIN. A similar approach is described in [528]. Notice that if only one switch output terminal is active at a time, as in ADM, IADM, or Kappa network, the switch bandwidth computed above must be reduced by a factor of $m$.

Since failed packets are subsequently retransmitted, the above formula overestimates the network bandwidth. However, the relatively small maximum error involved in this approximation (8%, for $m/p > 0.75$), has made possible an extensive use of the above formula for many packet-switched MINs. Improved regeneration techniques either use Markov chains, or appropriately increase the input request rate to account for resubmission of rejected requests [131, 429, 614, 716].

Bhuyan has provided a simple extension of this basic model to nonuniform memory traffic. This new model would model better memory traffic arising from parallel application programs. Let $m_0$ be the rate to hot-spot memory [67]. Since $m_0 + (k-1)m_1 + (n-k)m_2 = 1$, the memory bandwidth can be obtained using the following equation:

$$BW = m\left[1 - [1 - rm_0]\,[1 - rm_1]^{k-1}[1 - rm_2]^{n-k}\right].$$

Also, if $r = 1$, Bhandarkar used an exact analysis by modeling outstanding requests using Markov chain analysis to compute a more accurate empirical estimate of the bandwidth as $BW = m'(1-(1-1/m')^{p'})$, where $m' = \max(n, m)$ and $p' = \min(n, m)$ [64].

The acceptance probability for an arbitrary request is then given by:

$$P_A = \frac{\mathrm{BW}}{rp} = (m/rp)(1 - (1 - r/m)^p)\,. \tag{3.3}$$

Similar to the buffered networks discussed in Section 2.4 for single-buffered MINs, switches on a given network path are interdependent, while switches on different network paths are independent. Thus, one may obtain a set of equations for computing the bandwidth (and latency) of a unique path MIN. If the MIN consists of $a \times b$ crossbar switches (in each stage), the number of access requests forwarded to the next stage is given by $b[1-(1-r_0/b)^a]$, where $r_0$ is the rate of requests at the input of the first stage. Hence, in general, the recursive expression for the number of access requests passed from the $i^{th}$ stage to the $i+1$-st stage of an $n$-stage MIN is $r_i = 1-(1-r_{i-1}/b)^a$, where $i = 1, 2, \ldots, n$ and $r_0 = r$. The bandwidth is given by:

$$\text{BW} = b^n r_n \ , \tag{3.4}$$

since the total number of outputs is $b^n$. Since there are $a^n$ inputs, the probability of acceptance of a request is

$$P_A = \frac{b^n r_n}{a^n r} \ . \tag{3.5}$$

Performance measures for dynamic MINs in terms of bandwidth and probability of acceptance have been obtained in [106, 360, 528]. For an $m$-stage, unique path MIN configured with $k \times k$ switches, the acceptance probability is $P_A = r_m/r_0$, where $r_0$ is the input rate at the first stage and $r_m$ the corresponding rate after first the $m$ stages.

Kruskal and Snir asymptotically approximated $r_m$ with a closed form formula as, $r_m \sim \frac{2k}{(k-1)m}[1 - \frac{(k+1)\log m}{s(k-1)m} + \mathcal{O}(1/m)]$.[1] Furthermore, for a relatively small $r_0$, they have provided a good empirical approximation of $r_m \sim \frac{2k}{(k-1)m} + \frac{2k}{r_0}$. Thus, the probability that a message is successfully transmitted to destination is inversely proportional to the number of stages.

Cherkassky generalized the above empirical approximation for nonrectangular MINs configured with $s \times f$ switches ($s < f$), to compute $P_A = r_m f^m/(r_0 s^m)$ and $r_m = \frac{2s}{(s-1)\sum_{i=0}^{m-1}(\frac{f}{s})^i + \frac{2s}{r_0}(\frac{f}{s})^m}$ [106]. Notice that for large networks, i.e., when $f^m$ is large, $P_A \sim [1 + \frac{(s-1)p}{2(f-s)}]^{-1}$, which is a nonzero quantity. Extensions of Bhuyan's analytical nonuniform memory access models to MINs result in a set of seven nonlinear equations per each MIN stage [67].

The effect of reliability on synchronous MIN performance has been considered in [75]. For nonbanyan, redundant path networks, the presence or absence of messages at the inputs of any switch is dependent on the presence or absence of messages at the other inputs of the switch.

[1] This approximation is actually an upper bound.

For such networks, Sobalvarro proposed probabilistic techniques for computing the exact blocking probability, by considering the joint probability function of the load on channels that lead to output terminals [639]. However, since the number of equations increases with the number of communication channels entering a stage, an exact solution can be obtained only for very small irregular, or small symmetric networks. Thus, one promising direction might be to derive approximate solutions to these exact equations, possibly using simulated annealing, or Monte Carlo approximations [639].

The performance of *asynchronous packet-switched* MINs can be examined by assuming a random delay between injecting processor requests. In this case, processors are modeled as delay servers, and memory modules as FIFO queues, with buffers provided at the input of each output link of a switch. Packets are transmitted asynchronously, one at a time, on first-come first-serve basis, and either infinite queues [306], or finite queues are assumed [481].

In contrast to previous studies, using queuing models with deterministic service time, Mohapatra proved that a small buffer size provides performance equivalent to an infinite buffer only under light traffic, such as when we restrict input rate to one outstanding request per PE in the MIN. Multiple outstanding requests, e.g., possibly caused during nonblocking communication, increase network traffic, and then the buffer size needs to be large enough to mimic infinite buffer performance [481]. Furthermore, while large buffers are more suitable for throughput-oriented systems, smaller buffers are better when response time is more critical.

#### 3.4.6.2 An Extension from Packet-Switched Static MINs

In [4], the bandwidth and the probability of acceptance applied to a binary (static) $n$-cube describe a performance measure. It is obvious that such performance measures based on solving network equations, e.g., Kirchoff's law, are very useful in analyzing MIN performance as well. Thus, in order to generalize that approach to any static MIN we denote by $r$ the probability that a MIN node can generate an *internal request* for another node [704]. At each node, there can be only one internal request, and an arbitrary number of *external requests* (bounded from above by the number of incident links). Denote by $r_{\text{out}}$ the probability that a node makes a request, either internally, or that it forwards on an external request for memory access. Denote by $r_{\text{in}}$ the probability that a node receives an external request. Let $A$ denotes the percentage of the external requests forwarded on to another node. It follows that:

$$r_{\text{out}} = r + A\, r_{\text{in}}\, (1 - r)\,, \tag{3.6}$$

while the bandwidth in an $N$-node MIN is now:

$$\text{BW} = (1 - A)\, r_{\text{in}}\, N\,. \tag{3.7}$$

Assuming the regularity of the MIN (all nodes have the same degree $n$), and assuming the usage of all links, the probability that a node $y$ has not requested another node $x$ is $(1 - r_{\text{out}}/n)$. On the other hand, the probability that no node has requested $x$ is $(1 - r_{\text{out}}/n)^n$. Hence, the probability that $x$ has at least one external request is

$$r_{\text{in}} = 1 - (1 - r_{\text{out}}/n)^n\,. \tag{3.8}$$

The probability of a node generating an access request, in terms of its degree, and in terms of the probability of an internal request is given by a recursive expression:

$$r_{\text{out}} = r + A(1 - r)(1 - (1 - r_{\text{out}}/n)^n)\,, \tag{3.9}$$

where a convenient initial $r_{\text{out}}$ can be $r$.

## 3.5 Permutation and Nonblocking Networks

### 3.5.1 Beneš Permutation Network

*Rearrangeable networks*, also called *permutation networks*, realize edge-disjoint paths to admit any arbitrary permutation. Input permutation requests are assumed to arrive in a batch. Each batch of connection requests is handled independently of previous ones, by establishing a new set of edge-disjoint paths which realizes these requests. This may involve tearing down and rebuilding any of the existing connection paths.

A conflict-free permutation connection assignment which is realized by a circuit-switched network is called an *admissible permutation*. The class of all admissible permutations is of interest to the designer of parallel algorithms and networks. Several families of admissible permutations have been studied, e.g., BPC, Omega, affine, and linear; these classes contain common communication patterns, such as shifts and bitonic (see Chapter 2).

A permutation network is called *cost-optimal* if it contains the smallest number of switches possible. Shannon's argument provides the minimum number of switches necessary to realize the symmetric group $\Sigma_N$ of all permutations on $N$ letters as $\log_2 N!$, or roughly $N log_2 N$ switches [615]. For references to exact formulas and interesting discussions on lower bounds see [548]. Notice that shifter networks which can arbitrarily shift data have the same asymptotic complexity in the size and depth as permutation networks.

A permutation network is called *path-length optimal* if the length of its longest path is as short as possible. If a permutation network is configured with switches of size $b \times b$, then since the number of nodes connected by paths of length $m$ can not be greater than $b^m$, and each input must reach all outputs, the minimum path length is $\min m \geq \log_b N$.

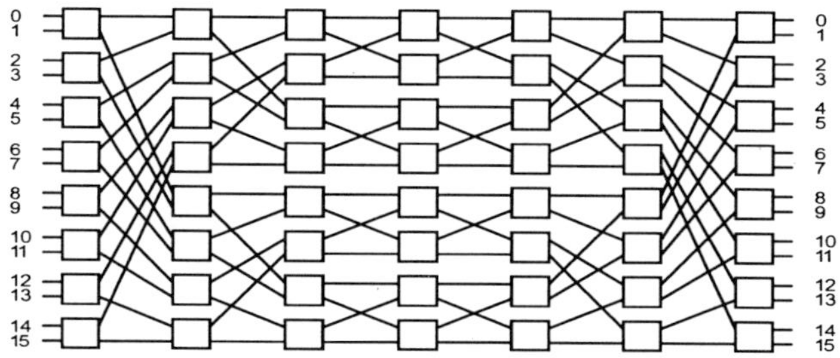


**FIGURE 3.20**
**Beneš permutation network for** $N = 16$

The Beneš permutation network takes its name from Beneš who discovered it in 1962 [53, 54]. In Figure 3.20 we show a 16-input, Beneš network, designed by placing in tandem a butterfly together with its inverse. Notice that the two identical intermediate levels (third and fourth for the network in Figure 3.20) can always be collapsed to a single level. As we show in the following section, this network can be recursively constructed from a more general three-stage Clos network. The algorithm for setting the network switches to realize a given permutation follows the recursive construction and performs $\mathcal{O}(\log N)$ steps. As shown in Figures 3.21, 3.22, and 3.23, each step requires: (a) realizing switch settings for the two atomic stages, i.e., the first and last network stages

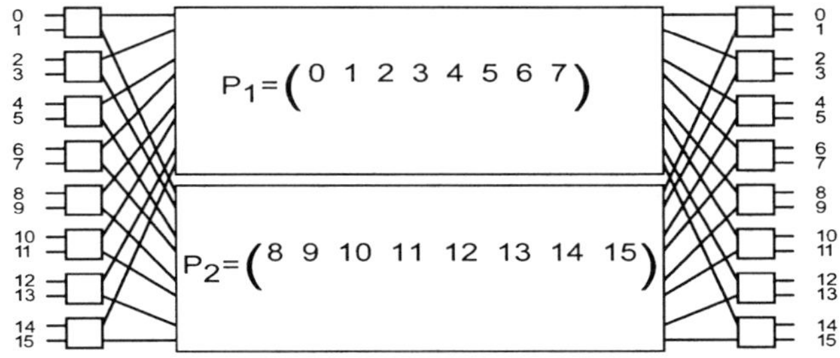


$$P = \begin{pmatrix} 0 & 1 & 2 & 3 & 4 & 5 & 6 & 7 & 8 & 9 & 10 & 11 & 12 & 13 & 14 & 15 \\ 5 & 15 & 12 & 7 & 3 & 0 & 8 & 11 & 9 & 6 & 13 & 14 & 10 & 2 & 1 & 4 \end{pmatrix}$$

**FIGURE 3.21**
**Switch settings on a Beneš network (step 1)**

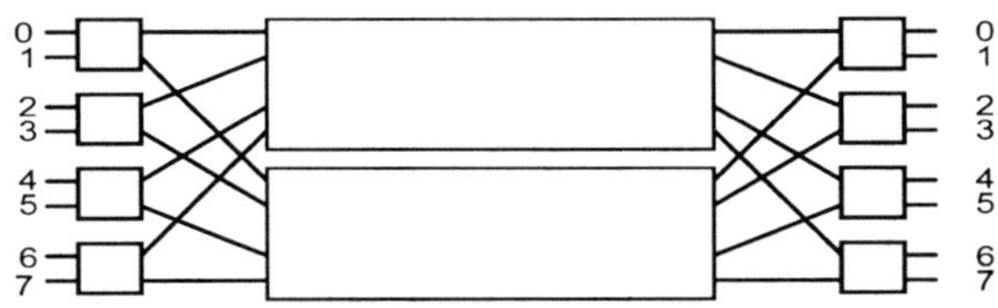

$$P_1 = \begin{pmatrix} 0 & 1 & 2 & 3 & 4 & 5 & 6 & 7 \\ 2 & 6 & 1 & 4 & 3 & 7 & 5 & 0 \end{pmatrix}$$

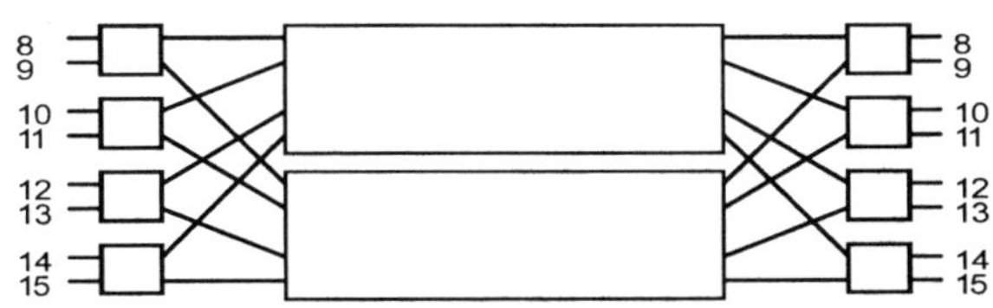

$$P_2 = \begin{pmatrix} 8 & 9 & 10 & 11 & 12 & 13 & 14 & 15 \\ 15 & 11 & 8 & 13 & 12 & 14 & 9 & 10 \end{pmatrix}$$

**FIGURE 3.22**
**Switch settings on a Beneš network (step 2)**

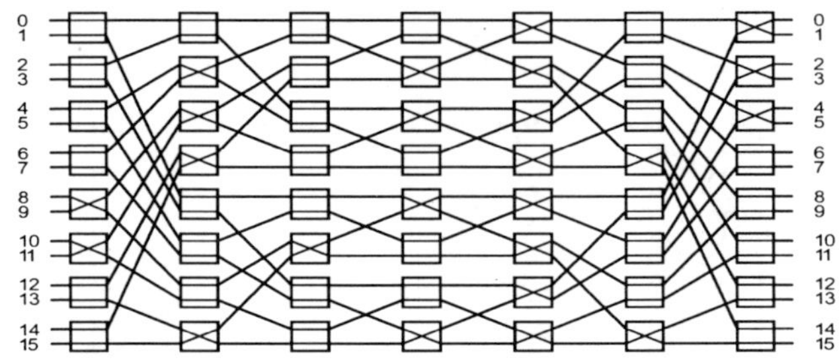


$$P = \begin{pmatrix} 0 & 1 & 2 & 3 & 4 & 5 & 6 & 7 & 8 & 9 & 10 & 11 & 12 & 13 & 14 & 15 \\ 5 & 15 & 12 & 7 & 3 & 0 & 8 & 11 & 9 & 6 & 13 & 14 & 10 & 2 & 1 & 4 \end{pmatrix}$$

**FIGURE 3.23**
**Switch settings on a Beneš network (final step 3)**

(see Figure 3.26), and (b) obtaining a new permutation to be realized by the intermediate stage using the recursive construction [698].

The Beneš network is topologically equivalent to several other networks, since the (inverse) butterfly can be replaced with an (inverse) Omega, baseline, reverse-shuffle, indirect binary cube, or other bidelta MINs. Two such networks are shown in Figure 3.24.

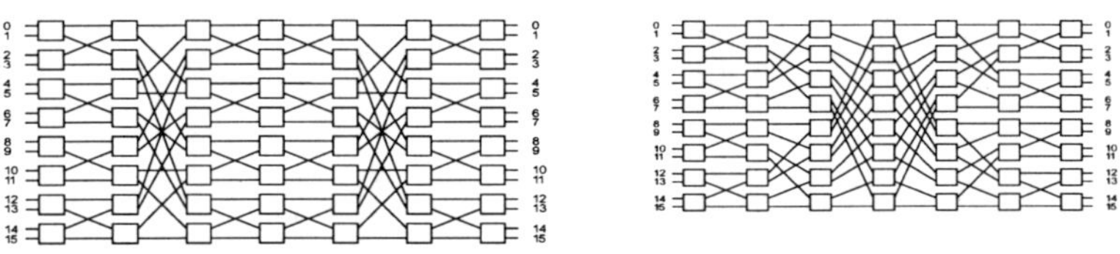

**FIGURE 3.24**
**MINs topologically equivalent to Beneš network**

Finally, an interesting question is whether an $N = 2^n$-node, $2n-1$-stage shuffle-exchange network is rearrangeable. This MIN is formed by concatenating two Omega networks and collapsing the two identical intermediate levels to one. Initially, progress in this problem was slow. At first, Parker showed that $3n$ shuffle-exchange stages are sufficient for rearrangeability [524]. Wu and Feng improved this to $3n-1$ stages [707], and later various researchers improved this to $3n-3$ stages [288, 347]. Then, using a difficult case by case analysis, Varma and Raghavendra showed that it is possible to realize all permutations on an 8-node, 5-stage shuffle-exchange network [567]. They, also subsequently proved that, if a $2^r$ node, $2\log r - 1$ stage shuffle-exchange network is rearrangeable, then a size $N > 2^r$, $3\log_2 N - (r+1)$-stage shuffle-exchange network is also rearrangeable, thus improving the previous best result marginally [687]. A breakthrough came when Abdennadher and Feng proved that the $N$-node, $2\log N - 1$-stage shuffle-exchange network is always rearrangeable [1]. Simpler routing algorithms for Omega-Omega rearrangeability were given in [613].

### 3.5.2 Other Permutation Networks

Self-routing permutation networks with slightly less cost than Batcher's sorting network can be constructed by using a number of 1-sorter devices [346]. A 1-sorter network can classify messages into two classes: high and low destinations. A self-routing 1-sorter uses a separate ranking circuit, which is implemented as a binary tree consisting of adder slices, followed by an indirect binary cube based on $2 \times 2$ communication switches [346]. The ranker network computes the rank of a given message by counting messages which must move to the top output port (in left to right order). It can be proved that the indirect binary cube can self-concentrate any of its inputs into consecutive outputs using ranking information only. Thus, self-routing control of the 1-sorter network is straightforward.

Counting elementary devices, such as binary adders and comparators, the cost of the 1-sorter with $N = 2\,n$ inputs is $\mathcal{O}(n\ \log_2 n)$. This cost can be further reduced to $\mathcal{O}(n)$ by using extensive time multiplexing [302]. Furthermore, it is simple to construct a self-routing $s$-sorter, which can separate $(s\ n)$ inputs into $s$ blocks of $n$ messages each. The cost of the $s$-sorter is $\mathcal{O}(n\ s\ \log_2 s)$.

### 3.5.3 The Clos Family of Networks

Clos networks were first described by Clos in 1950s [122]. The family of Clos networks of size $N$, which as a special case contains the Beneš permutation network, is constructed as follows (see Figure 3.25).

- $N/m$ input switches of size $m \times m'$ are used at stage I.
- $m'$ switches of size $\frac{N}{m} \times \frac{N}{m}$ are used at intermediate stage II.
- $N/m$ switches of size $m' \times m$ are used at stage III.
- The $j^{th}$ output link of every switch in stages I and II is connected to the $j^{th}$ switch at the following stage.

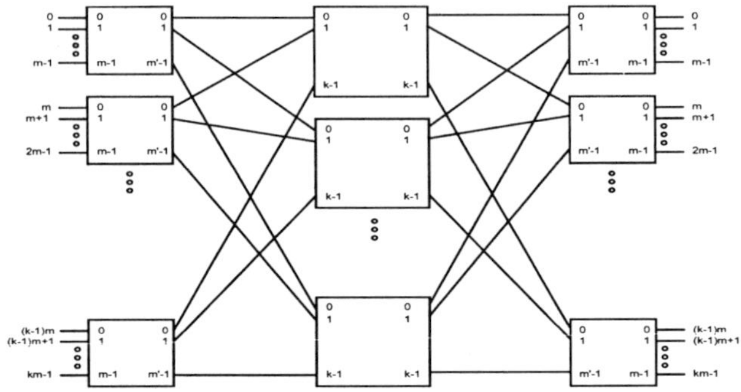


**FIGURE 3.25**
**A general, three-stage Clos network ($k = N/m$)**

As shown in Figure 3.26, Clos networks can be recursively constructed by allowing for a recursive intermediate stage. This extension leads to $2t+1$-stage Clos networks ($t \geq 1$) with just $\mathcal{O}(N^{1+1/t})$ crosspoints. The optimal choice for $m$ satisfies $N = \mathcal{O}(m^{t+1})$. This number of crosspoints is smaller than the $N^2$ crosspoints required by the $N \times N$ crossbar for $N \geq 36$ [122].

Clos networks provide a framework for designing blocking, rearrangeable, and also nonblocking MINs, by directly relating combinatorial power to the relative values of $m$ and $m'$.

### 3.5.3.1 Blocking Clos Networks

***LEMMA 3.1***
*Any three-stage Clos network with $m' < m$ is blocking.*

**PROOF** The proof follows easily from the observation that at the first stage some inputs may be lost. ∎

By assuming that all first stage Clos switches are set to pass, each internal switch setting corresponds to a distinct permutation which may be self-routed, i.e., routed based on destination tag only. Thus, the class

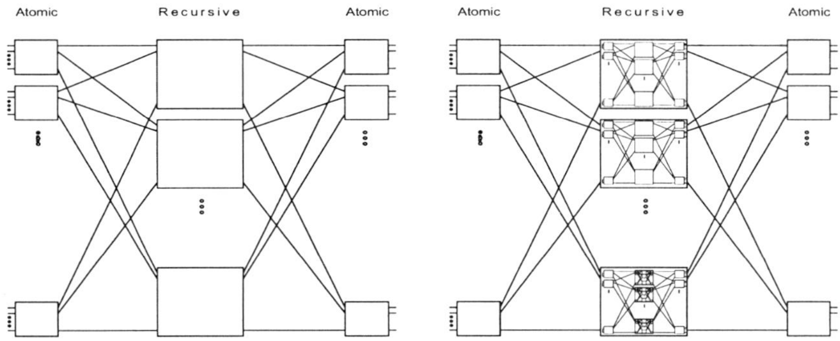


**FIGURE 3.26**
**Unfolding the recursive construction of Clos network**

of self-routed permutations is at least $m' \cdot (m/m')^{m'}$, which is minimum if $m' = \sqrt{m}$ [176].

### 3.5.3.2 Rearrangeable Clos Networks

***THEOREM 3.1***
*Any three-stage Clos network with $m \geq n$ is rearrangeable.*

**PROOF** The proof, due to Slepian (1952, unpublished) and Duguid (1959, technical report), uses Hall's theorem [247] and induction on network size to show that: (a) a single intermediate stage switch can always be routed (basis step), (b) routing on a general Clos network can be reduced to routing on a Clos subnetwork which consists of only the intermediate, nonatomic stage (induction step). ∎

The algorithm for setting the switches of a rearrangeable Clos network consists of a basic loop [515]. For binary networks, let the *dual* of a $2 \times 2$ switch input (output) be the other input to (respectively , output from) that switch. Then, the loop goes as follows:

- If all inputs have been routed, then exit. Otherwise, choose an unrouted request and set the first stage switch arbitrarily.

- *Continue loop*: Set the middle and last stage switches to realize the unrouted request. For the dual of the output just routed: set the middle stage switch (back towards inputs). If the first stage switch is already set, then go to start of the loop. Otherwise, consider the dual of the input, and backtrack again to the continue loop.

The above loop realizes a general permutation pattern in $\mathcal{O}(N)$ time. The time complexity of the above simple looping algorithm is at least as good as any routing algorithm which can be obtained from matching or edge-coloring [95, 290]. However, only special permutation patterns can be realized in $\mathcal{O}(\log N)$ time.

Notice that we can remove one input switch in a rearrangeable Clos network, by shortcuting the corresponding input-output links, without affecting its combinatorial power. For example, the Clos network shown in Figure 3.27 is rearrangeable, since $m' = m = 2$ [698].

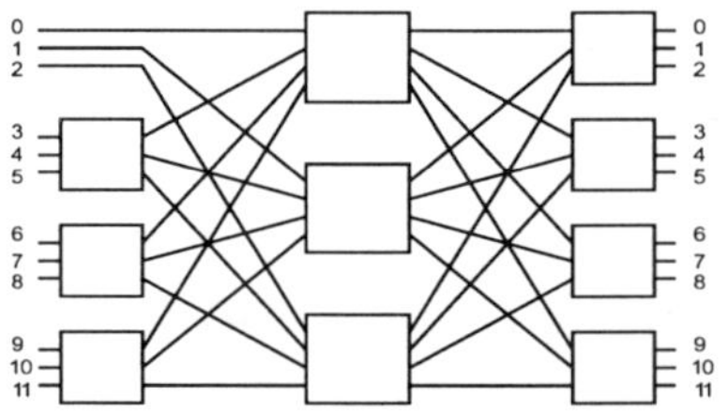


**FIGURE 3.27**
**A three-stage, reduced cost, rearrangeable Clos network**

The Beneš permutation network can be obtained from Clos networks by selecting a) $N = m \cdot r$, b) $m' = m$, and c) for the binary case, $m = 2$. Similar to Clos networks, the intermediate stage of the Beneš permutation network can be expanded recursively. In this case, the number of switches becomes $\mathcal{O}(N\ logN)$, which is asymptotically optimal. This is almost twice the cost of an Omega network, but much less than that of a nonrecursive Clos network.

#### 3.5.3.3 Strict-Sense Nonblocking Clos Networks

***THEOREM 3.2***
*Any three-stage Clos network with $m' \geq 2m - 1$ is strict-sense nonblocking.*

**PROOF** The "necessary" part considers finding a route for a request under worst case conditions. Since the request may be blocked by $m-1$ requests in the input links and also $m-1$ requests at the output links of an intermediate stage switch, a total of $2m-1$ intermediate stage switches are necessary to create at least one edge-disjoint path to the desired destination. The "sufficient" part follows from a straightforward argument which considers equally all available routing paths. ∎

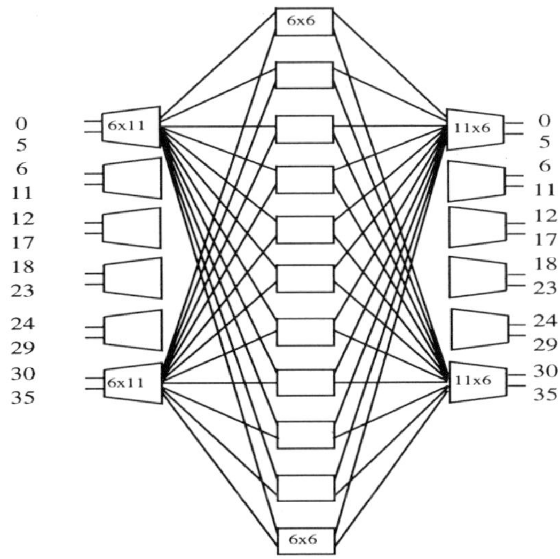


**FIGURE 3.28**
**A three-stage Clos network ($N = 36$, $m = 6$, and $m' = 11$) with some symmetric links omitted**

The number of crosspoints of a strict-sense nonblocking Clos network is less than that of a crossbar, but more than the $\mathcal{O}(N \log N)$ of the banyan. An example of a strict-sense nonblocking Clos network with $N = 36$, $n = 6$, and $m = 11$ is shown in Figure 3.28.

### 3.5.4 Other Nonblocking Networks

The Cantor strict-sense nonblocking network of size $N$ is constructed by stacking together $Q = \log_2 N$ identical $N$ input, binary Beneš permutation networks, each configured with square switches [94]. In addition, there are two multiplexing/demultiplexing stages providing the input-output connections to/from each of the $Q$ Beneš networks, i.e., the first stage consists of a column of $N$ $1 \times Q$ multiplexing switches, while the last stage consists of a column of $N$ $Q \times 1$ demultiplexing switches. Each of these additional stages distributes its links to all $Q$ Beneš networks. Deterministic and probabilistic routing on Cantor nonblocking networks has been considered [430]. Similar extensions to rearrangeable and non-blocking networks based on stacking (and cascading) nonbinary Beneš permutation networks have been considered in [390, 620, 621].

While efficient sorting networks can also be used for nonblocking routing, the randomly wired multistage structure of the multibutterfly with optimal $\mathcal{O}(N \log N)$ cost and $\mathcal{O}(\log N)$ depth and nice fault tolerant properties [45, 679, 407] has inspired researchers to further examine randomly wired networks for approximate nonblocking routing in parallel systems [451]. Empirical probabilistic results for large randomly wired circuit-switched networks are encouraging [436].

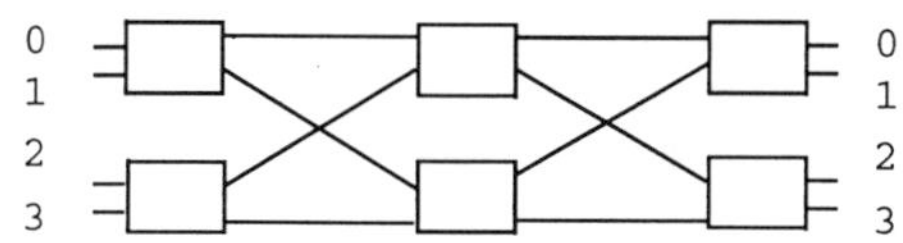


**FIGURE 3.29**
**An example of a nonblocking non-Clos network**

In a multirate MIN, we assume that each link has a capacity of one packet, while each connection request is associated with a bandwidth weight, usually in $[0, 1]$. Conditions for achieving rearrangeability and nonblocking properties, as well as routing algorithms on multirate Clos-type networks are discussed in [119, 179, 220, 469, 470]. Notice that *not* all nonblocking networks belong to the Clos family. A simple example is shown in Figure 3.29.

## 3.6 Generalized Connectors

A generalized connector with $m$ input and $n$ output ports has the following properties:

- Each output port can independently be selected by any input port, i.e., output destinations for several (and even for all) connection requests may be identical, and

- All connection requests can be set up so that no network link carries more than one signal.

Generalized connectors are useful for multiparty calling in telephone networks, network partitioning, and multicasting in parallel and distributed systems.

A *strict-sense nonblocking generalized connector* (also called strict-sense nonblocking broadcasting network) can provide a path for any connection request among free ports without disturbing existing connection paths; if several paths are available, any path may be selected without being concerned for future connections. Notice that a crossbar is a strict-sense nonblocking generalized connector with non-optimal cost.

A *wide-sense nonblocking generalized connector* provides a connection path among free ports if (in general) several existing requests are rerouted; thus, a control algorithm provides the new path selections based on the existing connections.

A *rearrangeable generalized connector* can realize any static connection request from any input to any free output by rearranging existing I/O paths.

Obviously, a multicasting single-stage crossbar switch is the simplest form of a generalized connector. Less expensive generalized connectors can be constructed using a three-stage Clos network equipped with broadcasting switches [290, 463, 715]. In a broadcast switch, a single input may connect to both outputs. As shown in Figure 3.25 (with unicasting switches (crossbars) replaced by broadcasting switches), the general three-stage generalized connector is constructed as follows.

- $N/m$ broadcast switches of size $m \times m'$ are used at stage I.
- $m'$ broadcast switches of size $\frac{N}{m} \times \frac{N}{m}$ are used at intermediate (recursive) stage II.
- $N/m$ broadcast switches of size $m' \times m$ are used at stage III.
- The $j^{th}$ output link of every switch in stages I and II is connected to the $j^{th}$ switch at the following stage.

The most important properties of Clos-type generalized connectors relate their combinatorial power to the relative value of $m$ and $m'$ [715].

### Rearrangeable Generalized Connector

**THEOREM 3.3**
*Any three-stage Clos network with $m' \geq N$ is a rearrangeable generalized connector.*

### Wide-Sense Nonblocking Generalized Connector

**THEOREM 3.4**
*Any three-stage Clos network with $m' > \min\{(m-1)[x + (N/m)^x]\}$, where $1 \leq x \leq \min\{m-1, N/m\}$ is a wide-sense nonblocking generalized connector.*

The choice of $x$ determines cost and performance. The crosspoint cost of a (2t+1)-stage network is $\mathcal{O}(N^{\frac{t+2}{t+1}} \frac{\log N}{\log\log N}^{\frac{t+2}{2} - \frac{1}{t+1}})$; $m' = \mathcal{O}(\frac{m \log N/m}{\log\log N/m})$ minimizes the cost. A new request can be routed in $\mathcal{O}(N)$, while a full connection assignment can be routed in $\mathcal{O}(m\ N)$ time.

### Strict-Sense Nonblocking Generalized Connector

**THEOREM 3.5**
*Any three-stage Clos network with $m \geq (N + n - 1)$ is a strict-sense nonblocking generalized connector.*

A different design of rearrangeable generalized connectors is based on a network realized by connecting together three different subnetworks (see Figure 3.30) [511, 662]. Thompson referred to the first two subnetworks, described below, together as a generalizer network.

- A *packing network* routes active input requests to consecutive outputs, preserving order. If self-routing is required, then this network can be implemented as a sorting network. Otherwise an inverse Omega network can be used; notice that an Omega network can realize any spread connection assignment.

- A *copying network* routes consecutive inputs to multiple output links, preserving order. This network can be implemented using an Omega network equipped with broadcasting switches.

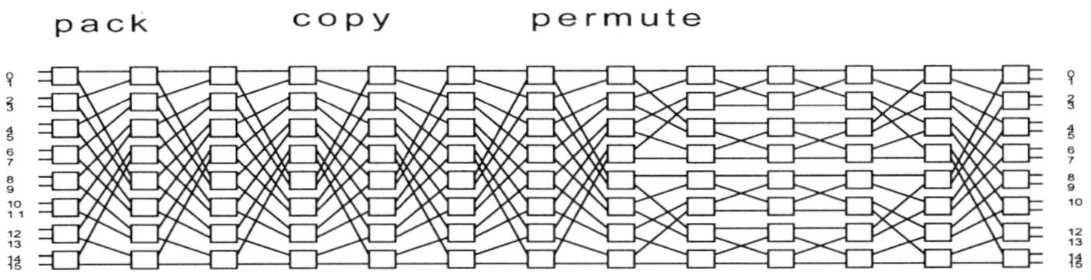


**FIGURE 3.30**
**Generalized connector: pack, copy, and permute networks**

- A *permutation network* permutes all messages to their desired destinations. For example, Beneš permutation network may be used. The setup time for switches in this network is obviously the overall network's bottleneck.

- An identity permutation is used for interconnecting the three subnetworks.

Following the above construction, Thompson's generalized connector with $3 \times 3$ switches uses a minimum of $7.6N \log_2 N$ switches. Chung and Wong have provided a generalizer configured with $2 \times 2$ switches (see Figure 3.31) which has smaller depth, thus reducing the cost of a generalized connector to $5.8N \log_2 N$ [118].

Similar to Beneš permutation network, identical intermediate stages between the three subnetworks of the generalized connector can be collapsed. Thus, the generalized connector with $N = 2^n$ inputs shown in Figure 3.31, requires $(4n - 3)\ 2^{n-1}$ switches of size $2 \times 2$ [715].

Nassimi and Sahni use a different design which consists of three subnetworks. The first subnetwork combines multicasting with copying packets, the second network concentrates inputs, while recursion is exploited in the third subnetwork [503]:

- A *generalizer network* which makes a copy of each input for each *concentrator network*. This network can be implemented with a $b$-ary tree, for any given integer $b > 1$.

- $2^m$ *concentrator networks*, which concentrate active inputs to outputs. This network can be obtained from a general $N$ input, $N$ output concentrator network by removing any switches leading to unused outputs [662].

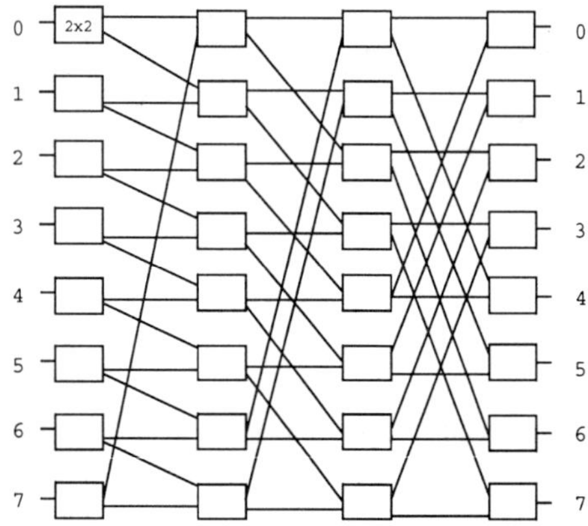


**FIGURE 3.31**
**Generalizer network: packing and copying network**

- $2^m$ smaller *generalized connector* networks, each with $N/2^m$ inputs and outputs.

- An identity permutation is used for interconnecting the three subnetworks.

Nassimi and Sahni's generalized connector has a total of $N = 2^n$ inputs, $\mathcal{O}(k \ N^{1+1/k} \ \log N)$ $2 \times 2$ switches, $\mathcal{O}(k \ \log N)$ depth, and $\mathcal{O}(k \ \log N)$ routing time, for any $1 \leq k \leq \log N$. Similar ideas have been used in Lee and Oruc's construction which uses distributors, concentrators (built recursively using odd-even splitters), and smaller generalized connectors to provide a generalized connector with $N = 2^n$ inputs, $\mathcal{O}(N \ \log N)$ $2 \times 2$ switches, $\mathcal{O}(\log^2 N)$ depth, and $\mathcal{O}(k \ \log N)$ routing time [392]. Further lower bounds on generalized connectors, including expander constructions, are given in [172].

Finally, notice that Zegura has presented the first analytic model for evaluating the blocking probability in buffered generalized connectors which use the basic *series* and *parallel network design operators* [720].

An extension to *self-routing generalized connectors* may use a *prefix network*, which (a) creates a dummy packet for all inputs that do not request for a path, (b) computes a parallel scan on $N$ items, and (c) uses a sorting network to route each input request to a different output. For descriptions of these special purpose networks refer to Section 3.8.

## 3.7 Fault Tolerance, Reliability, and Diagnosis

### 3.7.1 Fault Tolerance and Reliability

Reliability of a multiprocessor system is defined as the ability to support a certain quality of service in a specified operational environment. It can be viewed as a form of fault tolerance enabling the system to survive loss of network components (nodes and links).

In general, component faults may be classified in three groups:

1. *Transient faults* which are nonrecurring and they persist for a short period of time.

2. *Intermittent faults* which are recurring and they last for a finite period of time.

3. *Permanent faults* which cause an irreversible faulty condition of the network.

In practice, apart from reconfigurability, two other metrics of faulty networks are useful: *reliability* and *availability*.

Reliability ($Rel(t)$) is defined as the probability that the system (in our case, a MIN) is operational during a time interval $(0, t)$. Reliability is related to connectivity and scalability; an unreliable system is definitely not extendable. This reliability metric is important for mission-critical systems as it assumes that repairs of systems are impossible.

On the other hand, sometimes in the event of failure repairs are feasible. Then, we define the availability ($Ava(t)$) as the average fraction of time the system (MIN) was operational in time interval $(0, t)$.

Reliability and availability are both based on two-state characterization of components, i.e., each network component is either operating or faulty, and the entire network is either operating or faulty. Usually for a single component the probability of failure, called *component unreliability* is $1 - e^{-t/\mu}$, where $\mu$ is the mean time between failures for the component.

*Performability* measures network operation in the presence of faults. Thus, performance-related reliability measures deal with degraded network operation, e.g., increased congestion, packet latency, and distance to destination when there is no loss (or limited loss) of system connectivity. Performability can be represented in an elegant way if certain restrictions are made. For example, if one assumes bistate (unreliable) network links $E$, and reliable vertices $V$, with known probability $Pr[S]$ of each state $S \subseteq E$, and known value $\Psi(S)$ of some performance measure $\Psi$, the expected performance measure can be expressed in the form

$$Perf(V, E) = \sum_{S \subseteq E} Pr[S]\Psi(S) .$$

Similar expression can be found for the reliability of a network under the assumption of bistate independent component failures. It is known that calculation of general network performability measures is a computationally intensive process in a majority of nontrivial cases. Approximate reliability models which are based on system decomposition into PEs, memory modules, and switches, and which do not rely on exact computationally expensive probabilistic Markov chains, are discussed in [157]. Other reliability estimation techniques have been proposed [123]. They involve either combinatorial transformations on networks, or Monte Carlo simulation techniques. One interesting problem is assessment of the reliability of MINs by conversion of MINs to a series of channel graphs. A *channel graph* is any set of all input-output paths in a MIN [255]. In other words, a MIN $V = V_0 \cup V_1 \cup \ldots \cup V_{k-1}$ with $V_0 = \{s\}$ (the *source*), and $V_{k-1} = \{t\}$ (the *terminal*) is called a channel graph (see Figure 3.32).

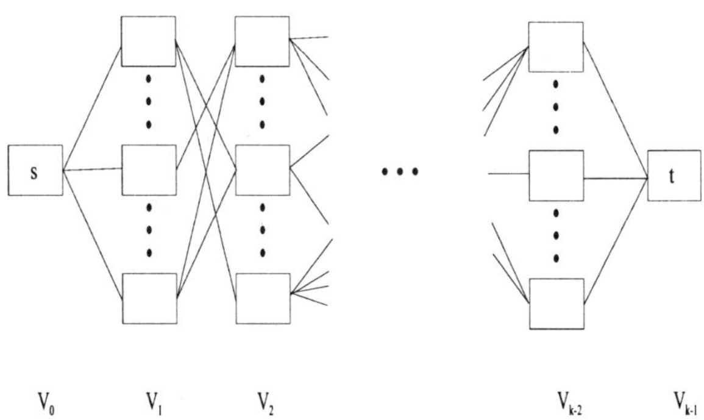


**FIGURE 3.32**
**A $k$-stage channel graph**

In [124, 255, 348, 349], a number of efficient methods for computing lower and upper bounds and Monte Carlo estimation of two-terminal reliability for channel graphs are presented. The model assumes (a fairly restrictive) reliable network vertices, and bistate, independent link failures. All of the methods used are computationally efficient, and do not exhibit the exponential growth in time required by known exact algorithms.

As MINs are collections (of mutually nondisjoint) channel graphs, one could hope that these bounding techniques can be applied to reliability assessment of the interconnection network. Lee proposed taking the average input-output probability as an estimate of the reliability of the whole MIN [391]. This approach has been extensively used in [76, 77, 86, 295, 341, 688].

For large MINs, regularity can be used to discover that many of the channel graphs in the network are in fact isomorphic. In this case, we calculate or bound reliability for a small number of channel graphs. In case of irregular MINs, we have to calculate (or bound) the reliability of every input-output channel graph. Since the reliability bound calculations are efficient for a single channel graph, they are efficient for the bounding of average probabilities of input-output connection assignments of all channel graphs in a MIN. However, it is still computationally intensive to compute (upper and lower) bounds for every input-output pair.

For upper bounds one cannot avoid the computation of a bound for each source-terminal pair. For lower bounds, one can bound the average input-output connection probability without calculation of a bound for any specific input and output [255].

We now define the *shifting transformation* on MINs. Let $G = (V, E)$ be a $k$-stage MIN with stages $V_0, \ldots, V_{k-1}$. $G$ may have multiple arcs (directed links) between the same pair of vertices. Denote by $p_{x,y}$ the operational probability of some arc from $x$ to $y$. When there is no arc from $x$ to $y$, $p_{x,y} = 0$, and in case of multiarcs from $x$ to $y$, $p_{x,y}$ is the probability that at least one of these operates. Let $w \in V_{j-1}$, and $z \in V_{j+1}$ be arbitrary vertices in stages $V_{j-1}$ and $V_{j+1}$, respectively. For $x, y \in V_j$ with $1 \leq j \leq k-2$, $shift(G, x, y)$ is the MIN obtained by:

(1) Deletion of all arcs from $w$ to $x$ and $y$, and addition of a new arc $(w, x)$ having operational probability $\min\{p_{w,x}, p_{w,y}\}$, and a new arc $(w, y)$ having operational probability $\max\{p_{w,x}, p_{w,y}\}$.

(2) Deletion of all arcs to $z$ from $x$ and $y$, and addition a new arc $(x, z)$ having operational probability $\max\{p_{x,z}, p_{y,z}\}$ and a new arc $(y, z)$ having operational probability $\min\{p_{x,z}, p_{y,z}\}$.

Arcs whose operation probability is zero need not be retained in this transformation. Note that this transformation locally changes the MIN in such a way that the least reliable arcs entering $x$ and $y$ are paired

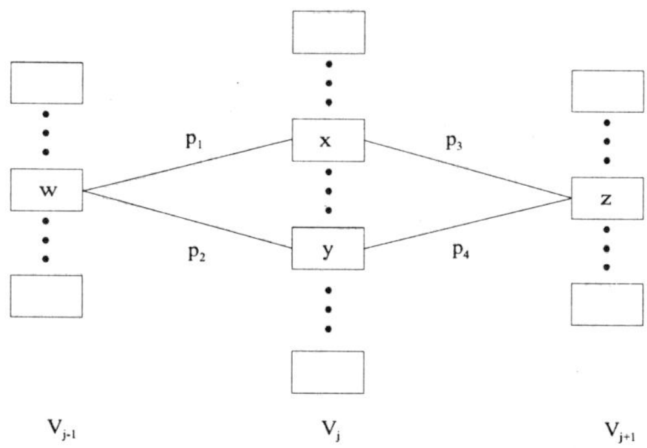


**FIGURE 3.33**
**Channel graph before shifting**

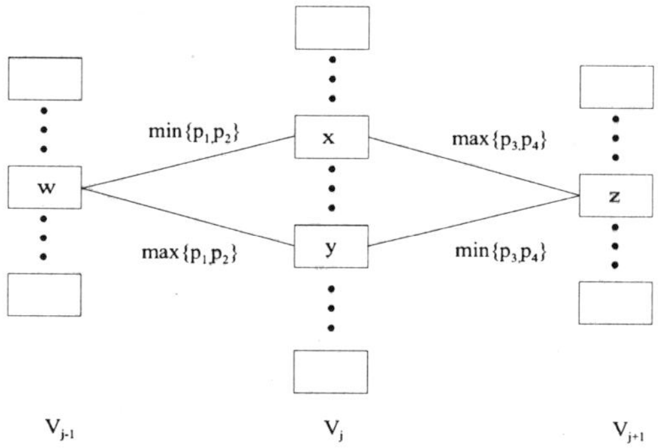


**FIGURE 3.34**
**Channel graph after shifting**

with the most reliable ways exiting (see Figure 3.33 and Figure 3.34). The proof that $shift(G,x,y) \leq Rel(V,E)$, i.e., that the $shift(G,x,y)$ is a lower bound for the reliability of MIN $G$, can be found in [255].

We claim that on average, shifting in a MIN makes the input-output channel graphs no more reliable than before the shift. In order to prove this, we convert the $k$-stage MIN $G$ to a $k+2$-stage channel graph $\widehat{G}$ by adding a source connected to each vertex in $V_0$, and a terminal connected from each vertex of $V_{k-1}$. We assign to each added edge operation probability $\tilde{p}$. Notice that first shifting in $G$, and then forming the channel graph yields the same result as first forming the channel graph and then carrying out the shift. In order to see what does shifting do to $\widehat{G}$, we express the source-to-terminal reliability of $\widehat{G}$ as a polynomial $\text{Rel}(\tilde{p})$. If $n$ denotes the total number of inputs and outputs in MIN $G$, one can write $\mathrm{Rel}(\tilde{p})$ as:

$$\mathrm{Rel}(\tilde{p}) = \sum_{i=0}^{n} E_i \, \tilde{p}^i (1 - \tilde{p})^{n-i} , \tag{3.10}$$

where $E_i$ is the expected number of input-output pairs communicating, given that the total number of inputs available plus outputs available is $i$. It follows that $E_0 = E_1 = 0$. Moreover, $E_2$ is exactly the sum of the source-terminal connection probabilities in all channel graphs of $G$. For $\tilde{p}$ sufficiently near 0, $E_2\tilde{p}^2(1-\tilde{p})^{n-2} \gg E_i\tilde{p}^i(1-\tilde{p})^{n-i}$ for all $3 \leq i \leq n$. Thus, on average, the reliability of input-output channel graphs (after the shift) has decreased.

Using this we can transform any MIN into a MIN where no more effective shiftings are possible, while the process will not increase the average source-terminal connection probability. We have to find a bound on this average given the ordered MIN. In [255] a lower bound on the probability that an input can reach at least $j$ outputs was developed, by applying the methods to a MIN with a single source but any number of outputs. Thus after shifting, we can determine for each input a lower bound on the expected number of outputs that it can reach, by a single application of bounding algorithm. By repeating this for each input, we obtain a lower bound on the expected number of input-output pairs that can communicate. This represents the average source-terminal connection probability multiplied by the number of input-output pairs. Thus we have a computationally efficient method for obtaining lower bounds on the average source-terminal connection probability of an interconnection network.

A second measure of the reliability of MINs is the probability of an input communicating with all of the outputs (*broadcast*) [688]. As the shifting can be used to bound from below the expected number of outputs reachable from a fixed input, we get a lower bound on the probability of reaching any number of outputs. Therefore, we obtain a lower bound on the probability of reaching all outputs. Upper bounds on the broadcast probability can be also obtained as direct extensions of the methods discussed in [255]. Colbourn et al. examine the extension of the method of consecutive cuts to reliability problems of MINs with a single input but many outputs, and an efficiently computable upper bound is given [125].

### 3.7.2 Fault Diagnosis Models

Fault diagnosis of MINs has been studied extensively. With static faults (of fixed location, and permanent duration) one may use an offline diagnostic procedure.

Feng and Wu considered MINs consisting of $2 \times 2$ crossbar switches with two valid states (pass or cross). Since any of the 16 possible input states of the $2 \times 2$ switch may be matched to 16 possible states (and not to just the two valid states), there are $16^2$ possible state combinations in which a potentially faulty switch can behave. By examining these state combinations we can detect any stuck-at-cross or stuck-at-pass error, and any stuck-at-zero or stuck-at-one switch fault with just two input patterns $(x_1, x_2) = (1, 0)$, and $(x_1, x_2) = (0, 1)$. Furthermore, this switch testability property leads to simple fault detection and location for the baseline multistage network, provided that a single fault (switch or link) is present [198]. A hardware implementation of this detection algorithm in a dynamic setup is discussed in [422].

Huang and Lombardi extended these results and obtained a constant number of rounds (the maximum is 12) for full network diagnosis, providing also exact location procedures with reduced computational complexity [289]. Many other authors have improved and extended these results and models. For instance, multiple switch-only faults can be detected and located for baseline MINs, provided that (a) any output terminal is connected to at least one processor (called dynamic full access (DFA) property), and (b) permanent faulty values appear at the output of faulty switches. The tests are similar as before, but now $2(1 + log_2 N)$ test rounds are necessary [445].

If one considers a model of a network with four valid states (pass, cross, and also upper/lower broadcast) the location problem becomes much harder, since there are now $16^4$ state combinations. Feng and Zhang proved that for a baseline network with broadcast switches, both link and switch faults can be detected, but they can not be exactly located. Also, $6(2 + log_2 N)$ tests may be needed for detecting multiple faults [199].

Agrawal and Leu extended the DFA network property to hold if every processor can communicate with every other processor in a finite number of passes through the network, routing the data through intermediate PEs if necessary [15]. The DFA property is computed by the reachability matrix $R$, an $N \times N$ matrix with $R_{ij} = 1$, iff PE$_j$ is reachable from PE$_i$, and $R_{ij} = 0$, otherwise. Since individual stages of a MIN are basic permutations $A_i$, $R = \prod_{i=1}^{L} A_i$, where $L$ is the number of stages in the

MIN. The reachability after $k$ passes is indicated by $R^k$. It can also be proved that if there exists an $i$, such that $R^i = R^{i+1} \neq I$ (where $I$ is the identity matrix), then the DFA property does not hold. It is usually interesting to consider the number of passes $\lambda$ that make $R^{\lambda}_{ij} = 1$; $\lambda$ indicates the path length between $i$ and $j$.

In the case of multiple faults, complicated tests sufficient for 3-pass and $(log_2 N - 2)$ pass Omega network have been derived by Varma and Raghavendra [688]. Test sequences of minimal length for fault detection on the Omega network have been considered by Thanawastien et al. [661].

A wrapped baseline-like unique path network with $f \times f$ switches arranged in $L$ stages, and with up to $f - 1$ stuck-at faults, switch, or link faults can be diagnosed using rerouting on $f$ Hamiltonian, edge-disjoint cycles [457]. An example is the diffusion machine (see Figure 3.35), which uses an $f = 3$, $L = 2$ bidirectional generalized baseline network.

Extra stage, or link networks increase reliability by providing multiple paths to any destination. As an example, an extra stage butterfly (or a topologically equivalent indirect binary cube network), which is formed by adding an extra stage along with a number of multiplexers and demultiplexers, can tolerate one link, or one node fault [7]. For a summary of diagnosis techniques on these networks see [8, 676]. Bounds and approximations for reconfigured butterfly networks and augmented shuffle-exchange are discussed in [365, 675]

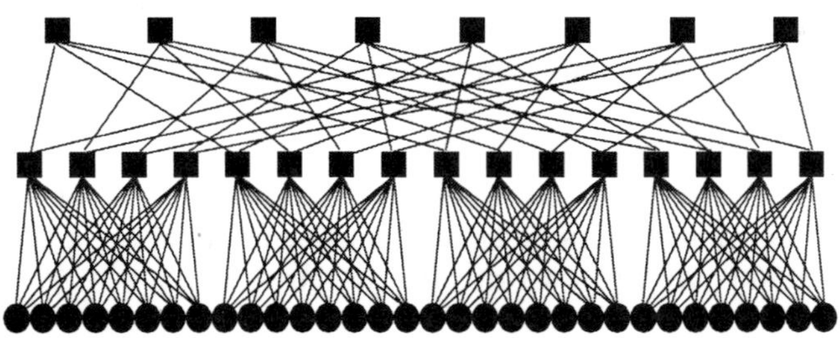

**FIGURE 3.35**
**Diffusion machine: generalized baseline MIN with $N = 32$ PEs (shown as circles)**

For nonrectangular $(s, f, L)$ regular SW-banyans, based on $L$ stages of $s \times f$ switches ($f > s$) with a single node/edge fault (exclusive of PEs), the maximum number of PEs (corresponding to input terminals) which are connected to all output terminals is $s^L - s^{L-1}$, if $s > 1$.

Similarly, the maximum number of output terminals connected to all PEs is $f^L - f^{L-1}$, if $f > 1$ [222]. A member of $(s, f, L)$ networks is the 4-ary butterfly, whose fault tolerant properties are considered further in [619].

Relatively efficient methods for testing packet switches for control and routing faults have also been explored in [437]. The symmetry of this approach guarantees $\mathcal{O}(f^2 2^f)$ testing complexity for a unicast packet communication switch $f \times f$ [488]. The overhead of testing an $N$-node unique path multistage network is only $\log N$ times, since up to $N$ parallel paths can be tested together from different PEs [488].

## 3.8 Special Purpose Networks

In addition to communication networks, several other multistage networks assume a dual communication and computation role in accomplishing certain data processing tasks, such as sorting, combining, and distributed counting.

### 3.8.1 Sorting Networks

*Sorting networks* sort arbitrary input data identified based on a key (and not a destination tag) by rearranging elements in a multistage network. Sorting networks use sequences of "increasing" and "decreasing" comparator elements; these elements propagate packets based on their relative key values (see Figure 3.36).

A uniform sorting network connects columns of comparators together with permutations as in a multistage routing network (see Figure 3.37). Sorting circuits are used as switching networks, e.g., in the AT&T Starlite switch.

Testing that a parallel sorting algorithm (or a sorting network) is correct can use the 0/1 sorting principle (or the zero-one lemma) [343]. These lemmas state that if an oblivious comparison-exchange sorting algorithm (or a sorting network) sorts all inputs of 0's and 1's correctly, then it sorts all inputs correctly.

The most popular sorting network is Batcher's *bitonic sorting network* [46]. This network has roughly $\log N$ times more switches than an $N$ input banyan network. A sequence of numbers is bitonic if it has at

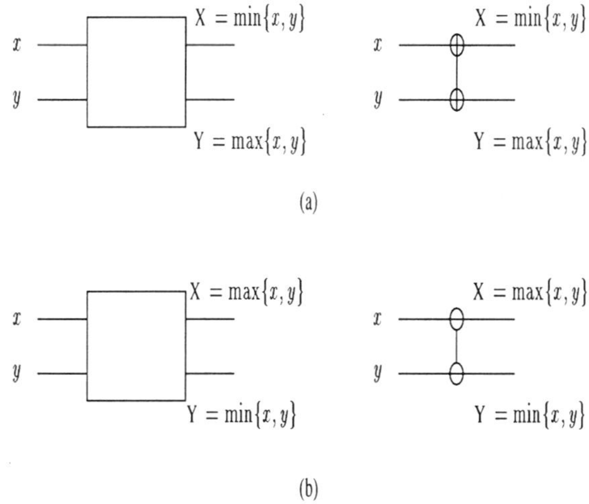


**FIGURE 3.36**
**Increasing (a) and decreasing (b) comparators**

most one local maximum and one local minimum; the two endpoints of the sequence may also wrap around. For example, 2, 3, 7, 6, 0, 1 and 8, 9, 2, 1, 0, 4 are both bitonic integer sequences. It turns out that sorting is quite simple if the initial input datum is a bitonic integer sequence. Assuming a (possibly padded with dummy values) bitonic sequence of length $N = 2^n$, the "special case" sorting algorithm proceeds as follows.

- The list is divided into half,
- Elements at similar positions at each sublist compare and swap values, and
- This process is repeated with each newly obtained sublist; notice that both sublists are bitonic, and each list has elements that are relatively sorted in respect to each other.

Notice that $n$ stages of all increasing comparators connected together with exchange permutations on bit $i$, $(0 \leq i < n)$ can realize the above bitonic sorting algorithm. An example for $N = 2^4$ is shown in Figure 3.38.

The next step is how to generate a bitonic sequence. Since the initial $N$ data elements can be considered as $N/2$ bitonic sequences of size two

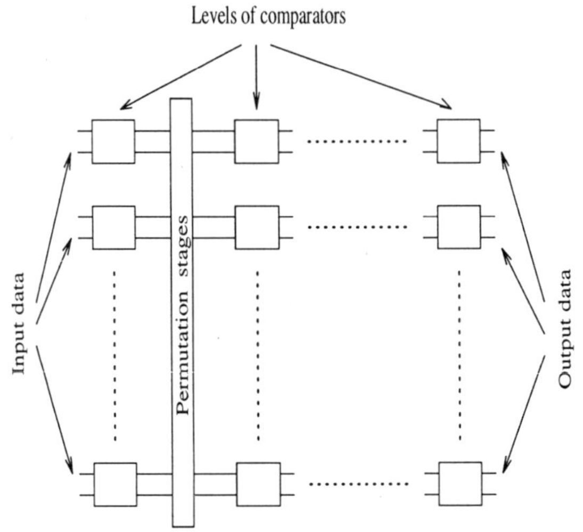


**FIGURE 3.37**
**Sorting network using comparators**

each, the task is reduced to recursively constructing a bitonic sequence of size $N$, from two bitonic sequences of half that size. As shown in Figure 3.39, this can be done by sorting individual bitonic sequences of increasing size: initially two, then four, and finally eight elements.

Combining the generation of a bitonic sequence with its sorting counterpart, we obtain a regular bitonic sorting network with $n(n+1)/2$ stages of comparator elements. The network repeatedly merges bitonic sequences of increasing length. The bitonic sorting network for $N = 32$ elements is shown in Figure 3.40; comparators are shown here with vertical lines. The depth of the circuit matches its time complexity, reducing the sequential sorting of $\mathcal{O}(N \lg N)$ to $\mathcal{O}(\lg^2 N)$.

Batcher also provided another simple sorting networks called *odd-even mergesort*. Starting with sublists $a$ and $b$ (e.g., each of even size $2l$), sorted in increasing order, the algorithm proceeds as follows.

- The half size lists $c = \{a_0,\ b_0, a_2,\ b_2,\ \ldots,\ a_{2l-2},\ b_{2l-2}\}$, and $d = \{a_1,\ b_1, a_3,\ b_3,\ \ldots,\ a_{2l-1},\ b_{2l-1}\}$ are merged recursively, and

- The lists $c$ and $d$ are combined into a new sorted list $e$ by assigning $e_0 = c_0$, $e_{4l-1} = d_{2l-1}$, and $e_{2i-1} = \min(c_{i+1}, d_i)$ and $e_{2i} = \max(c_{i+1}, d_i)$, $1 \leq i < 2l$.

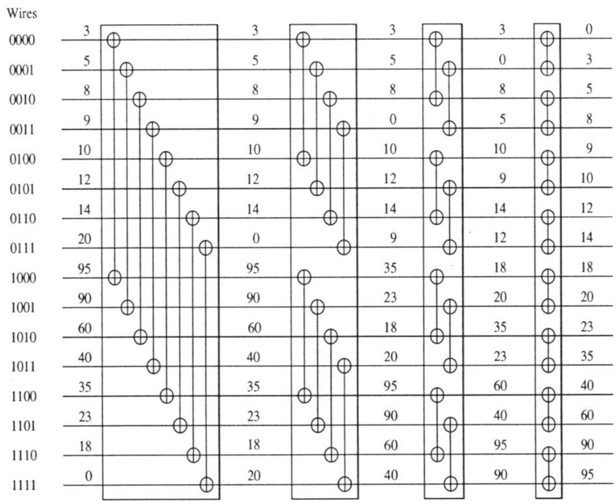


**FIGURE 3.38**
**Sorting a bitonic sequence of 16 elements $(3, 5, \ldots, 0)$ [364]**

The network directly constructed using the above construction is not regular. As we can see from the $e$ computation, some comparators are not needed. Hence, now some nonadjacent comparators are connected, unlike what one normally implies from Figure 3.37. Although odd-even mergesort saves on the number of comparators over a bitonic network (~15% for a 1024-node sorter), asymptotically the size of the network is still $n^2N/4 + \mathcal{O}(nN)$, and the depth is also $\mathcal{O}(\lg^2 N)$. The irregularity and complexity of the odd-even mergesort network has attibuted to its lesser popularity for sorting applications and constructing nonblocking ATM networks.

Stone first noticed that indices of every pair of keys that enter a comparator at any step of the bitonic sorting network differ in a single bit in their binary representations [649]. Thus, the bitonic sorting network can be considered as a sequence of $\log N$ stages, with the $i$th stage consisting of $i$ steps, and step $i$ inputs differing in their least significant bit being compared. Since shuffling is equivalent to a left shift, a sequence of perfect shuffle permutations could be used between the comparators of any two adjacent switch stages. Input lines can be ordered before each comparison-exchange step by shuffling them as many times as required by the bitonic sort algorithm. Thus, this modified sorting network has $\log N$ stages, with each stage consisting of $\log N$ steps. At each step, the

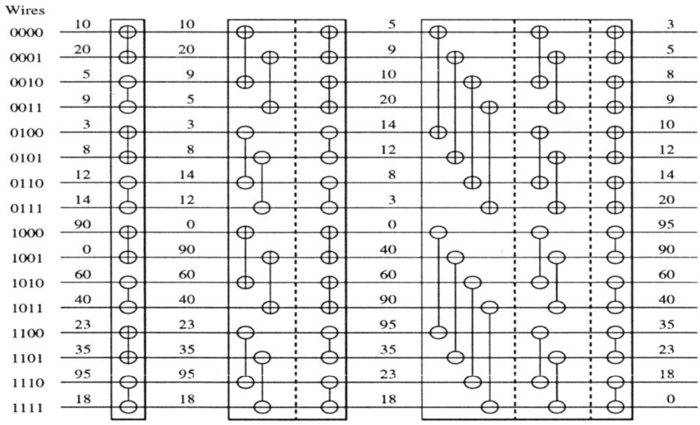


**FIGURE 3.39**
**Generating a bitonic sequence from a random sequence of 16 elements [364]**

output lines are shuffled before they enter the next rank of comparators. The comparators associated with the first $i$ steps of the $i$th stage do not exchange their inputs. Their only use is to shuffle their input lines.

Exploiting the above idea, Dowd et al. introduced a *balanced periodic sorting network*, of size $N$, nd period $p > 2$. This network is a cascade of $p$ identical blocks, where output $i$ of each block is fed as input $i$ of the next block, $0 < i < N - 1$. A periodic sorter is attractive since it offers the possibility of using a single block, with the outputs of the block feeding back to the inputs. By recirculating the data through such a block $p$ times, the numbers are sorted. Obviously this uniform sorting network, called the balanced periodic sorting network, has a much simpler VLSI implementation [177].

Theoretical design issues of periodic sorting networks (with variant period) using chains of sorted items is discussed in [48]. Periodic sorting networks with constant period, and also a method for converting any nonperiodic sorting comparator network to a periodic one are discussed in [375, 376].

Expander graphs became increasingly important in search for asymptotically optimal permutation routing and sorting networks [314]. The AKS sorting network [18] and the multibutterfly are the only two known

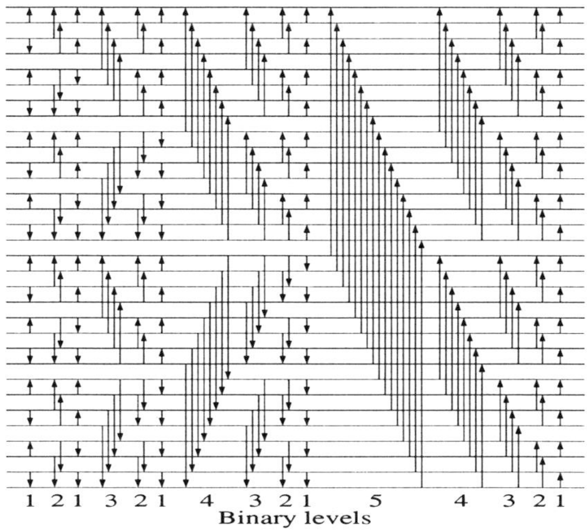


**FIGURE 3.40**
**Bitonic sorting network for 32 inputs**

bounded-degree networks which can sort in $\mathcal{O}(\log N)$ bit steps with deterministic algorithms [453, 550]. The recent proof that multibutterflies can sort is based on a load 1, congestion 1, and dilation 2 embedding of an $N$-node AKS network onto a $\frac{3N}{2}$-node degree-8 multibutterfly [453]. Hence, multibutterfly networks are at least as powerful as AKS. Both the AKS, and the multibutterfly network can solve the $(N, N, k_1, k_2)$-routing problem in optimal $\mathcal{O}(k_1 + k_2 + \log N)$ time.

The multibutterfly discovered by Bassalygo and Pinsker [45] has a simple randomly wired multistage structure; it consists of butterfly networks merged together after randomly permuting switches at each level. Upfal discovered that this network can deterministically permute its inputs in $\mathcal{O}(\log N)$ time [679]. Furthermore, Arora et al. [32] proved that back-to-back multibutterflies (called Beneš networks) are strict-sense nonblocking, i.e., any unused connections between any input-output network pairs can be self-routed in $\mathcal{O}(\log N)$ time using the circuit switching model. The multibutterfly is also robust against faults; optimal online and offline routing algorithms on multibutterflies are based on greedy "grouping" schemes [226, 407]. The multibutterfly has sparked an interest in randomly wired networks for packet routing in parallel systems [436, 451].

AKS networks are point-to-point networks with an irregular structure which does not admit good embeddings on nonexpander, regular topologies [141]. Although significant reductions on the size and depth of the AKS sorting networks have been made [19, 120, 529], the AKS sorters remain impractical compared to classical $\mathcal{O}(\log^2 N)$ delay, bitonic sorting networks [46]. For example, there exists a class of the AKS sorting networks with depth $1830 \log_2 N - 58657$ and size $N \geq 2^{78}$ [120]; the smallest such network still has hundreds of times the depth of a $2^{78}$-node bitonic sorting network. Improved lower bounds on the minimum size and depth of comparison-based sorting networks are provided in [315]; a $(3.27 - o(1)) \log N$ lower bound on the depth is given. Small sorting networks of optimal depth are known for $n \leq 10$, and of optimal size for $n \leq 8$ [523]. Sorting networks with depth $\sim 4.5 \log_2 N$ which can sort a permutation with high probability are discussed in [405].

The feasibility of high capacity VLSI sorters depends on an efficient layout of the sorting network. VLSI complexity bounds for sorting networks have been investigated in [70, 71, 128, 419, 664]. Using general results from [664], Even et al. proved that the grid area required by a size $N$ sorting network is at least $(N-1)^2/2$ [189]. Muthukrishnan et al. illustrated complicated layouts (also with knock-knees, i.e., routing path corners) of bitonic sorting which achieve a grid area less than $1.39N^2$ [491].

For large sorting networks fault tolerance has also been studied. In the presence of independent random faults at comparators, fault tolerant sorting networks which can sort either correctly, or correctly with high probability have been considered [34, 403, 404, 405, 406, 448].

### 3.8.2 Combining Networks

*Combining networks* combine atomic and read/write requests directed to the same memory unit and location into a single memory request. Combining is based on an operation code which is included in the packet format. This opcode usually represents an associative mathematical or Boolean operation, which is usually identical for all concurrently active combining requests.

Combining is useful for nonuniform communication patterns which create hot-spots either at certain memory modules, or at certain network switches. The former case happens with shared variables in scan and reduction operations, and also in implementations of synchronization mechanisms such as barriers, binary locks, pointers to shared lists or queues, and condition variables (semaphores). The latter case arises with special source-destination patterns which result in some form of routing collision within the network.

When hot-spot locations are known at compile time, *software combining* can be used effectively by spreading accesses to an asynchronous tree of variables [233, 471, 477, 606]. Although most PEs may only ascend a few levels on the tree, all must wait for the whole tree to be ascended, and then descended. This makes software combining time depend on the number of participating PEs, usually as $\log N$. In addition, for a multistage network, each communication step (on the binary tree of processes) requires $\log N$ transit time through the network, thus a software combining operation takes $\log^2 N$. In contrast, *hardware combining*, usually implemented with binary trees, takes only $\log N$ time.

An example for Fetch&Add operations executed on a combining tree is shown in Figure 3.41. Tree nodes accept the values sent initially from PEs (assigned to tree leaves, marked with a '*'), and push the combined result up. Finally, a reverse process starts, and the final result at the root is decombined, providing each PE with a different value, as required by atomicity constraints.

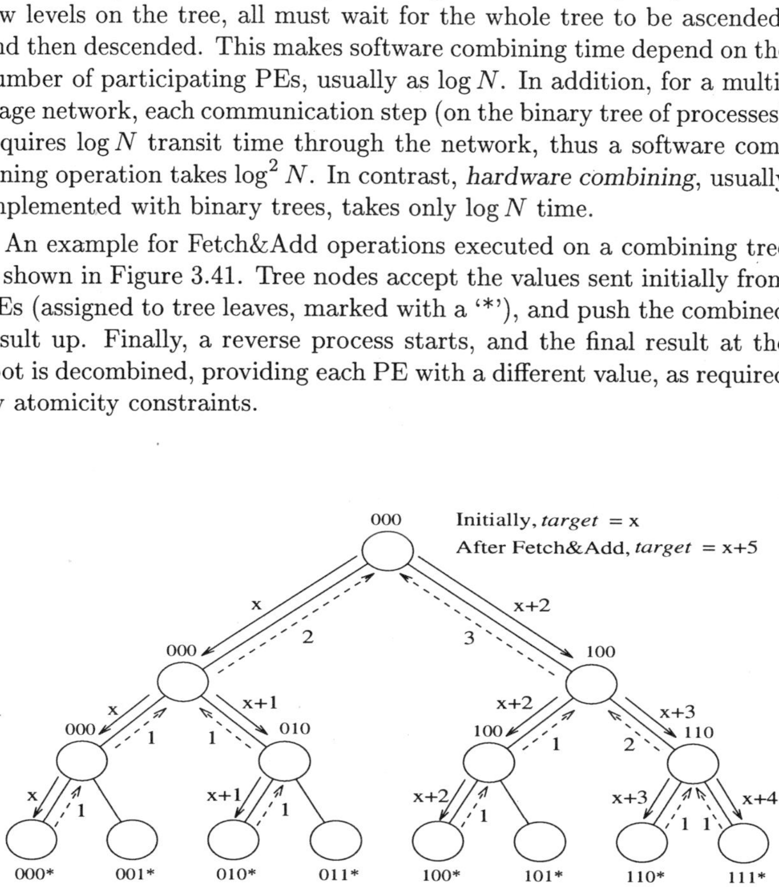


**FIGURE 3.41**
**Combining Fetch&Adds on an 8-node complete binary tree [364]**

Hot-spots cause increased memory and network contention resulting in considerable network latency and the so called *tree saturation effect*; full buffers at later stages of the network cause buffers at previous stages to fill, and therefore both random and hot-spot requests are being delayed. For example, this event happens during vector reads from a single PE with fine interleaving. With only 0.125% of the input traffic heading to hot memory on a 1000-node unique path network, network bandwidth is reduced to half! Combining alleviates tree saturation effects by reducing the rate at which requests arrive at the hot memory module, when requests access the same memory location. Another experimental feedback-based approach effectively controls the input rate of requests, by monitoring queue lengths and providing "fuzzy" thresholds (called damping) for admission traffic control. This method is particularly helpful with hot requests that can not be combined [608].

Network saturation is affected by the existence of hot-spots. A continuous traffic model considering a single hot-spot on a unique path multistage network configured with switches implementing output queuing has been considered by Pfister and Norton [539]. The following parameters are defined.

**N** - network size

**r** - random input rate applied at each input port

**h** - percentage of traffic departing from the hot-spot port, $0 \leq h \leq 1$

The total traffic at the hot-spot port is

$$r\,(1-h) + rhN\ .$$

Since messages are removed from a last stage switch with a rate of at most one message per cycle, the maximum hot-spot traffic $h_s$ can be obtained by setting the above quantity equal to one. Thus,

$$h_s = \min\left\{\frac{1-r}{r\,(N-1)},\ 1\right\}\ .$$

Notice that for a network with output queuing, the saturation rate depends on the network size. As the network size increases, it takes the network longer to recover, and the hot-spot saturation rate decreases accordingly. Thus, the asymptotic network performance is severely degraded by hot-spot traffic unless some sort of message combining is used. Also notice, that more sophisticated queue designs, e.g., using virtual circuits, may improve the access latency for messages not directed to the hot module in the short term, but eventually tree saturation will occur.

Previous simulations and Markov models, for both finite and infinite buffer size, on a $d$-way ($d = 2, 3$) combining Omega network show vast improvements in throughput and latency over the single-buffered packet-switched Omega network for hot-spot traffic [165, 391, 393, 394, 439, 482]. Systolic combining switches contain combining/decombining logic based on cyclic shifters, parallel comparators, memory, and sophisticated controllers are provided in [164]. Fault tolerant combining networks based on extra stage, $4 \times 4$ combining switches have been proposed [41]. Tzeng proposes separation of the routing and combining network, and uses a tree as the combining structure [672, 673]. His design reduces the cost of the $N$-node combining tree by a $\log_2 N$ factor; it uses the same tree to combine messages directed to different hot-spot modules, by loading level ($\log_2 N - i$) in the combining tree with hot-spot messages from level $i$ of the underlying Omega network.

Previously proposed combining switches do not combine messages directed to different memory locations or modules. Such messages can still cause link conflicts named *nonuniform traffic spots* (NUTS), for common data patterns, such as the bit reversal permutation on the Omega multistage network. In some situations, up to $\sqrt{N}$ messages may contend for a link in the middle stages of an $N$-node multistage network, possibly causing tree saturation.

Tzeng's solution to the saturation problem for NUTS uses multiple buffers at each input port (crosspoint switching), instead of using more powerful switches which dynamically assign available buffer memory space to requests. Outstanding requests are delayed and reattempt advancement to next stages whenever the corresponding buffer becomes free and the link becomes available [674]. Another solution to tree saturation effects for NUTS uses (a) Batcher's sorting network to group hot-spot requests, followed by (b) a concentrator configured as a double-width Omega network to combine the requests, and (c) a distributor configured as a double-width inverse Omega network to route the combined requests [25].

Other solutions provide multiple network paths, limit the number of hot messages per switch [163], or use diverting switches with retransmission schemes [388]. In the latter case, retransmissions always occurs from a different source processor.

### 3.8.3 Prefix Circuits

The parallel prefix computation (also called parallel scan) can be illustrated as follows. Initially, let vector $x = (x_0,\ x_1,\ \ldots\ x_{p-1})$, where $x_i$ is assigned to processor $i$. The prefix of $x$ produces a new vector $y = (y_0,\ y_1,\ \ldots\ y_{p-1})$, where $y_i = y_0 + y_1 + \ldots + y_i$, and $y_i$ is assigned to processor $i$.

*Prefix circuits* realize a parallel scan computation on $N$ items. Alike sorting circuits, they provide edge-disjoint routes for each prefix output. Constructions usually exploit recursion which halts on a basic building block, e.g., a $2 \times 2$ prefix circuit. One can easily spot similarities between prefix circuits and scan communication patterns presented in Chapter 2.

Prefix forms the kernel to many operations such as polynomial evaluation, partial sums, recurrence solution, tridiagonal systems, etc. Simple patterns (as shown in Figure 2.9) may be used to implement scan operation. For example, MPI has "scan" collective communication routines that provide this computation. Similar operations also exist in HPF, CMF, and CMC*.

Snir's prefix circuit [637] follows the classical divide and conquer pattern, and for even $N$, Ladner and Fischer's prefix circuit [380] essentially realizes a recursive broadcast pattern. The size of the prefix circuit relates to the network cost, while its depth is a measure of communication time [72, 73, 270, 380, 637]. Like scan operations, prefix is very useful for parallel computations such as partial sums, linear recurrences, digital filtering, various graph problems, parallel sorting, load balancing, and scheduling [359]. A comprehensive survey of parallel algorithms based on prefix is given in [383].

### 3.8.4 Balancing and Counting Networks

*Balancing networks* are acyclic circuits consisting of balancers, just as sorting networks are acyclic circuits consisting of comparators. Whereas a comparator receives a single number on each input and outputs the numbers in sorted order, a balancer receives a sequence of tokens along each input wire asynchronously and alternates in sending them to the top or bottom outputs. Thus, a balancer balances the number of tokens on its output wires.

*Counting networks* provide distributed mechanisms for implementing Fetch&Increment, by avoiding the traffic associated with accessing a single shared memory location. They also help reduce bottlenecks for applications which access a centralized counter, such as locks, consumer/producer, or barrier synchronization. By implementing counters as counting networks we avoid the traffic associated with accessing a single (possibly virtual) shared memory location.

Counting networks form a special case of balancing networks. Let $y_i$ represent the number of tokens delivered at the $i$th network output, $0 \le i \le N-1$. An $N$ input counting network is a balancing network, with the additional property that for any input sequence, $0 \le y_i - y_j \le 0$, $0 \le i < j < N$.

In Figure 3.42 (a) we show a sequence of tokens passing through a balancer. The number of tokens per input-output wire is provided in Figure 3.42 (b). Finally, in Figure 3.42 (c) we show a sequence of tokens passing through a counting network [342]. The individual balancing steps at each network stage can be implemented using Compare&Swap operations on different shared memory locations. The final value at the last network stage is the incremented counter value, as can be seen in Figure 3.42.

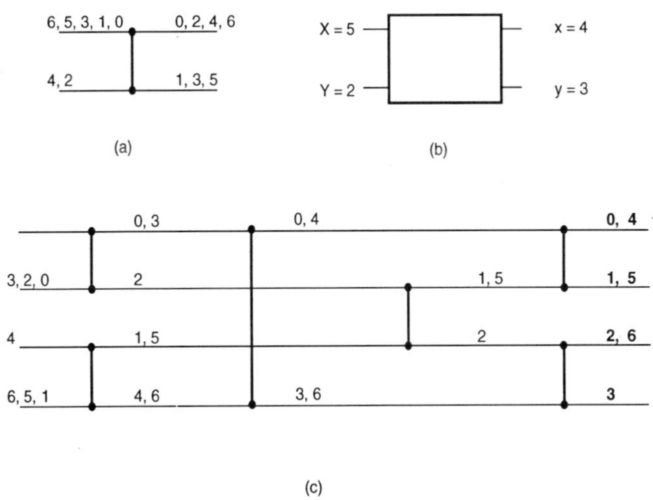


**FIGURE 3.42**
**Examples of embedding graph $G$ onto graph $H$**

Tokens are labeled with distinct numbers, which may indicate order of arrival. In respect to counting, it can easily be proven that assigning the value $i\,n + j$ to the $i$th token from input wire $j$ results in the tokens receiving distinct integer values in $[0, L]$, where $L$ is the total number of tokens input to the network.

The depth of counting networks is $\mathcal{O}(\log N)$ for random constructions, or $\mathcal{O}(\log N\ \log^* N)$ for deterministic constructions [342]. However, the most interesting constructions for implementations seem to be two $\mathcal{O}(\log^2 N)$ depth constructions [33, 265]. These are isomorphic to Batcher's bitonic sorting network [46], and to the balanced periodic sorting network [177]. Fan-in two, $\mathcal{O}(\log^2 N)$-depth counting networks are practical for implementing low-level synchronization primitives [265]. For counting networks with fan-in greater than two, see [16]. Network testing for proving counting capability is complicated [92].

### 3.8.5 Interconnection Networks in Cryptology

Interconnection networks have an interesting property that makes them interesting for cryptology, i.e., they provide the possibility to specify and perform permutations at the same time. Secure cryptosystems, such as S-boxes [540, 541] or private key block ciphers [74, 557], can possibly be constructed using Boolean functions, and the Beneš or other interconnection networks [540, 541, 557]. Such constructions can be made secure, e.g., if one can construct pseudorandom permutation generators from pseudorandom Boolean function generators [63, 74].

Furthermore, insertion networks may be used to realize special permutations which move a given item to an arbitrary position in a list, while shifting the remaining items accordingly [159].

## 3.9 Shared Memory Algorithms

Ranade has presented an efficient algorithm for simulating a PRAM on a butterfly network which supports distributed shared memory. A computation step of the PRAM can be emulated in near-optimal $\log N$ steps on the butterfly with high probability, where $N$ is the number of processors. The emulation uses message combining, and a complicated routing algorithm.

## 3.10 VLSI Layout and Customization

Few results have focused on the VLSI complexity of multistage networks, with the exception of butterflies and Clos networks.

The VLSI layout of the butterfly network has (a) 2/3 $N^2$ area, if the order of I/O terminals is fixed, or (b) $N^2/2$ area, if the order of I/O terminals can be permuted [491]; this result improves on previous results for banyan networks [705]. Furthermore, a direct customization of a butterfly network into modules results in a complete graph intranetwork, which has the smallest pin count, but also the highest number of bundles for any given bisection width [569]. Trade-offs between pin count and number of bundles can be obtained by using column partitioning and shuffling rows within each subnetwork to form a smaller butterfly; in this case, the intranetwork can be designed as an $n$-hop network.

Chien and Oruc provided optimal designs with a given fixed pin count using various recursive decompositions of Clos-type rearrangeable and nonblocking networks based on Beneš networks [108]. Assuming constant pin-out count, VLSI implementations of banyan networks which use the largest possible $k \times k$ switches that fit onto a single integrated circuit are discussed in [215].

Notice that constant pin-out constraints imply that the throughput of an $4 \times 4$ switch is twice as much that of an $8 \times 8$ switch. An extension to unique path, multistage networks of switches (e.g., a $64 \times 64$ network configured as a three-stage butterfly network with 16, $4 \times 4$ unicast switches per stage, or a two-stage network with 8, $8 \times 8$ unicast switches per stage) will yield similar results; saturation throughput is essentially determined by the first network stage [238].

## 3.11 Commercial Multistage Systems

Several multistage networks have been used for routing.

For example, the BBN TC2000 was a NUMA multiprocessor based on the butterfly network. It consisted of up to 128 PEs (Motorola 88100 CPU, including three 88200 chips) and used distributed shared memory. The network did not maintain cache consistency, thus the operating system placed a restriction on caching read only data.

The IBM SP2 used an Omega multistage network for interconnecting PEs into an MIMD system [638].

The IBM GF11 parallel computer was an SIMD architecture with 576 PEs connected together using a three-stage rearrangeable Beneš network. The network supported broadcasting, and partial multicasting. The GF11, like the control network of the TMC CM-5, used offline circuit switching by first simulating the parallel program on a sequential host PE. Switch configurations were then loaded as part of the program startup. The target application of the GF11 supercomputer was the numerical evaluation of quantum chromodynamics problems; it is possible that offline circuit switching may also work well for other scientific and engineering applications.

The only special purpose networks which have been built as part of a parallel interconnection networks are sorting and combining networks.

In most cases, a bitonic sorting network has been used because of its simplicity in providing nonblocking routing, e.g., in the AT&T Starlite switch.

Because of the limitation that only requests to the same memory location can be combined, and the additional hardware complexity for combining/decombining messages, combining networks are not popular. However, they have sometimes been considered in MPP systems.

An independent combining network was designed for the NYU Ultracomputer for executing Fetch_Φ atomic operations. Combining was also initially considered for the IBM RP3 prototype. However, the cost of a dedicated combining network is deemed high; for the NYU Ultracomputer, it amounted to six times the cost of an Omega network, on which both the NYU Ultracomputer and IBM's RP3 prototype were based.

On the TMC CM-5, a separate control network which used a complete binary tree topology with system components located at the leaves, provided global or segmented operations, such as broadcasting, synchronization, reduction, and parallel scan [416]. Several FIFO buffers at the network interface perform the appropriate combining functions. The control network is designed for low node-to-neighbor latency and high-bandwidth long distance transmissions and provides good scalability. Thus, reduction and scans on the CM-5 are performed very efficiently, since they rely on fast vector processors (for local reduction/scan) and the hardware combining ability of the control network. Performance for the provided operators (SUM, MAX, MIN, bitwise logical OR, bitwise logical XOR, and router-done) differs very little. Notice that communication completion is signaled using bitwise logical OR.

Asynchronous eureka operations signal special conditions and exceptions, such as a new SUM, or a 0/1 transition in the value of a variable at a PE. Barriers and eureka (AND, OR) events are implemented on the CM-5 and Cray T3E using hardware combining trees. These parallel operations use special buffers and high-priority packets over the three-dimensional torus data network to reduce cost and hardware complexity [607].

# Chapter 4

# Hypercube Networks

In the following chapters we consider general undirected graphs, especially hypercubes and meshes, but also star graphs and related networks. We concentrate on packet routing algorithms, for which theoretical results are richer. Although architectural issues, such as virtual shared memory, prefetching, multithreading, and cache coherence play a prominent role in the design of massively parallel systems, numerous efficient platform-independent parallel solutions are derived directly from either hypercube or mesh based algorithms. Unless otherwise specified, we assume full duplex (bidirectional) links, and the constant model for node to neighbor communication. For detailed description and analysis of some important communication algorithms contained here, the reader is referred to [401]. System issues are discussed in [137, 294]. Parallel algorithms, mainly for meshes and hypercubes, are covered in detail in [200, 300, 364, 561, 580]. Sorting algorithms are examined in [21, 114, 384, 587].

## 4.1 Topological Properties

The $n$-dimensional binary hypercube (also called $n$-cube) is a static interconnection networks with $N = 2^n$ processors, numbered $0, 1, \ldots, N-1$. A processor is addressed by a binary string $x = x_1\ x_2\ \ldots\ x_n$, where $x_i \in \{0, 1\}$, for all $1 \leq i \leq n$. Two processors with addresses $x$, $y$ are connected if their binary representations differ in exactly one bit. Thus, each node of the $n$-cube corresponds to a collapsed row in the $2^n$ node butterfly. Both the degree and diameter of the $n$-dimensional binary hypercube equal log $N$, while its bisection width is $N/2$, and the graph is $2n-1$-connected. The hypercube is not a Moore graph, since it is not an asymptotically optimal graph in terms of the relation of its diameter to the number of nodes. [1]

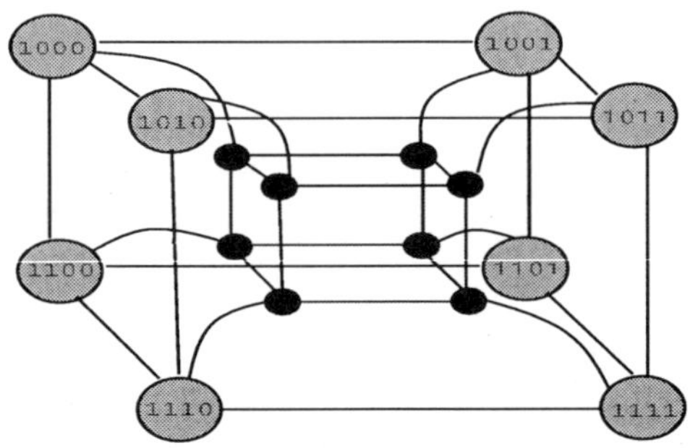


**FIGURE 4.1**
**A 4-cube partitioned into two 3-cubes**

The hypercube topology can be recursively partitioned into smaller hypercubes, hence divide and conquer algorithms can be efficiently implemented on hypercubes. For example, the $n$-cube can be decomposed into two $(n-1)$-cubes (see Figure 4.1 for $n = 4$), or one $(n-1)$-cube and two $(n-2)$-cubes. Partitioning is based on grouping together hypercube nodes with identical bit patterns at given dimensions. Furthermore, since the hypercube graph is vertex and edge symmetric, server-client type algorithms with different root nodes are isomorphic, and thus easily described.

The *greedy routing* algorithm communicates packets from source to destinations along shortest paths. Dimensions are examined left to right (or right to left) order. Packets which have not reached their final destination move by correcting each dimension for which the corresponding source and destination bits are different. Packet conflicts are commonly resolved with a farthest-first or random queuing discipline.

We proceed with a description of the greedy routing algorithm on the $n$-cube. The notation $x : (x_j \Longleftarrow \alpha)$ defines the node obtained by changing the $j$th digit in the binary representation of $x$ to $\alpha$. Let an arbitrary packet $\Psi$ originate at source node $x = x_1\, x_2\, \ldots\, x_n$ and head to destination node $x' = x'_1\, x'_2\, \ldots\, x'_n$.

[1]An example of a Moore graph is the Petersen graph, see Chapter 6.

*Greedy routing* ("cobegin ... coend" refers to all packets).
cobegin
while $x \neq x'$ do
transmit $\Psi$ from $x$ to $x : x_i \Longleftarrow x'_i$, where $x_i \neq x'_i$
coend

Other important topological characteristics of the hypercube are its large number of available node- and edge-disjoint paths, and its interesting embedding properties.

For example, two nodes at distance $h$ on the $n$-cube have $n$ node-disjoint paths of length $\leq n + 2$. These paths can be constructed by correcting an arbitrary dimension in the first step, and thereafter using the greedy algorithm. The $h$ paths consist of,

- $h$ node-disjoint shortest paths of length $h$, and
- $(n - h)$ node-disjoint paths of length $h + 2$.

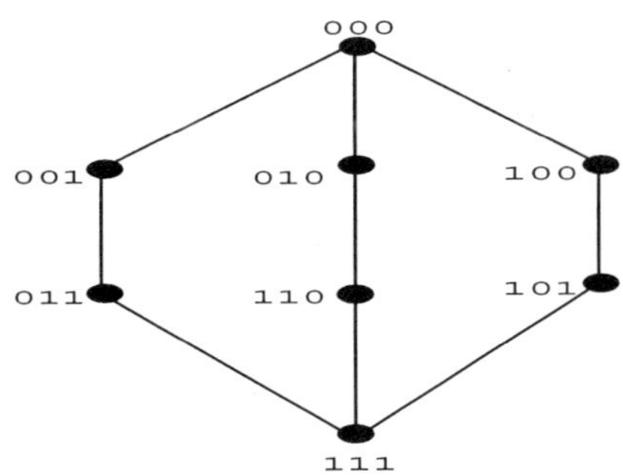


**FIGURE 4.2**
**Node-disjoint shortest paths between 000, 111 on 3-cube**

As an example, the node-disjoint shortest paths between node 000 and 111 on a 3-cube are shown in Figure 4.2. Another topological property is that, if sets $S = \{s_1, s_2, \ldots, s_n\}$ and $D = \{d_1, d_2, \ldots, d_n\}$ contain $n \geq 3$ distinct $n$-cube nodes, then there exist node-disjoint paths between $s_i$ and $d_i$, $1 \leq i \leq n$.

The general problem of deciding whether a general graph is a subgraph of the $n$-cube is NP-complete [12, 139]. General tree embeddings onto the $n$-cube are also NP-complete [697]. Gray codes are very useful for defining embedding relations of many symmetric graphs on the hypercube.

Gray codes have been described in a U.S. telephony patent in 1947, by Frank Gray at Bell Labs. The code had also been used by Baudot in telegraphy experiments in the 1870s. A binary reflected Gray code $G_n$ is an ordered cyclic sequence of nodes on the $n$-cube, such that two consecutive such strings differ in just one bit position. Furthermore, the first and the last strings are also neighbors. The sequence is recursively defined as follows.

- $G_1 = \{0, 1\}$ *for 1-cube*
- $G_i = \{0G_{i-1}, 1G_{i-1}^R\}, 1 \leq i \leq n.$

Hence, we have:

- $G_2 = \{0G_1, 1G_1^R\} = \{00, 01, 11, 10\}$
- $G_3 = \{0G_2, 1G_2^R\} = \{000, 001, 011, 010, 110, 111, 101, 100\}$

Notice that $G_n$ essentially defines a Hamiltonian cycle on the $n$-cube; this path contains all vertices exactly once, and the first and last vertices in the path are also neighbors. Gray codes assist in defining unit dilation and expansion embeddings onto the $n$-cube for certain important symmetric graphs, as follows.

- *Linear arrays of size* $2^k$, *when* $k \leq n$ *[593].*
- *The smallest* $n$*-cube onto which a linear array of size* $s_1 \times \ldots \times s_k$ *can be mapped has dimension* $n = \sum_{i=1}^{k} \lceil \log s_i \rceil$ *[97, 605].*
- *Rings of even size* $2^k$, *when* $k \leq n$ *[593].* There are no odd rings on the hypercube.
- *Rings of even size* $s$, *with* $4 \leq s \leq 2^n$ *[593].* Let $m = s/2$, $n = \lceil logs \rceil$ and $G_{n-1}[m]$ be the first $m$ elements of $G_{n-1}$. Unit dilation embedding is obtained by forming the ring: $\{0G_{n-1}[m], 1G_{n-1}^R[m]\}$
- $k$*-dimensional torus of size* $2^{s_1} \times \ldots \times 2^{s_k}$, *if* $s_1 + \ldots + s_k \leq n$. Also the $k$-dimensional torus of size $a_1 2^{s_1} \times \times \ldots \times a_k 2^{s_k}$ can be embedded with dilation one, and congestion $\prod_{i=0}^{k} a_i / \min_i a_i$ [97, 605].

While any two-dimensional mesh can be embedded onto a minimal expansion hypercube with dilation at most two, for three-dimensional meshes the dilation is bounded between 2 and 5 [97, 98]. Among the many grid embedding techniques [311], mesh to mesh reshaping (changing the number of elements per axis), dimension collapsing or splitting, compression, folding, graph decomposition are the most modular [272]; if graph $G_i$ can be embedded onto graph $H_i$ with expansion $e_i$, dilation $d_i$, and congestion $c_i$ ($i = 1, 2$), then $G_1 \times G_2$ can be embedded onto the target graph $H_1 \times H_2$ with expansion $e_1 \times e_2$, dilation $\max d_1, d_2$, and congestion $\max c_1, c_2$.

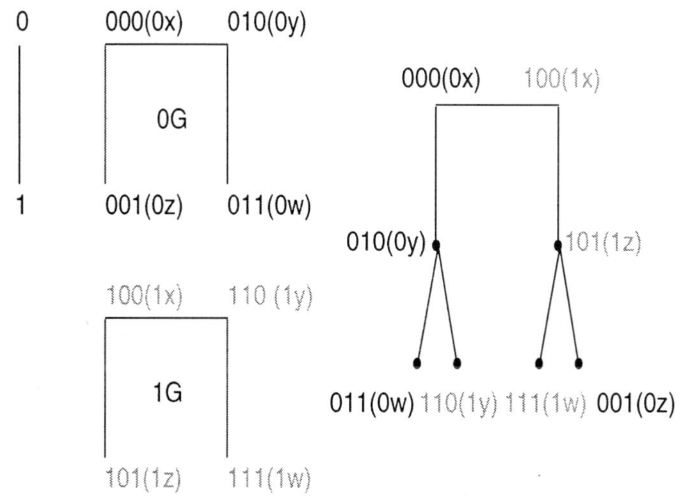


**FIGURE 4.3**
**Embedding a double-rooted binary tree on the $n$-cube, $n = 1, 2, 3$**

Tree embeddings have also been considered. The $n$-cube can emulate any bounded degree tree of $2^n$ nodes with $\mathcal{O}(1)$ dilation and congestion [66]. Unit dilation and expansion embedding of the complete binary tree with $2^n - 1$ nodes onto $n$-cube, $n \geq 3$ is *not* possible. The proof is interesting; we count the number of odd/even parity nodes on the $n$-cube and compare with that on the complete binary cube [401]. However, dilation one (expansion two) embedding of the complete binary tree of $2^n - 1$ nodes is possible on the $(n + 1)$-cube. Furthermore, a slightly different tree, the double-rooted complete binary tree, with $2^n$ nodes can be embedded with dilation one onto the $(n + 1)$-cube. This embedding technique is based on recursion and moves around in interesting ways tree sub-branches, as shown in Figure 4.3, for $n = 1, 2, 3$.

Asymptotically optimal embeddings exist for arbitrary binary trees with $k$ nodes, dynamic trees, meshes of trees, trees of meshes, and pyramids [401, 413, 442]. A tree of meshes is formed by replacing each node of a complete binary tree with a mesh and each edge with a set of edges linking together the meshes. A mesh of trees is formed by a grid with edges along each axis of the grid replaced with complete binary trees.

## 4.2 Hypercube Communication and Sorting

### 4.2.1 Online Permutation Routing

Theorem 2.3 specializes to an $\Omega(N^{1/2}/n)$ lower bound for oblivious permutation routing on the MIMD $n$-cube. By examining the bit reversal and matrix transposition permutations it is simple to show that greedy routing on the binary hypercube is not optimal within the class of deterministic oblivious algorithms, for MIMD communication.

***THEOREM 4.1***

*Greedy permutation routing on the $N$ node, $n$-cube takes $\mathcal{O}(N^{1/2})$ steps, for both SIMD and MIMD communications.*

Applying Theorem 2.3 on the hypercube, we obtain:

***COROLLARY 4.1***

*Deterministic permutation routing on a hypercube with $N$ nodes requires $\Theta(N^{1/2}/\log N)$ steps for oblivious algorithms with MIMD model.*

The $\Omega(N^{1/2}/n)$ lower bound for MIMD communication is tight since a deterministic oblivious algorithm, based on a Hamiltonian cycle decomposition of the hypercube, can route any permutation in $2N^{1/2}/(\log N - 2\log\log N + 2) + \log N/2 = \mathcal{O}(N^{1/2}/\log N)$ time [321]. The algorithm is based on many-one and one-many routing as explained below.

***THEOREM 4.2***

*There is an oblivious algorithm for routing of many-one and one-many relations on a hypercube with $N$ nodes in $N/(\log N - \log\log N) + \log N/2$ time [321].*

**PROOF** Using a classical result on a Hamiltonian decomposition of an (undirected) hypercube with $N$ nodes into $N$ Hamiltonian cycles, we are able to partition the algorithm for an oblivious many-one routing of an arbitrary permutation into two phases. In the first phase we correct the first $\log M$ bits and then the last $\log N - \log M$ ($M$ will be determined later). This can be accomplished by traversing a Hamiltonian cycle in whichever of two possible directions. Thus, the source-destination is $\leq M/2$ away, meaning that our routing problem can be solved in $M/2$ steps.

In the second phase we implement a one-many routing. We route $\geq N$ packets in an $N/M$ node hypercube $G$ with $\leq M$ packets at each node. If we decompose $G$ into $\log(N/M)$ edge-disjoint Hamiltonian cycles of the Hamiltonian decomposition of $G$, the packets will be evenly distributed among cycles. This can be done respecting the origins of packets, thus ensuring the oblivious routing. It follows that on each Hamiltonian cycle of length $N/M$ there are $\leq M/\log(N/M)$ packets which can be routed in time $\lceil (N/M) + M/\log(N/M) \rceil = N/\log(N/M)$. The total time needed for routing is $(M/2) + N/\log(N/M)$. The inverse (one-many) routing can be performed by reversing the two phases. The theorem follows if we take $M = \log N$. ∎

Thus, for oblivious permutation routing an $N$-hypercube one has:

***THEOREM 4.3***

*There is an oblivious algorithm for routing of arbitrary permutation on a hypercube with $N$ nodes in $\leq (2\sqrt{N}/(\log N - 2\log\log N + 2)) + (\log N)/2$ time [321].*

**PROOF** To route a permutation, we first correct the $\log(N/2)$ most significant bits of the address. To do that, we implement $\sqrt{N}$ independent many-one routings in an $\sqrt{N}$-hypercube. We then correct the $\log(N/2)$ least significant bits by performing $\sqrt{N}$ independent one-many routings in an $\sqrt{N}$-hypercube. Half of the packets may use one of the two edge-disjoint partitions in many-one routing. The second half may use the second partition. Kaklamanis et al. claimed that, similar to the proof of Theorem 4.2 one can save a factor of two in the first term, by routing the first half forward and the second half backward in the bit sequence. However, Hwang and Yao proved that this technique can not reduce the total number of steps by a factor of two, see [292]. ▮

For practical hypercubes of dimension $n \leq 14$, the above algorithm was initially considered as the most efficient. Small improvements have been made by adapting optimal routing methods (many-to-one and one-to-many) for small cubes. For $n \leq 7$, an arbitrary permutation can be online routed on the binary $n$-cube in just $n$ moves [293]. The proposed adaptive routing is *local* since a message does not need to know the other packet routes, only assuming that a packet at a node can detect whether an adjacent link is already occupied. For $n \leq 6$, optimal oblivious permutation routing algorithms based on uniformly constructed destination graphs, have been proposed in [234]. Dasgupta, Hwang, and Yao have recently extended the destination graph idea to incorporate delay schedules, proving optimal oblivious permutation routing for $n = 7, 8$, and providing the most efficient permutation algorithms known to date for $n \leq 14$ [158].

**Problem 3** *Is $n$-move local permutation routing on the MIMD $n$-cube possible for arbitrary $n$?*

***DEFINITION 4.1*** *A routing function $f$ with the property that messages from two different nodes cannot be sent to the same node is called an injective routing [378].*

All routing functions in this section are assumed to be injective.

***DEFINITION 4.2*** *A routing $f$ is called a semi-contraction if for all $i, j$ one has $|i - j| \geq |f(i) - f(j)|$. Here $|\cdot|$ denotes the integer difference of numbers.*

***LEMMA 4.1***

*At every hypercube node $n$, at any time, there is at most one message $m$ which wants to cross the $j^{th}$ dimension.*

**PROOF** By induction on $i$. Trivially, before the algorithm begins, there is at most one message at each node. For the inductive step, assume that the lemma is valid for all $i$, $0 \leq i < l$, and that a message is routed from node $j$ to $f(j)$ where $f$ is a semi-contraction. The messages are able to cross all dimensions $\leq l - 1$. As a consequence, the $l - 1$ destination low order bits are all zero. Therefore, we can partition the message destination address field in the following way: $\{x_{n-1} \cdots x_1\, x_0\}$ $= \{x_{n-1} \cdots x_{l+1}\, x_l\, 0 \cdots 0\}$. The bits $x_{n-1}, \ldots, x_{l+1}, x_l$ are the bits in the original message address, and we use the convention of writing the low order bits on the right. We claim that there can be no packet collision in the dimension $l$.

Assume that there exists a hypercube node $n$ with messages $m_1$ and $m_2$, both wanting to cross the dimension $l$, i.e., the last $l$ bits in both destination addresses will coincide; the $(l+1)^{st}$ (from the right) bit must be the same because of crossing, and the rest $l$ bits are both zeros anyway. The message destinations are different, so the difference between final destinations must be $\geq 2^{l+1}$. Since $f$ is a semi-contraction, the origins of $m_1$ and $m_2$ must differ for the same amount. It follows that some of the bits $x_{n-1}, \ldots, x_{l+1}$ for both messages are different, which is a contradiction with the assumption they both are at the same node $n$. ∎

***THEOREM 4.4***

*Semi-contractions (or semi-expansions) can be routed in bit time $\mathcal{O}(n)$, with $\mathcal{O}(n)$ bits of storage at each node of an SIMD $n$-cube [378].*

**PROOF** The hypercube dimensions are numbered $0, \ldots, n-1$.

```
Semi-contraction routing algorithm:
        for i = 0 to n − 1 do
            for each node n in parallel
                if  a message m at hypercube node n wants to cross
                    the ith dimension then
                send message m across the ith dimension
```

The algorithm terminates because of the Lemma 4.1 (since there are no packet collisions). ∎

Since packings (unpackings) are also semi-contractions (respectively, semi-expansions), we have:

***COROLLARY 4.2***
*Packings and unpackings can be routed on the SIMD $n$-cube in just $n$ steps.*

The above lemma also applies to the MIMD model, since this model is stronger. The methods of greedy routing (correcting dimensions either from right to left, or left to right) are also used to perform optimally general vector operations, such as odd-even merge and split, rotation, reversal, compression, and expansion [431]. Furthermore, for messages of size $\mathcal{O}(n)$, the greedy algorithm can be pipelined:

***LEMMA 4.2***
*Any semi-contraction, semi-expansion, or semi-broadcast can be routed on a pipelined MIMD $n$-cube in $\mathcal{O}(h+n)$ steps using queues of size $h$ [689].*

Partial permutation routing can use Batcher's sorting algorithm followed by an unpacking. Load balancing can also be based on simple operations of broadcast, concentrate, and distribute (which together form a semi-expansion) [301, 552, 580]. Thus, using synchronous greedy routing and load balancing after each routing step, JáJá and Ryu proved:

***THEOREM 4.5***
*$(N, N, k_1, k_2)$-routing can be solved on the pipelined MIMD $n$-cube in $\mathcal{O}(k_1 + k_2 n + n^2)$ time with $\mathcal{O}(k_2)$ queue size [301].*

Up to this point we have considered deterministic hypercube routing. A major theoretical breakthrough was the probabilistic permutation routing algorithm on the binary hypercube, proposed by Valiant [682] and extended to general graphs by Valiant and Brebner [685]. The two-phase algorithm consists of randomization (Phase I) and deterministic routing (Phase II). In Phase I, packets are sent to independent, randomly selected nodes through the network. The role of this phase is to reduce the difference between the average and the worst case performance. In Phase II, packets follow a shortest path to their final destinations. Dimensions are corrected either in an orderly [685] or random fashion [682].

***THEOREM 4.6***

*Probabilistic permutation routing on the MIMD $n$-cube runs in $\mathcal{O}(n)$ time using queues of size $\mathcal{O}(n)$, with high probability [682, 685].*

Probabilistic routing was later applied to multistage networks. Upfal showed how to route on the $\log N$-stage butterfly in $\mathcal{O}(\log N)$ steps using queues of size $\mathcal{O}(\log N)$ [678]. Pippenger was the first to show how to reduce the queue size to a constant, by probabilistic routing on a variant of the butterfly network in $\mathcal{O}(\log N)$ steps using constant size queues [549]. Valiant and Brebner extended the above theorem to routing $k$-relations on an arbitrary graph (Theorem 2.7) and proved that the two-phase strategy is optimal within the class of oblivious routing algorithms for graphs of asymptotically optimal diameter [683]. By introducing more randomness at each node (and some precomputation) Valiant reduced the propagation delay, effectively reducing the time delay of the randomized routing algorithm to $n + \mathcal{O}(n/\log n)$ steps using constant size queues, with high probability [684]. Ferreira and Grammatikakis improved the probability of routing all packets in asymptotically optimal time by routing packets in Phase I according to a random permutation, and not random destinations [201].

Valiant's algorithm on the binary hypercube requires $\mathcal{O}(N \log N)$ bits of randomness. Ranade showed how to route on the hypercube in $\mathcal{O}(\log N)$ time using only $\mathcal{O}(\log^2 N)$ random bits [577]. The effects of limited randomness and pseudorandomness on routing time were studied in [322, 536, 680].

### 4.2.2 Offline Permutation Routing

A permutation can be decomposed offline into a sequence of elementary transpositions, corresponding to the $2 \times 2$ crossbar settings of an appropriate size Beneš permutation network [53]. These transpositions can then be simulated in optimal time by embedding the Beneš network (with a single-input per each input terminal node) onto the binary hypercube. The mapping is one to one, since during each step data are exchanged between nodes differing in a single bit position.

**THEOREM 4.7**
*Any permutation can be offline routed on the n-cube in:*

1. $2n-1$ *steps using queues of size two, with SIMD model [662], or*
2. $2n-3$ *steps using queues of size two, with MIMD model [618].*

The latter algorithm provides a recursive construction based on Szymanski's result on realizing permutations on a circuit-switched 3-cube (see Lemma 4.12. No parallel algorithm for setting up switches in this model has been provided, while a sequential algorithm seems to require $\Omega(N)$ steps.

Serial setting of the switches takes $\mathcal{O}(N \log N)$ time, which is prohibitive [395, 396, 698]. The best parallel algorithm for computing the switch settings on the binary hypercube uses Cypher and Plaxton's sorting algorithm, improving the results of Nassimi and Sahni [502] and Lev, Pippenger and Valiant [423] as follows:

**LEMMA 4.3**
*Computing the switch settings for the* $2^n$ *node Beneš permutation network takes* $\mathcal{O}(n^3 \log n)$ *steps on the MIMD n-cube [140].*

All ($N$ log $N$!) *BPC* permutations can be realized on the $n$-cube in $n$ steps, using the routing algorithm of Nassimi and Sahni [501]. The algorithm is synchronous and follows individual cycles in the permutation. As an example, let $N = 16 = 2^4$, and $\pi(x) = x_3\ \bar{x}_1\ x_2\ \bar{x}_4$. Since $x_1 \Longrightarrow x_3 \Longrightarrow x_2 \Longrightarrow \bar{x}_1$, and $x_4 \Longrightarrow \bar{x}_4$, this permutation consists of a 3-cycle $(1,3,2)$, and a singleton cycle $(4)$. For each such cycle, and for all packets not in their final destinations, the corresponding source digits are changed one at a time. The *BPC* algorithm routes packets as follows:

$$0000 \Longrightarrow \text{wait} \Longrightarrow \text{wait} \Longrightarrow 0100 \Longrightarrow 0101$$
$$1000 \Longrightarrow 0000 \Longrightarrow \text{wait} \Longrightarrow \text{wait} \Longrightarrow 0001$$

**LEMMA 4.4**
*Any BPC permutation can be routed on the MIMD n-cube in just n steps with buffers of size two [501].*

Assuming the linear model, Johnsson and Ho showed how to route some *BPC* permutations (including the shuffle and matrix transposition) in optimal time for both the 1-port and $n$-port pipelined hypercube [309, 310]. The class of all permutations which can be routed optimally on the hypercube using the above cyclic scheme is wider than the BPC class [521]. Furthermore, it has been shown that cyclic shifts (a subclass of BPC) can be routed in $4n/3$ time on the $n$-cube without any local message buffers [518].

***LEMMA 4.5***
*Any Omega (or inverse Omega) permutation can be routed in n steps using synchronous greedy routing on the SIMD n-cube, correcting dimensions from left to right (respectively, right to left) [378].*

***LEMMA 4.6***
*Normal Ascend/Descend permutations can be performed in $\mathcal{O}(n)$ time with SIMD model [486]. Furthermore, $2^b$ Ascend/Descend computations on the n-dimensional binary hypercube take $\mathcal{O}(1)$ steps (using Gray-code) with MIMD model [495]. With SIMD model, $2^b$ Descend computations take $\mathcal{O}(n)$ steps, while $2^b$ Ascend computations take $\mathcal{O}(n^{1.5})$ steps with $\mathcal{O}(1)$ space per node [495, 496].*

Varvarigos and Bertsekas studied isotropic communication tasks which are symmetric with respect to any origin node. By considering also symmetric routing algorithms, one instance of isotropic tasks, the total exchange was reduced to a matrix decomposition problem, minimizing both completion time and average packet delay on the MIMD hypercube (and torus) [691, 694]. The hypercube definitions of isotropic tasks were based on XOR operations, while those for the torus used modular addition.

**Problem 4** *Classify isotropic and nearly isotropic tasks in symmetric topologies.*

### 4.2.3 Sorting Algorithms

Sorting is usually according to an implicit processor order, such as decimal order, or Gray-code ordering on the $n$-cube.

*Odd-even transposition* is one of the simplest parallel sorting algorithms. The $N$ elements are sorted in $N$ phases, each phase requiring

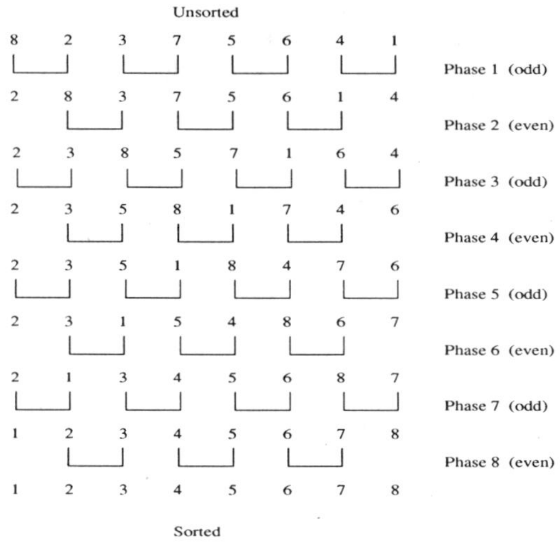


**FIGURE 4.4**
**Odd-even transposition sorting for $N = 8$ items [364]**

$N/2$ Compare&Swaps (see Figure 4.4) The Hamiltonian path consisting of the alternating odd/even phases can be directly mapped on the $n$-cube using Gray encoding.

In Batcher's *bitonic sorting* (or mergesort) network, a Compare&Swap takes place at each node. During a sequence of phases, data are exchanged between nodes differing in a single bit position. Using local node information, it can be decided whether the smaller or the larger data will be kept for further processing. Like the Beneš network, the bitonic sorting network (or mergesort) on $N = 2^n$ nodes can be easily embedded with dilation one onto the SIMD $n$-cube; links between adjacent stages of the sorting network map to edges of the hypercube one to one. The last four steps of bitonic sorting on 16 elements (Compare&Swap on dimensions $4, 3, 2, 1$) are shown in Figure 4.5.

Bitonic sorting can be performed on the hypercube with $\mathcal{O}(\log^2 N)$ time delay and queue size one [46]. It can be implemented more efficiently on the $n$-port pipelined hypercube; it takes $\mathcal{O}(\log^2 N/\text{loglog}\, N)$ time, with high probability [465].

*Parallel quicksort* can be implemented recursively. Step $i$, $1 \leq i \leq n$ consists of (a) selecting a pivot element, (b) broadcasting the pivot to each $n-i$-subcube, (c) partitioning local elements into two sets, (d) exchanging appropriate blocks between PEs whose labels differ in the

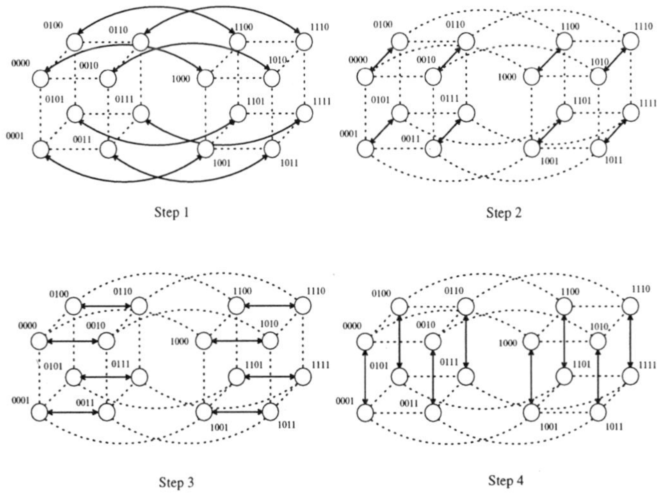


**FIGURE 4.5**
**Sorting a bitonic sequence - part of sorting on the $n$-cube [364]**

$i^{th}$ dimension, and (e) PEs with a 0 (1) in the $i^{th}$ bit retaining the elements smaller (respectively , larger) than the pivot. Poorly selected pivot elements would lead to sequences with different sizes causing severe degradation of the algorithm. With a uniform input data distribution, the median value constitutes a good pivot. Quicksort can be implemented on the $n$-port hypercube in $\mathcal{O}(\log^2 N/\log\log N)$ time, with high probability [701].

Cypher has shown that any embedding of the AKS $\mathcal{O}(\log N)$ depth sorting network requires $\Omega(\log^2 N)$ time on the shuffle-exchange and cube-connected cycles [141]. Furthermore, many efforts have produced lower bounds on the depth and size of shellsort and hypercubic sorting networks [553, 554, 555]; a lower bound on the depth of all hypercubic networks is $\mathcal{O}((\log^{1+\epsilon} N))$, for any constant $\epsilon > 0$ [553].

Recently, Cypher and Plaxton have given an involved $\mathcal{O}(\log N \log\log N)$ deterministic sorting algorithm (called sharesort) [144]. Since the constant involved is in the thousands, this algorithm, although asymptotically better, it is not practical for small dimension hypercubes ($n \leq 16$).

**Problem 5** *Is it possible to perform deterministic sorting on the MIMD $n$-cube in $\mathcal{O}(n)$ steps?*

We have concentrated here on the saturated (fine grain) case, where the number of data elements ($p$) matches the number of PEs ($N$). Most sorting techniques when $p < N$ use a fast $\mathcal{O}((p/N)\ \log\ (p/N))$ local quicksort or heapsort and replace the Compare&Swaps with block based Compare&Splits. At first, data blocks are swapped between PEs. Then, these blocks are merged and each PE retains its appropriate half of the merged block. For sorting techniques when $p < N$ refer to [145, 231, 465, 689].

### 4.2.4 Global Communications

Global communication schemes on hypercubes have been extensively considered. Broadcasting can be easily implemented, since the hypercube belongs to the class of broadcasting graphs, i.e., in a broadcasting graph, the number of nodes possessing the necessary information can be doubled after each communication step. Thus, hypercube broadcasting is implemented based on an embedded binary spanning tree. The algorithm can be used on an SIMD hypercube, since dimensions are considered in sequential order [308, 650]. Assuming the constant delay model, broadcasting takes an optimal $n$ steps on the SIMD $n$-cube.

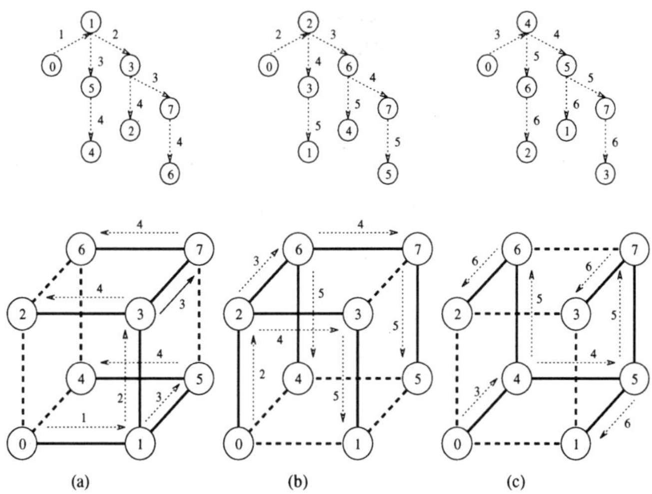


**FIGURE 4.6**
**Broadcast on $n$-cube using $n$ edge-disjoint spanning trees [364]**

Many optimal global communication algorithms for the hypercube are based on binomial spanning trees. A one-node binomial tree is the node itself. A $k$ node binomial tree is obtained from two $k/2$ node binomial trees by joining (with an edge) the roots of the two trees. Binomial trees can be easily constructed for the hypercube. Assuming the linear model, the time complexity can be improved by splitting the packet (taken as a no cost operation) into $n$ smaller sub-packets which are first routed to nodes: $0\ldots01$, $0\ldots010$, $\ldots$, and $10\ldots0$. From these nodes packets are routed farther using $n$ edge-disjoint symmetric binomial spanning trees which are constructed using circular shifts [210, 308]. This technique is illustrated in Figure 4.6. Optimal time can be achieved by further splitting of sub-packets into groups and exploiting pipelining on the binomial spanning tree graph [650].

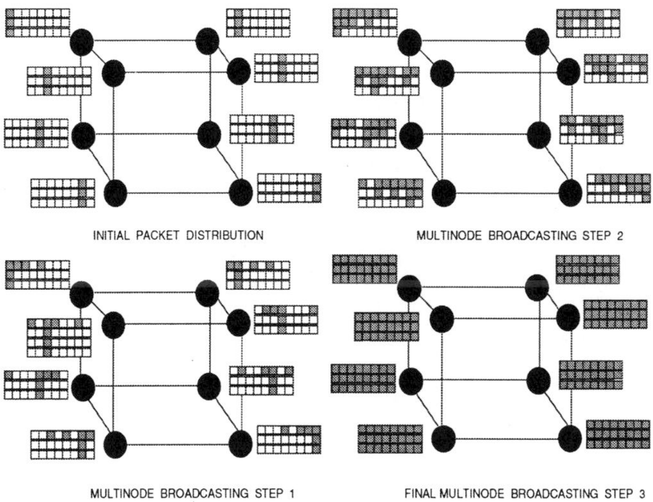


**FIGURE 4.7**
**Multinode broadcast on $n$-cube by using $n$ edge-disjoint spanning trees**

Multinode broadcasting can be implemented using simultaneous single-node broadcasts from each hypercube node; in this case, pipelining can not be used. The time complexity of each step approximately doubles, since each node must now forward twice the information. The information exchanged during each step is shown in Figure 4.7. The algorithm achieves $\mathcal{O}(\lceil (N-1)/n \rceil)$ delay which is optimal for the MIMD $n$-cube [62, 308]. A similar algorithm is optimal for the SIMD hypercube [308].

Single node scatter may use spanning trees used for broadcasting; by giving priority to nodes farthest away the algorithm achieves optimal $\mathcal{O}(\lceil (N-1)/n \rceil)$ time on the MIMD $n$-cube [62]. For an optimal number of packet transmissions a new perfectly balanced spanning tree has been introduced [62]. For the SIMD hypercube, similar techniques achieve optimal complexity [308, 594]. Varvarigos and Bertsekas have also considered the effect of a random packet size on the performance of multinode broadcasting. While for the hypercube only small latency variations were shown, the situation is much different for a ring [692].

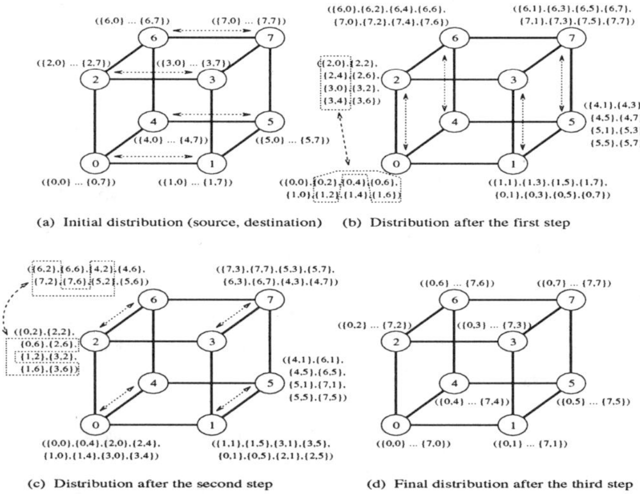


(a) Initial distribution (source, destination) (b) Distribution after the first step

(c) Distribution after the second step (d) Final distribution after the third step

**FIGURE 4.8**
**Total exchange on $n$-cube [364]**

As shown in Figure 4.8, total exchange algorithms are similar to multinode broadcasting ones; only the packet contents and sizes really differ [133]. Furthermore, since a lot of information must be exchanged, simpler techniques based on routing packets along Hamiltonian circuits on the $n$-cube are also competitive. A simple algorithm which achieves an optimal $\mathcal{O}(\lceil N/2 \rceil)$ delay for the MIMD $n$-cube is given in [62]. This recursive algorithm is based on three overlapped phases: (a) total exchange on each $(n-1)$-dimensional sub-cube, (b) packet exchange between corresponding $(n-1)$-dimensional sub-cube nodes, and (c) total exchange of the packets received in phase 2 on each $(n-1)$-dimensional sub-cube. Optimal SIMD algorithms for this problem are much simpler [594].

### 4.2.5 Global Communications with Linear Model

Pipelining is a way to improve broadcast as an example of intensive hypercube communication. In essence, pipelining applied to a message means splitting the message into pieces and sending them in waves across the hypercube, in order to make more nodes busy at the same time [650].

Let $\tau$ represent the packet transfer rate, and let $\beta$ denote the startup and termination time. We assume that it takes $\tau l + \beta$ time for sending a packet of length $l$ from a hypercube node to its neighbor.

Denote by $C_k$ the subcube containing all hypercube nodes which are distance $k$ from node 0.

In an $n$-dimensional hypercube, $\oplus_k(m)$ represents the flipping of the $k^{th}$ bit in binary representation of node $m$ (i.e., the bitwise $XOR$ of $m$ and $2^k$).

Denote by $\overset{i}{\hookleftarrow}(m)$ the left circular shift of the $n$-bit representation of node $m$ by $i$ bits. Here, $i \in \{0, \ldots, n-1\}$.

#### 4.2.5.1 Broadcast

*Algorithm:* BROADCAST 1 *(Recursive Doubling)*

The broadcast is done in $n$ stages, $0, \ldots, n-1$. In $k^{th}$ stage, hypercube nodes $0, \ldots, 2^k - 1$ send the message to nodes $\oplus_k(0), \ldots, \oplus_k(2^k - 1)$ respectively and concurrently. Notice that the entire algorithm requires time $n(\tau l + \beta)$ for a packet of length $l$ .

Since BROADCAST 1 does not take the advantage of the concurrent read/write channels of the link bound hypercube, we can improve the algorithm by working simultaneously on $n$ subproblems along separate dimensions.

*Algorithm:* BROADCAST 2

Partition the message into $n$ equal packets $P_0, \ldots, P_{n-1}$ of size $l/n$. In $k^{th}$ stage, nodes $\overset{i}{\hookleftarrow}(0), \ldots, \overset{i}{\hookleftarrow}(2^k - 1)$ send packet $P_i$ to nodes

$\oplus_{(i+k) \bmod n}(\overset{i}{\hookleftarrow}(0)), \ldots, \oplus_{(i+k) \bmod n}(\overset{i}{\hookleftarrow}(2^k-1))$ respectively, for each $i \in \{0, \ldots, n-1\}$. Notice that each stage of the algorithm takes $\tau(l/n)+\beta$ time, and that the total time is $\tau l + n\beta$.

In spite of the improved performance, in BROADCAST 2 half of hypercube links never get used, and large proportion of used links are almost always idle. In order to rectify this, we attempt to make all links always busy sending packets to free nodes.

*Algorithm:* BROADCAST 3

There are $n+1$ stages, numbered $0, 1, \ldots, n$. Messages are partitioned into packets as in BROADCAST 2. In $(k-1)^{st}$ stage, each node in $C_k$ receives $k$ distinct packets from $C_{k-1}$. In $k^{th}$ stage, the nodes in hypercube sector $C_k$ send each only one packet along $n$ outgoing links. In $(k+1)^{st}$ stage, each node in $C_k$ receives $n-k$ remaining packets from $C_{k+1}$. Since there are now $n+1$ stages, the total time for this algorithm is equal to:

$$(n+1)\left(\tau\frac{l}{n}+\beta\right) = \frac{n+1}{n}\tau l + (n+1)\beta \, .$$

This hypercube broadcasting algorithm can be improved so that the last stage is eliminated by sending the second packets wave immediately after the first one [650].

*Algorithm:* BROADCAST 4

In first stage, node 0 relabels packets $P_0, \ldots, P_{n-1}$ as $Q_{n-1}, Q_0, Q_1, \ldots, Q_{n-2}$, respectively. From node 0 we implement again BROADCAST 3 (for $n-1$ stages only), but now using $Q_0, Q_1, \ldots, Q_{n-1}$ instead of $P_0, P_1, \ldots, P_{n-1}$, respectively. Each node in $C_{n-1}$ will now receive every packet. The saving in time results in total running time of $\tau l + n\beta$.

We now apply the pipelining in order to get the following:

*Algorithm:* BROADCAST 5

At first, we pipeline BROADCAST 3 (i.e., we divide the message into $g$ groups $0, 1, \ldots, g-1$ of equal length $l/g$, and then apply separate BROADCAST 3 algorithms to the first $g-1$ groups). After that, we execute the BROADCAST 4 algorithm for the last group. No node should be broadcasting on more than one stage at a time. This means that we have $n+g-1$ stages, numbered by $0, 1, \ldots, n+g-2$. In $k^{th}$ stage, each node in $C_j$, $k-g+1 \leq j \leq k$, executes the $j^{th}$ stage of BROADCAST 3 for $(k-j)^{th}$ group.

Since each of the $n+g-1$ stages takes $\tau(\frac{l}{ng}) + \beta$ time, the whole BROADCAST 5 takes:

$$(n+g-1)\left(\tau\frac{l}{ng} + \beta\right) = \frac{(n+g-1)\tau}{ng}l + (n+g-1)\beta$$

time. This function (of $g$) is minimized when:

$$g = \begin{cases} 1\,, & n = 1 \\ \sqrt{\frac{(n-1)\tau}{n\beta}}\, l^{1/2}\,, & n \geq 2\,, \end{cases}$$

which yields (minimal) final time for broadcasting of:

$$\begin{cases} \tau l + \beta\,, & n = 1 \\ \frac{\tau}{n}l + 2\sqrt{\frac{(n-1)\beta\tau}{n}}\, l^{1/2} + (n-1)\beta\,, & n \geq 2\,. \end{cases}$$

This represents a significant improvement on time in [308] of

$$\frac{\tau}{n}l + 2\sqrt{\beta\tau}\, l^{1/2} + n\beta\,.$$

***THEOREM 4.8***
*For packets of fixed size,* BROADCAST 5 *is optimal.*

**PROOF** Let $s$ denote the (fixed) packet size. Each hypercube node must receive $\geq l/s$ of these packets, and $\geq 1$ such packets will wait $((l/(ns)) - 1)(\tau s + \beta)$ time for a free link before leaving node 0. This is because only $n$ packets can be sent out at a time from node 0. In order to reach node $2^n - 1$, this packet must traverse $n$ links, which takes time $n(\tau s + \beta)$, yielding the total minimum time of $(n + (l/(ns)) - 1)(\tau s + \beta)$. Since BROADCAST 5 takes exactly this time for $g = l/(ns)$ the proof follows. ∎

### 4.2.5.2 Scatter

In this communication mode, a hypercube node sends a different message to each of the remaining nodes. Various authors use different names for this type of hypercube communication [594, 650].

*Algorithm:* SCATTER (DISTRIBUTE 2 *in [650]*)

For fixed $n$ let

$$k = \min\left\{k \mid n\,\frac{(n-1)^k - 1}{n-2} \geq 2^n - 1\right\},$$

and let $r$ be the solution of the equation

$$n\frac{r^k - 1}{r - 1} = 2^n - 1 .$$

One can take any $r'$, such that $r \leq r' < n - 1$ (when $r$ is an irrational number). Starting from node 0, there will be $k$ waves of packets:

1. $k^{th}$ wave will contain messages for $C_1$ such that each outgoing link from 0 carries a message of size $l$.

2. $(k-1)^{st}$ wave starts from 0 with packets of size $rl$ along each outgoing link with all messages for $C_2$, and in case of large $n$, with portions of the messages for $C_3$.

3. $(k-2)^{nd}$ wave will start with packets having size $r^2l$ with the rest of the messages for $C_3$, all messages for $C_4$ (and in case of large $n$, with portions of the messages for $C_5$).

Each packet wave starts with packets $r$ times larger that the following wave, containing messages for a set of further $C_i$'s. After a wave reaches all the nodes in $C_1$, it is partitioned into wavelets, one for each $C_i$ within main packet wave.

If $B(C_i, C_j)$ denotes the bandwidth from subcube $C_i$ to subcube $C_j$, one has $B(C_1, C_2) = (n-1)B(\{0\}, C_1)$. Since $r \leq n-1$, for a sufficiently large message length $l$, each packet wave leaves $C_1$ before next wave arrives. It is easy to show that for sufficiently large $l$, all wavelets reach their destination by the time the last wave reaches $C_1$. It follows that the total time for the algorithm is determined by the time it takes for the node 0 to send out all $k$ waves, equal to:

$$\frac{(2^n - 1)\tau}{l} + k\beta \sim \frac{(2^n - 1)\tau}{l} + \frac{n}{\log_2 n}\beta .$$

We see that one cannot get a lower bound by adding the bandwidth lower bound to the startup lower bound.

#### 4.2.5.3 Total Exchange

In this communication mode, each hypercube node wants to send (receive) a different message to (from) each other node, respectively.

| *Algorithm:* TOTAL EXCHANGE |
|---|

Since every node starts with $(2^n - 1)l$ items which have to be sent to all other nodes, in $k^{th}$ stage the node starts sending messages to the $\binom{n}{k}$ nodes at distance $k$. In $k^{th}$ stage, a total of $2^n \sum_{j=0}^{k} \binom{n}{j}$ messages are being worked on. Because of the symmetry, each of the $2^n n$ links sends a packet of size:

$$\sum_{j=0}^{k} \frac{\binom{n}{j}}{n} l = \sum_{j=0}^{k} \frac{\binom{n-1}{j}}{n-j} l ,$$

the total time is

$$\sum_{k=0}^{n-1} \left[ \tau \sum_{j=0}^{k} \frac{\binom{n-1}{j}}{n-j} l + \beta \right] = \sum_{j=0}^{n-1} \sum_{k=j}^{n-1} \frac{\binom{n-1}{j}}{n-j} l + n\beta = \sum_{j=0}^{n-1} \binom{n-1}{j} \tau l + n\beta$$
$$= 2^{n-1} \tau l + n\beta .$$

We see an improvement of the time of $2^{n-1} n\tau l + n\beta$ from [594].

### 4.2.6 Dynamic Routing Problems

Every node of the $n$-cube independently generates packets with rate $\lambda$. Each packet's destination is chosen randomly, with each node at distance $r$ being assigned an *a priori* probability $p^r(1-p)^{n-r}$. Thus, if $p = 0.5$ the distribution of the destinations is uniform, while if $p < 0.5$ the distribution favors shorter (localized) distances. Greedy routing (Section 3.8) is stable if the load factor satisfies $\lambda p < 1$. This is also the best possible region for stability.

***THEOREM 4.9***

*If packets are generated at each node with rate $\lambda$, and destinations are selected randomly with each node at distance $r$ having a probability $p^r(1-p)^{n-r}$, the average packet delay $T$ on the MIMD $n$-cube satisfies*

$$np + \frac{n\lambda p}{2\,(1-\lambda p)} \le T \le \frac{np}{1-p},$$

*for any $\lambda p < 1$ [645].*

Approximate models for *dynamic routing* aim at giving intuitive support for simulation results. Abraham and Padmanabhan's analysis for the buffered and unbuffered (single packet buffered) binary hypercube is based on the approximately correct assumption that each packet moves independently of other packets [4]. This model was later modified to consider locality of references [616]. Stability conditions and approximate delay models, when packaging several PEs at each hypercube node, have also been investigated [116]. Dynamic routing on the hypercube outperforms the star graph and alternating-group graph [236]. For vertex-transitive graphs, the maximum generating rate prior to saturation depends on the degree and average node distance.

*Dynamic broadcasting* has also been extensively considered. In this problem, packet broadcasts are generated at random instants at each node. The problem also arizes in practice, when overlapping global communication with local computation, e.g., when solving first order recurrence equations, $X_{i+1} = A \cdot X_i$, where $X$ is distributed to different nodes. Although dynamic broadcasting may be based on spanning trees [644, 643], an approach based on developing efficient partial multinode broadcasting algorithms (used as subroutines) is more general and achieves the widest range of stability with asymptotically optimal latency [693, 690].

Routing $N$ packets from distinct nodes to *independently chosen random destinations* corresponds exactly to Phase I of the probabilistic routing algorithm. Therefore, greedy routing can be used to achieve $\mathcal{O}(n)$ delay, with high probability. In this scenario, deflection routing also achieves $\mathcal{O}(n)$ delay, with high probability [191].

Another interesting problem concerns the maximum distance $h_{max}$ over all (source, random destination) pairs. While if $h_{max} \leq 2$, all $N = 2^n$ packets with distinct source nodes can be optimally routed on the MIMD $n$-cube, there is only one general (nontight) bound on the time complexity for all $h_{max} = 3, 4, \ldots, n$.

### 4.2.7 Fault Tolerant Routing

**$P$-faulty model.** Assuming dynamic node failures and focusing on a single packet route, ignoring packet congestion, Gordon and Stout proved:

***LEMMA 4.7***

*If processors have no local knowledge, the probability that a single message is successfully routed to its antipodal destination on the n-cube, using shortest path routing, is* $(1-p)^{n-1}$*, and converges to zero as n approaches infinity. If processors have local knowledge, the corresponding probability is* $\prod_{i=2}^{n}(1-p^i)$*, and converges to nonzero as n approaches infinity [229].*

**Worst Case Model.** Bounds on the hypercube diameter for a fixed number of faults are provided in [512, 667]. Other known results for this model are given in:

***LEMMA 4.8***

*By maliciously placing* $\mathcal{O}(n)$ *faulty nodes on the n-cube, the hypercube can be disconnected into components of size* $\mathcal{O}(N/n^{1/2})$ *[229].*

***THEOREM 4.10***

*Any shortest path routing on the n-cube is* $(3, n-1)$*-tolerant [88].*

***LEMMA 4.9***

*Greedy shortest path routing on the n-cube is* $(2, n-1)$*-tolerant [88].*

Also, if each node has at least one nonfaulty neighbor, there is a routing on the $n$-cube which is $(4, 2n-3)$-tolerant [611].

Chen and Shin first considered routing directly on the faulty hypercube. Let's first assume that each node can determine the status of its own communication links (live or faulty). Algorithm $A_1$ tries to move a packet closer to its destination. If all possible moves are blocked, then the packet is sent along spare dimension $i$, chosen so that dimension $i$ has not been used before as a spare dimension.

***THEOREM 4.11***

*Algorithm* $A_1$ *routes a packet on the n-cube, with* $f < n$ *faulty links (or* $f < n$ *faulty nodes), via an optimal path, with high probability. The expected path length is* $n + \mathcal{O}(n/N)$ *(respectively,* $n + \mathcal{O}(n^3/N)$*) [103].*

Now, we assume that each node can determine the status not only of its own communication links, but also of its neighboring components.

Algorithm $A_2$ works exactly as $A_1$, except that it checks if a move is blocked one more hop away.

**THEOREM 4.12**

*Algorithm $A_2$ routes a packet, from its source $u$ to its destination $v$, on the $n$-cube with $f < n$ faulty links (or $f < n$ faulty nodes) via a path of length at most $H(u, v) + 2$, where $H$ represents the Hamming distance [103].*

In the presence of more than $n - 1$ faults, shortest path routing can be obtained using network delay tables. Since these tables are sometimes difficult to maintain and update, an alternative technique based on depth-first search *(DFS)* routing has been proposed. The *DFS* algorithm tries to move a packet closer to its destination. If all such moves are blocked, then an alternate path is considered. If no alternate path exists, backtracking is enforced. To avoid message looping, the *DFS* algorithm never visits a node twice, unless backtracking is enforced.

**THEOREM 4.13**

*DFS routes a packet on the $n$-cube via an optimal path, with high probability, provided that a live path exists [104].*

Hayes and Lee considered fault tolerant routing and broadcasting schemes, when each node has no more than $k$ faulty nodes in its $k$-neighborhood [398]. They proposed a routing algorithm $A_1'$ which works as follows. A packet currently at distance $r$ from its destination examines $r$ node-disjoint shortest paths, formed by use of cyclical shifts. The first path without any faults in its first $k$ nodes is taken. If no such path is found, the $n - r$ nonminimal paths of length $r + 2$ are checked, and one of the remaining dimensions is shifted.

**THEOREM 4.14**

*If every nonfaulty node has at most $k$ faulty nodes in its $k$-neighborhood ($k \leq 2$), then algorithm $A_1'$ routes a packet on the $n$-cube via an optimal path [398].*

Algorithm $A_1'$ is not optimal for $k \geq 3$. Suppose that another algorithm $A_2'$ works as $A_1'$, but searches for a feasible path, not only within the $r$ node-disjoint paths, but within all shortest paths. Then, we have:

**THEOREM 4.15**
*Algorithm $A'_2$ routes a packet on the n-cube via an optimal path, provided that every nonfaulty node has at most k faulty nodes in its k-neighborhood ($k \leq n-1$) [398].*

Finally, algorithm $A'_3$ works as $A'_2$, but prefers to route along the shortest path to safe nodes, which by definition are nonfaulty nodes with at most one faulty or unsafe nearest neighbor. All other nodes are unsafe or faulty.

**THEOREM 4.16**
*If the number of faulty nodes is $f < \lceil n/2 \rceil$, algorithm $A'_3$ routes a packet on the n-cube, from its source u to its destination v, via a path of length at most $H(u,v) + 2$ [398].*

**Problem 6** *Compare algorithms $A_1$ through $A'_3$ in terms of their performance and resources needed for their implementation.*

Leighton, Maggs, and Sitaraman considered the fault tolerance of *normal hypercube algorithms*. Communication in these algorithms is always performed by correcting dimensions in consecutive order (1-port communication); furthermore, wraparound of dimensions 0 and $n$ is also allowed.

**THEOREM 4.17**
*An N node faulty hypercube, with $N^{1-\varepsilon}$ worst case faults (for any constant $\varepsilon > 0$), can emulate any normal algorithm running on the N node hypercube with only constant slowdown [411].*

Similar results have been obtained for any algorithm (not necessarily normal) running on the wrapped butterfly and the shuffle-exchange graph [411].

**IDA Model.** Rabin's information dispersal approach (IDA) can be applied to the $N$ node, $n$-cube to tolerate $N/n$ link failures [447, 563]. Hastad et al. improved this result by employing a better choice of the parallel routing paths. The $n$ pieces of a message are first sent to the neighbors of the node which generated the packet. Then, these pieces are routed along parallel paths to the $n$ neighbors of a random intermediate node. From there, they are routed along parallel paths to the neighbors of the intended destination, and finally to the destination itself.

**THEOREM 4.18**

*Let all components of an MIMD $n$-cube fail independently with probability $1/(cn)$ (for a sufficiently large $c$), before or during a permutation routing. Then, assuming that the packet's source and destination nodes are nonfaulty, the routing will be completed with high probability in $\mathcal{O}(n)$ steps using queues of size $\mathcal{O}(n)$ [258].*

**PROOF** The bistate failure of the $i^{th}$ hypercube component is a binary random variable with the moment generating function

$$M_i(\lambda) = \frac{1}{cn}e^{\lambda} + \left(1 - \frac{1}{cn}\right) .$$

It follows that the moment generating function for the sum of random variables for all $(2n+3)n$ components is

$$M(\lambda) \leq \left(1 + \frac{e^{\lambda} - 1}{cn}\right)^{(2n+3)n} \leq e^{\frac{(e^{\lambda}-1)(2n+3)}{c}} .$$

Now, one can bound the failure of $\geq n/8$ hypercube components by

$$e^{\frac{(e^{\lambda}-1)(2n+3)}{c} - \frac{\lambda n}{8}} .$$

If we choose $\lambda = \ln \frac{c}{16}$, we get the upper bound on failure probability to be $(e^{1/8}(\frac{16}{c})^{1/8})^n$ which can be made arbitrarily small by increasing the constant $c$. ∎

This theorem is based on an improvement [258] of Rabin's [563] fault tolerant routing in presence of $\mathcal{O}(\frac{1}{n^2})$ failing nodes in a hypercube. In Rabin's algorithm, packets are partitioned into $n$ pieces which are routed twice along node-disjoint paths to the final destination. The algorithm from [258] uses a more uniform path selection; $n$ pieces are first sent to the neighbors of the source, and then (along parallel paths) to the neighbors of a random, intermediate hypercube node. The pieces are then routed (in parallel) to the neighbors of the destination, and finally, to the destination itself.

For $s$ and $t$, two nodes of a hypercube, denote by $p_i(s,t)$ the Valiant-Brebner path ([685]) from $s^i$ to $t^i$ (used in one phase), and let $P(s,t) = \{p_i(s,t) | 1 \leq i \leq n\}$ denote the set of all such paths. Varying $s$ over all nodes in hypercube, denote by $\mathcal{F}$ the family of all paths in $N$ sets $P(s,t)$, where target $t$ is chosen randomly and uniformly. Note that for a hypercube node $u$, the neighbor across the $i^{th}$ dimension is denoted by $u^i$.

***LEMMA 4.10***

*For all hypercube nodes $u$, and for all $j$, where $0 \leq j \leq n$, it is highly probable that $u$ is the $j^{th}$ node along only $\mathcal{O}(n)$ paths in the family $\mathcal{F}$ [258].*

**PROOF** Let $u$ denote the $j^{th}$ node along the path $p_i(s,t)$ from $s$ to $t$ in the hypercube. Then, $u^i = t_1 t_2 \ldots t_j s_{j+1} \ldots s_n$. We consider now two possible cases:

1. If $i \leq j$, then $s_{j+1} \ldots s_n = u_{j+1} \ldots u_n$. There are exactly $2^j$ such nodes $s$ in an $n$-hypercube. If one of $s$'s chooses a $t$ such that for some $i \leq j$, one has $t_1 \ldots t_{i-1} \bar{t}_i t_{i+1} \ldots t_j = s_1 \ldots s_j$, then there is only one path $p_i(s,t)$ with $u$ being the $j^{th}$ node. The other case is when there will be no such paths. It follows that for each of $2^j$ nodes, the probability of one such path is $j/2^j$.

2. If $i > j$, one has $s_{j+1} \ldots s_{i-1} \bar{s}_i s_{i+1} \ldots s_n = u_{j+1} \ldots u_n$. Such a condition is satisfied by exactly $(n-j)2^j$ nodes of hypercube. Choose $t$ now in such way that $t_1 \ldots t_j = u_1 \ldots u_j$. As a consequence, the probability of $u$ being the $j^{th}$ node along only one such path is $1/2^j$.

Out of those two cases it follows that one has to consider the sum $X$ of $2^j$ binary random variables, each of probability $j/2^j$ of having value one, and $(n-j)2^j$ binary random variables, each of probability $1/2^j$ of having value one. The moment generating function $M(\lambda)$ for $X$ satisfies the following inequality:

$$
\begin{aligned}
M(\lambda) &= \left(\frac{je^{\lambda}}{2^j} + 1 - \frac{j}{2^j}\right)^{2^j} \left(\frac{e^{\lambda}}{2^j} + 1 - \frac{1}{2^j}\right)^{(n-j)2^j} \\
&= \left[1 + \frac{(e^{\lambda}-1)j}{2^j}\right]^{2^j} \left[1 + \frac{(e^{\lambda}-1)}{2^j}\right]^{(n-j)2^j} \\
&\leq e^{(e^{\lambda}-1)j} e^{(e^{\lambda}-1)(n-j)} = e^{n(e^{\lambda}-1)} .
\end{aligned}
$$

It follows that

$$Pr[X \geq an] \leq e^{n(e^\lambda - 1)} e^{-an\lambda} = \left(e^{e^\lambda - a\lambda - 1}\right)^n .$$

Now, for $\lambda = \ln a$, one has $Pr[X \geq an] \leq \left[e^{a(1-\ln a)-1}\right]^n$. The expression on the RHS converges to zero as $a$ increases, hence the proof of the lemma follows. ∎

The basic outcome of Lemma 4.10 is noncrossing, with high probability of no more than $an$ packet pieces at each hypercube node. Since packets traversing hypercube links can come from two directions, $\leq 2an$ pieces cross any links during routing. If all hypercube components have the capacity of holding $2a$ original packets, we can conclude that with high probability there is no need for buffering and no packet pieces will wait in queues.

Since there are $\leq (2n+3)n$ different hypercube components incident to packet pieces from $s$ during the first routing phase, it would be valuable to have a bound on the number of encounters of the packet pieces routes:

***LEMMA 4.11***

*For $u \neq s, t$ in a hypercube, the number of possible paths crossings at $u$ in $P(s,t)$ is $\leq 2$.*

**PROOF** We count the nodes along $p_i(s,t)$ starting from $s^i$ (which is $0^{th}$ node on the path). Assume that for $i < j$, paths $p_i(s,t)$ and $p_j(s,t)$ cross in $k^{th}$ and $l^{th}$ node, respectively. Then, $t_1^i t_2^i \ldots t_k^i s_{k+1}^i \ldots s_n^i = t_1^j t_2^j \ldots t_l^j s_{l+1}^j \ldots s_n^j$. Here, $s_m^{m'} = s_m$ and $t_m^{m'} = t_m$, for $m \neq m'$.

1. For $k, l \leq j$ one has $s_j^j = s_j^i$, which is impossible.

2. If, $k, l \geq i$ one has $t_i^j =_i^i$, which is impossible.

3. If, $k < i$ and $l > j$, or $l < i$ and $k > j$, it follows that $t_i = \overline{s}_i$, $t_j = \overline{s}_j$, and $t_q = s_q$ for $i < q < j$. It follows that for such $q$'s, $p_q(s,t)$ cannot cross $u$, which is again a contradiction.

The proof of lemma follows. ∎

The topic of fault tolerant global communication algorithms on the hypercube has attracted an immense interest.

**Problem 7** *Do any constant degree networks share the above property?*

Many fault tolerant algorithms have been proposed for global hypercube communication and especially broadcasting [90, 209, 214, 273, 274, 397, 57, 533, 576, 711]. Most algorithms exploit the multiplicity of hypercube paths by extensively using packet replication. For example, in one particular technique, fault tolerant broadcasting can be accomplished by starting recursive doubling from each neighbor of the broadcast (root) PE. A greedy routing algorithm can tolerate $n - 1$ destructive faults, since there are $n$ disjoint paths on the $n$-cube through which a PE can receive information. Byzantine faults involving the malicious corruption of delivered messages (including omitting, rerouting etc) further limit the number of faults that can be tolerated to $\lfloor n/3 \rfloor$ (see [576]).

Using analytical models for small size systems [719], and simulations for large systems, it can be observed that hypercube system reliability, computed as the probability of a nondisconnected system under random failures (without repairs), is much better than the corresponding reliabilities of the mesh, tree, and ring. It also approaches that of a fully connected graph. However, considering cost vs. efficiency issues for dynamic routing on faulty graphs using reachability arguments to compute the average latency, we find that the worse graphs, after the fully connected graph, are ring, tree, mesh, and finally, hypercube [68, 156]. Furthermore, when considering dynamic routing with system repairs, tree deteriorates the most, while mesh, hypercube, and fully connected have similar reliabilities [68, 156].

Notice that under random faults, the mean time between routing failures may be computed using Markov chains. Each Markov state represents different distances from the packet's final destination, with one additional state representing system failure. State transitions depend on packet service rates, and also the probability of failure for each choice made by the routing algorithm, which may also involve packet priorities.

Work has also been reported on fault diagnosis on $n$-cube, by defining limits on the number of faults for isolation of nodes. The design of an optimal minimum traffic diagnostic algorithm on the hypercube is NP-complete [223]. It has been proved that the $n$-cube ($n \geq 3$) is precisely one-step $n$-fault diagnosable, while the $n$-cube ($n \geq 4$) is pessimistically one step $(2n - 2)$-fault diagnosable [223, 334]. Based on the theoretical foundation of system diagnosis for general graphs [121, 148, 714], complete diagnosis algorithms for the hypercube are provided in [654],

### 4.2.8 Cut-Through, Wormhole, Circuit Switching, and Hot-Potato Routing

Fast bit-serial algorithms are very difficult to construct. This is particularly true for the $n$-cube since the head of the packet may be far from its tail, preventing up to $\mathcal{O}(n)$ links to be used, and raising the issue of deadlock [153]. Bit-serial algorithms on the hypercube are often based on properties of more general routing schemes, e.g., edge-disjoint routing properties of sorting networks imply efficient schemes for wormhole, circuit-switched, and hot-potato routing. In particular, Aiello et al. provided a randomized algorithm for bit-serial routing on the $n$-cube. The algorithm consists of embedding a probabilistic two-phase routing algorithm, which runs on an $n$-dilated butterfly of $n2^n$ nodes, onto the $n$-cube. The embedding is based on 1-error correcting codes and achieves constant load and congestion and $\mathcal{O}(\log N)$ dilation. Leading address bits are deleted from the header as the packet moves on. The algorithm works in the wormhole, cut-through, or circuit-switching mode.

***THEOREM 4.19***

*For packets $b$ bits long, any permutation can be routed on the $n$-cube in $\mathcal{O}(b + \log N)$ bit steps using constant size queues, with high probability [17].*

An $n$-cube is *circuit-switching rearrangeable* if for any arbitrary permutation there exist $2^n$ edge-disjoint paths connecting all (source, destination) pairs.

***LEMMA 4.12***

*The $n$-cube ($n \leq 3$) is circuit-switching rearrangeable [655]. If all paths are restricted to minimal lengths, the $n$-cube is not circuit-switching rearrangeable [446].*

Circuit switching hypercubes can not perform better than store-forward ones, since they emulate optimal store-forward communication algorithms and incur higher startup costs.

Dynamic hot-potato routing on the $n$-cube has good empirical performance. Simulations and approximate models have been considered in [240, 241, 283].

## 4.3 Hypercube Embeddings and Algorithms

We provide a glimpse at hypercube algorithm design to illustrate the use of regular communication patterns. A general reference to parallel algorithms is [364].

Basic operations, such as packet sorting based on source or destination addresses with semi-contraction (and for random access read also semi-expansion), can be combined together to implement in time proportional to sorting, random access concurrent read, random access write, as well as some forms of concurrent write on hypercubes. The initial proof due to Nassimi and Sahni used basic rank, concentrate, and distribute operations to implement the equivalent semi-expansion and semi-contraction operations [499, 580]. Thus, it is obvious that any PRAM algorithm can be simulated on the $n$-cube with an overhead equal to the time complexity of the best $n$-cube sorting algorithm.

Efficient parallel algorithms for load balancing, polynomial evaluation, parallel reductions, recurrence relations, and tridiagonal systems can be structured around efficient scan (and scatter) operations [301, 485, 487].

Detecting application termination can also be based on an efficient collective Boolean AND reduction operation. Notice that using dilation one embeddings of the divide and conquer communication pattern (see Figure 2.9), prefix operations (scans) can be implemented in $n$ steps on the $n$-cube; other communication patterns which can be used to implement scans are more expensive: $2\ n$ steps for recursive doubling, $4\ n$ for odd-even reduction, and $\mathcal{O}(n^2)$ for recursive broadcasting on the MIMD $n$-cube [486].

Figure 4.9 illustrates possible partitions of a matrix to a number of PEs. By examining these matrix allocation schemes, we see that parallel algorithms for many problems, such as matrix transposition, matrix multiplication, Gaussian elimination, LU-decomposition, all-pairs-shortest path, parallel searching, and computing connected components require one-to-many broadcasting [303, 713]. These matrix algorithms are also used as functions in larger numerical problems.

An example of parallel matrix transposition which is based on total exchange and local only data moves (when matrix is allocated using horizontal stripes) is illustrated in Figure 4.10.

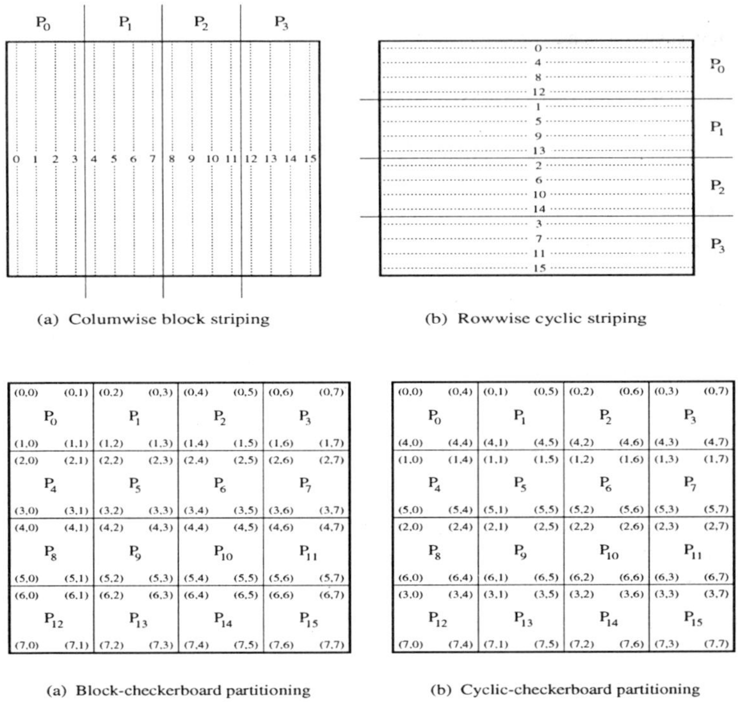

**FIGURE 4.9**
**Matrix partitioning techniques [364]**

Similarly, Figure 4.11 shows an example of matrix-vector multiplication using multinode broadcast and local only computation (when matrix is allocated using horizontal stripes).

An example of matrix multiplication using shift operations and local only computation (when matrix is allocated using block-checkerboard partitionings) is illustrated in Figure 4.12. Notice the submatrices moved to each processor at the end of each step. Multiplying them at each step, and summing the corresponding results provides the parallel multiplication algorithm.

Now, consider solving the system $A{\cdot}x = b$, where $A_{m \times m}$ is tridiagonal, and $m = 2^n - 1$. For example,

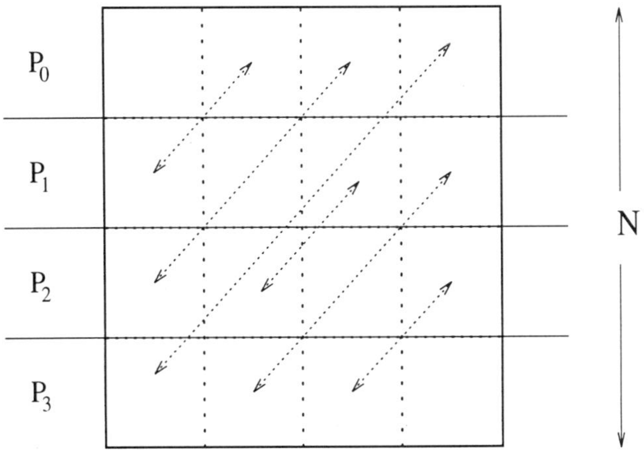


**FIGURE 4.10**
**Matrix transposition using total exchange communication [364]**

$$A = \begin{bmatrix} b_1 & c_1 & 0 & 0 & 0\,0 \ldots 0 & 0 & 0 \\ a_2 & b_2 & c_2 & 0 & 0\,0 \ldots 0 & 0 & 0 \\ 0 & a_3 & b_3 & c_3 & 0\,0 \ldots 0 & 0 & 0 \\ . & . & . & . & . \; . \; \ldots . & . & . \\ 0 & 0 & 0 & 0 & 0\,0 \ldots a_{m-1} & b_{m-1} & c_{m-1} \\ 0 & 0 & 0 & 0 & 0\,0 \ldots 0 & b_m & c_m \end{bmatrix}.$$

Balanced cyclic reduction can be applied during the Gaussian elimination phase; the backward substitution phase is similar to balanced cyclic reduction, except that arrows are reversed. The algorithm repeatedly considers triads of equations, eliminating $x_{i-1}$ and $x_{i+1}$, and relating $x_{i-2}$, $x_i$, and $x_{i+2}$. An example for $N = 15$ equations is shown in Figure 4.13.

Notice that the binary hypercube does not admit triangles! Also, in any mapping of cyclic reduction on the $n$-cube, at least $n/2$ steps involve communication between processors at distance two or more. We can easily find that a binary-reflected Gray code allocation works optimally for this problem on the $n$-cube.

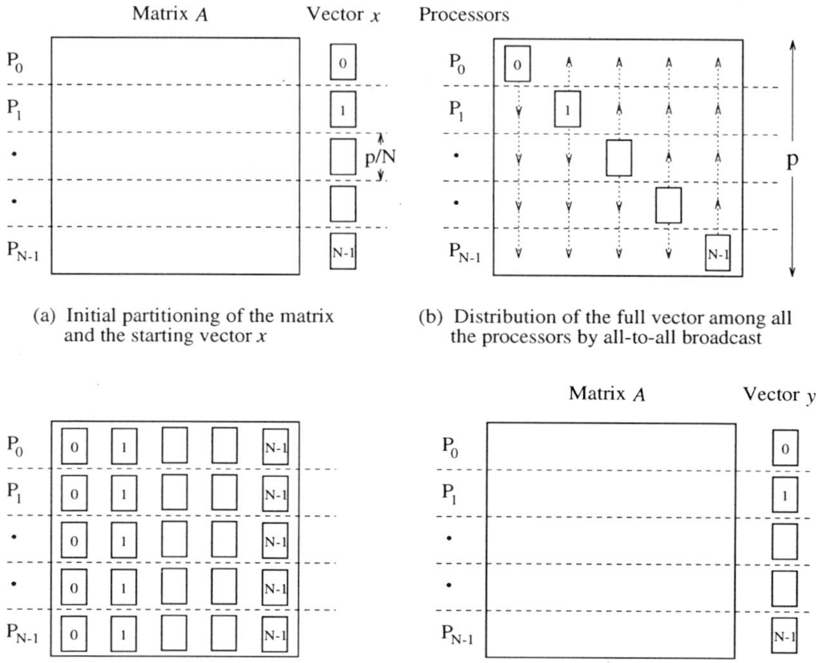


(a) Initial partitioning of the matrix and the starting vector $x$

(b) Distribution of the full vector among all the processors by all-to-all broadcast

(c) Entire vector distributed to each processor after the broadcast

(d) Final distribution of the matrix and the result vector $y$

**FIGURE 4.11**
**Matrix-vector multiplication using all-to-all broadcast [364]**

Node-to-neighbor communication is needed to implement TSP, or image segmentation using the simulated annealing approach [196]. Nondiagonal stencils for solving partial differential equations can be mapped to the hypercube using embeddings of the $d$-dimensional grid (Gray code in each dimension).

Geometric triangulation, convex hull, computing closest neighbors, and parallel algorithms for database and dictionary machines can be based on sorting and load balancing [514, 200, 648]. Furthermore, efficient implementations of priority queues and network simulation can be based on sorting ideas, see [239].

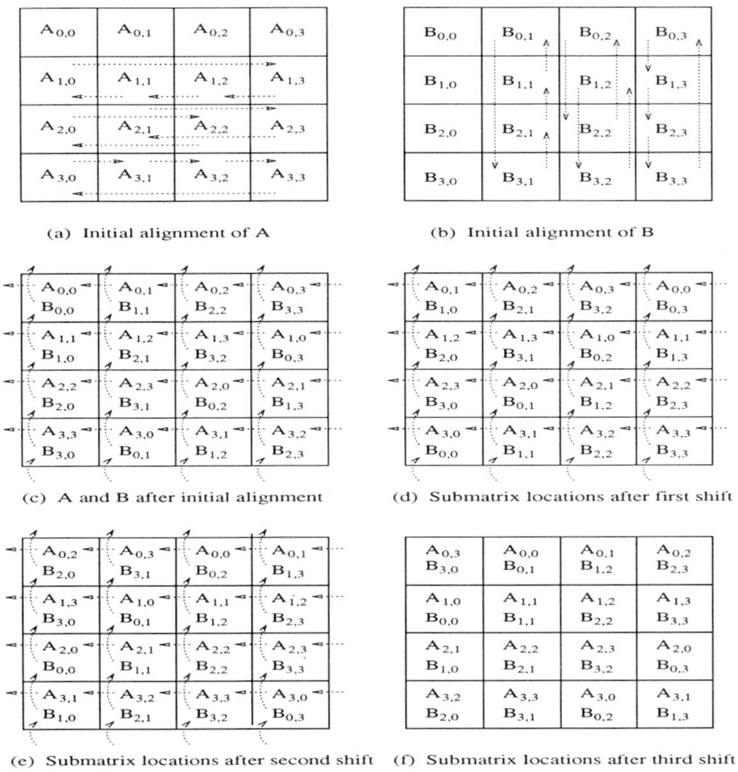

FIGURE 4.12
Matrix multiplication using shifts [364]

A hypercube algorithm is called *normal*, if at any step, all active communication links associate with a single dimension, with consecutive dimensions used in consecutive steps. Most hypercube algorithms we have described are normal. Hypercubic networks, such as the butterfly, the cube-connected cycles, and the shuffle-exchange simulate any normal hypercube algorithm with constant slowdown.

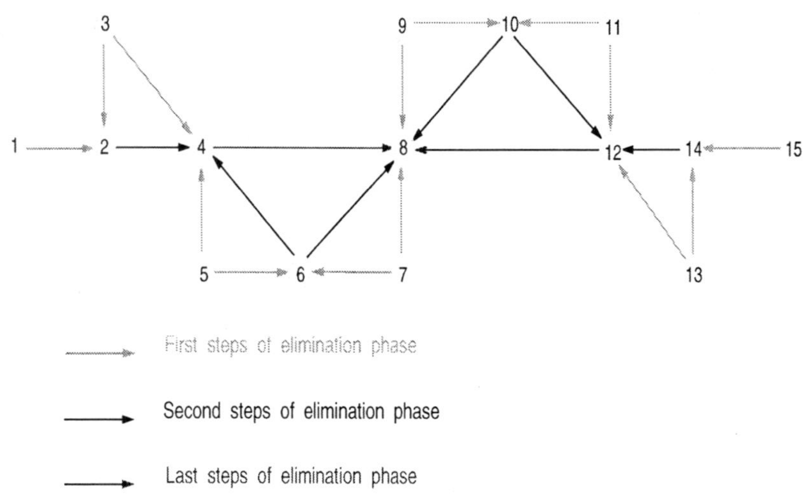


**FIGURE 4.13**
**Task graph allocation for solving tridiagonal system using cyclic reduction (elimination phase)**

## 4.4 Allocation and Scheduling Strategies

With multitasking operating systems on the $n$-cube, users dynamically request subcubes. Upon program completion, subcubes are released to the operating system and become available for future requests.

In general, requests may or may not be granted depending on availability. A request for a $k$-dimensional subcube may be denied even if there are $2^k$ processors available, due to hypercube fragmentation depending on the allocation/release pattern.

Any good allocation strategy assigning jobs to a set of processors must satisfy two objectives,

- *Maximize processor utilization*, and
- *Minimize system fragmentation* [192].

Detecting availability of a subcube of required size, and merging released subcubes of small sizes into a larger subcube is NP-complete.

Thus, several allocation heuristics on the $n$-cube, which avoid exhaustive search, have been explored.

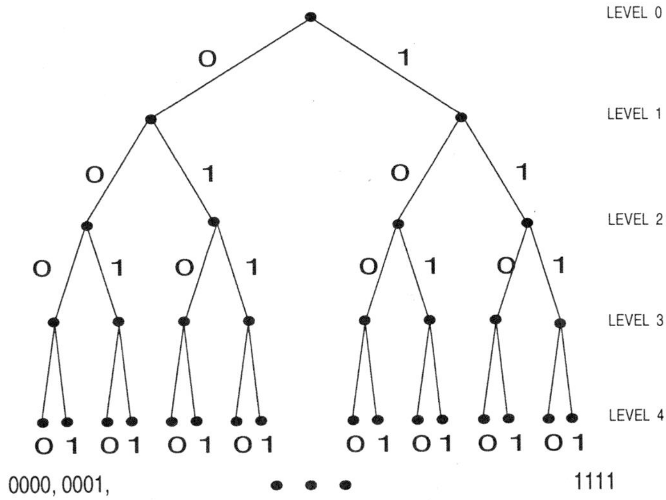


**FIGURE 4.14**
**Binary spanning tree for buddy allocation on the 4-cube**

With the *buddy allocation* strategy, when an incoming request needs a $k$-subcube, the level $n - k$ of a binary tree with $2^n$ hypercube nodes as leaves is examined (see Figure 4.14). Tree nodes at this level are searched in left to right order (but not cyclically), and the first free $k$-subcube is allocated [105]. With this technique, up to $2^{n-k}$ different subcubes are examined. Subcubes formed by nonconsecutive dimensions may still not be recognized, e.g., the 2-subcube consisting of nodes $\{1000, 1001, 1100, 1101\}$ shown in Figure 4.14.

The buddy strategy is *statically optimal*: it can accommodate any batch input request sequence if and only if the total number of requested PEs is less than $N = 2^n$.

With a *binary reflected Gray code allocation* strategy, when an incoming request needs a $k$-subcube, level $n - k + 1$ of a binary tree (encoded with Gray code) is examined (see Figure 4.15). Tree nodes at this level are searched circularly, in left to right order, for two adjacent free nodes [105].

The number of $k$-subcubes recognizable is $2^{n-k+1}$, twice that of the buddy strategy.

Allocation based on Gray code is also statically optimal and generally outperforms the buddy strategy. It packs requests more densely (with larger contiguous regions).

Like the buddy strategy, the binary reflected Gray code strategy can not recognize all free subcubes, e.g., the $\{0011, 0010, 1010, 1011\}$ 2-subcube shown in Figure 4.15. Notice that this cube would still be recognizable by a $\{2, 1, 3, 4\}$ bit-permuted Gray code; this code can be derived from the standard Gray code sequence by permuting the corresponding dimensions, or equivalently permuting bits at the leaf nodes in the tree.

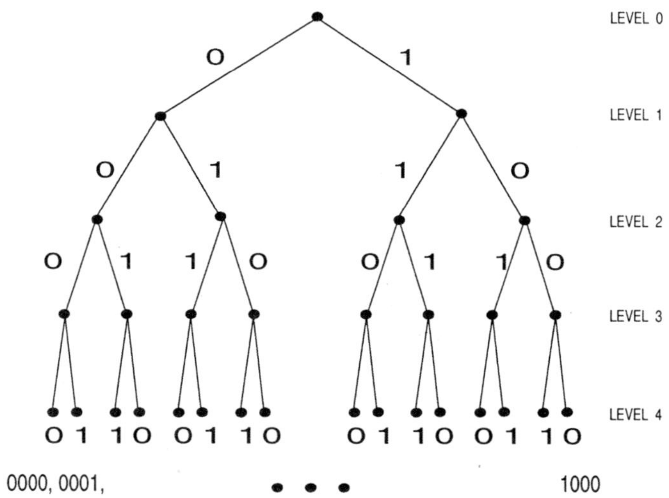


**FIGURE 4.15**
**Binary spanning tree for Gray code allocation on the 4-cube**

Since each permutation corresponds not only to a different allocation strategy, but also to a different set of recognizable subcubes, processor utilization on a multitasking $n$-cube can be significantly improved by increasing the number of Gray codes in the binary reflected Gray code allocation scheme. Notice that, only $C(n, n/2)$ Gray code sequences (out of a maximum of $n!$) are sufficient to recognize any free subcube [105].

The limitation of examining only part of the available subcubes causes fault tolerance degradation. For example, with the buddy system 4 worst case (or 24.5 on the average case) faults are enough to make all examined subcubes faulty, while normally 8 worst case (or 8.3 on the average case) faults would have been needed; relevant upper and lower bounds are discussed in [232]. The method of using permutations of Gray code has been analytically studied in [441].

Other algebraic techniques achieve full subcube recognition ability in polynomial time [117, 185, 338, 340, 483, 570, 617]. Intuitively their polynomiality depends on sophisticated descriptors. However, the extraction of these functions from normal node representation makes these algorithms behave in exponential manner [642].

Static allocation with faults has also been considered. Abraham and Padamanabhan considered how much of the static partitioning ability of the hypercube is lost when random node or link failures occur on the hypercube. The mean time between failures is also computed analytically by integrating appropriate probability functions [5]. Alam and Melhem considered spare allocation in fault tolerant hypercubes. Through the use of spanning subgraphs of the hypercube, they computed close approximations to the hypercube reliability functions for small or large node failure probabilities [22].

In *dynamic allocation*, job scheduling algorithms select jobs to be allocated, either using a *first come first serve* (FCFS) protocol, or by searching a waiting queue (usually implemented as a priority queue) for an optimal candidate. Once the jobs have been selected for execution, allocation techniques for the $n$-cube are commonly based on,

- Showing the existence of a free subcube of requested dimension,
- Detecting the addresses of the free subcubes, and
- Choosing a subcube from the set of candidates which results in smaller fragmentation.

While the FCFS scheduling protocol is simple, there are two major drawbacks with this approach. First, it is likely that an incoming job with a request for a large cube is blocked, until some existing jobs finish execution and free PEs to form a large enough cube. During this waiting period, any new arriving requests, even for smaller cubes, are queued. Another problem is that FCFS scheduling creates fragmentation, and thus makes allocation of succeeding jobs difficult. Two nonreal-time scheduling algorithms, *scan* and *lazy scheduling*, try to remove these problems of FCFS scheduling.

In the *scan scheduling* algorithm, a queue is maintained for each possible cube dimension. The queues are served similar to the c-scan algorithm used in disk scheduling [635]. This eliminates the blocking problem associated with the FCFS scheme. However, scan uses a greedy algorithm, since when processing a queue, scan attempts to allocate jobs whenever an appropriate subcube is available; this results in fragmentation, especially when service time variability for requests happens to be large.

*Lazy scheduling* removes FCFS limitations by maintaining separate queues for different job sizes, and delaying job allocation until no other jobs of the same dimension are running on the system. Thus, jobs wait for existing subcubes, rather than acquiring a new subcube and possibly causing additional fragmentation. In order to avoid discrimination against any job size, job waiting time is controlled using *dynamically selected queue size threshold values.* Notice that if a static (fixed) subcube partitioning to the queues was used, then all jobs of a particular dimension would wait for a *single* subcube executing in the system, i.e., one job running on an $(n-1)$-cube, one on $(n-2)$-cube, ..., one on 1-cube, and two on 0-cubes. However, the usefulness of this static partitioning approach is limited to a uniform job size distribution. In lazy scheduling, each queue threshold is initialized to zero, and subsequently incremented with each subcube allocation, and decremented with each subcube deallocation. Thus, for every subcube allocated in the system, there can be another job waiting to acquire it. If the queue size exceeds the corresponding threshold, then the scheduling algorithm tries to allocate to the queue a new subcube on an FCFS basis. This creates more servers for queues with a large number of requests to corresponding subcubes. To avoid starvation of requests, especially for larger subcubes, another threshold value for the maximum queuing delay that a job can tolerate is established; whenever the job waiting time exceeds this threshold, the job gets priority over all other jobs. This threshold can either be predefined, e.g., using system load characteristics upon job submission, or computed dynamically. A predefined threshold value, as used in the lazy scheduling algorithm, is very useful in imposing a deadline for job completion [479].

Simulation studies indicate that lazy scheduling outperforms FCFS and scan scheduling, especially for varying workloads, in terms of many performance indicators such as job completion time, average queuing delay, system utilization, throughput, and overall system power (throughput over delay) [479]. The same study exhibits the minimal effect of complete subcube recognition allocation strategies, thus pointing out the usefulness of a simple buddy allocation strategy together with a good scheduling mechanism [358, 479]. The proposed lazy scheduling scheme with buddy allocation has $\mathcal{O}(n)$ complexity on the $n$-cube.

With *real-time scheduling* (RT), in addition to a system dimensional requirement, each job is associated with a computation time and a deadline. The goal of real-time scheduling is to determine whether all jobs can complete their processing before their fixed deadlines, and to find an efficient schedule. Determining a schedule such that all jobs meet before their respective fixed deadlines on a hypercube system, when preemption is not allowed, is an NP-complete problem. Hence, we concentrate on heuristic scheduling algorithms for nonpreemptable real-time jobs. Kwon and Chwa have provided approximate nonpreemptive scheduling algorithms for independent jobs on the hypercube which achieve a factor of at most 1.875 from optimal [379]. Other real-time scheduling algorithms on the hypercube which are considered competitive are *buddy/RT*, *stacking/RT*, and *deferred earliest deadline first*.

In the *buddy/RT* algorithm all jobs are scheduled upon arrival using the FCFS discipline [357]. For each job, the buddy/RT algorithm tries to find the smallest $j, 0 \leq j \leq 2^{n-k}$, such that all PEs in the subcube $j2^k, (j+1)2^k - 1$ have earlier available times than the latest start time of the given job. If no such $j$ exists, then, the job can't be completed before its deadline; thus, if its latest starting time is larger than a worst case rescheduling time, a second phase is initiated which reschedules all jobs, whether currently executing or waiting.

The *stacking/RT* algorithm tries to minimize fragmentation by avoiding unnecessary coalesces and splits of subcubes [357]. This approach is based on stacking equal-sized jobs, and instead of choosing the first available subcube, selecting the earliest available subcube from all subcubes that allow the deadline to be met with the least number of splits or coalesces.

Mohapatra, Yu, and Das proposed a real-time scheduling algorithm, called *deferred earliest deadline first* [483]. The main idea of this algorithm is to defer the scheduling decision as late as possible so that a set of jobs is scheduled at a time (using optimal static allocation). Simulation results indicate that the DEDF algorithm performs significantly better than both the buddy/RT, and the stacking algorithm for a wide range of workloads. Most of the scheduling algorithms discussed in this section can also be extended to mesh interconnects and other parallel system topologies.

Several point-to-point hypercube multicomputers supporting SIMD or MIMD computation models have been developed, such as the T-series by FPS, the Thinking Machines Connection Machine CM-1 and CM-2, the Intel iPSC series, the NCube, and many others. Processors used range from bit-serial to transputers to Sparcs. For issues concerning the design of communication networks in actual parallel systems refer to [284, 402, 416, 568].

## 4.5 VLSI Layout and Customization

A major task in system design consists of the reliable integrations of thousands to millions of chips encountered with parallel systems. In network design, memory usually dominates chip area and may not scale linearly, even in VLSI.

Although, density and chip yield play an important role, the communication or processing speed is often determined by the worst case signal delay. Thus, the VLSI design problem may be stated as follows: for a given bisection bandwidth and pin count (between 256-512), maximize the clock speed and wire length within the chip.

Several signal delay models have been proposed.

- *Capacitive model*: the signal delay of a wire of length $l$ is proportional to $logl$.
- *Resistive model*: the signal delay of a wire of length $l$ is proportional to $l^2$. This can be reduced to $l$ if repeaters are provided, or if the speed of light is the limiting factor.
- *Constant delay model*: the signal delay is determined by parameters other than the wire length.

Ranade and Johnsson examined planar layouts for the $N = 2^{2n}$ node binary hypercube with the following interesting results [579].

- In a Gray code based $2^p \times 2^p$ mesh layout, the processor at position $(x, y), 0 \leq x, y \leq 2^p$ is labeled with the concatenated string $Gray(x) \cdot Gray(y)$. This layout results in different lengths of channels along a given dimension. Independent of the signal delay model, this layout is good for certain communication patterns only, like FFT.
- In a row-major based $2^p \times 2^p$ mesh layout, the processor at position $(x, y), 0 \leq x, y \leq 2^p$ is labeled $x2^p + y$. Independent of the

signal delay model, row-major labeling results in a smaller layout area. It also provides similar wire length along a given hypercube dimension.

Physical constraints imply limitations on the communication bandwidth and latency between nodes. Thus, the assumption of unit node-to-neighbor communication time is unrealistic. When a network, such as hypercube, is embedded in three-dimensional space, wires in some dimensions of the network are much longer than wires in other dimensions. Hence, for a synchronous system with a single bit on a given wire at a given time, the network clock cycle is related to the length of the longest wire. Therefore, networks with higher dimensionality will have a slower clock.

It has been shown that the VLSI layout for hypercube requires $\mathcal{O}(N^2)$ area [559]. If we assume an $\mathcal{O}(1)$ layout cost per PE, the *area efficiency*, defined as the ratio of the area occupied by the PEs to the total chip area, is just $\mathcal{O}(1/N)$ for the $N$ node hypercube. Furthermore, assuming a constant steady state power dissipation, a small area efficiency leads to an increased power efficiency, per unit area. While the efficiency of the hypercube layout is not as good as some other regular topologies, e.g., meshes, the many redundant paths provide nice fault tolerance properties.

Packaging constraints are also important, since board to board communication is usually one to two orders of magnitude slower. Customization studies considered a single level packaging hierarchy for interconnection topologies and focused on LSI complexity, rather than VLSI design issues. In these studies, different network topologies are constrained to have either the same bisection width, or the same pin count by appropriately varying their channel widths. At first, Dally, assuming a constant bisection width, proved that it is better to allocate a small number of high bandwidth channels, such as two-dimensional mesh [150]; a constant bisection width implies that a channel for a two-dimensional mesh has $\mathcal{O}(\sqrt{N})$ times more wires than a channel on the $N$ node binary hypercube. In contrast to Dally's study, Abraham and Padmanabhan, assuming a constant pin count constraint, proved that low dimensional networks, such as the hypercube, are favored [5].

A two-level LSI hierarchy, e.g., cluster-based hierarchical hypercubes or meshes, with either a constant bisection width or pin count constraint, may actually enhance performance [284, 285]. Hsu and Yew also showed that the relative performance of clustered networks depends not only on the chosen technological constraint, but also on system configuration and packet size; for a large packet size, or large number of clusters a clustered hypercube architecture is favored [285].

## 4.6 Commercial Hypercube Systems

Hypercube multicomputer systems were pioneered at universities. In the early 1980s Charles Seitz and researchers at California Institute of Technology built the first real MIMD point-to-point hypercube system [444, 610]. The initial term was homogeneous concurrent machine, but later it was called the *Caltech Cosmic Cube*. The system featured Intel 8086 CPUs, including 8087 floating co-processor, and bidirectional asynchronous links. Each node provided 512 Kbytes of local memory, and 8 I/O ports: 7 ports were used to form the 7-dimensional hypercube, while the 8th was used for an Ethernet connection from each node to the host. Store-forward routing was implemented, with an average latency in the order of few milliseconds. An independent host was provided for spawining processes. The user programs involved the process model, with nonblocking virtual communication channels established between processes using sends and receives. A test probe into the messaging facility was also provided. Lock variables were used for synchronization. There was no limit to the number of processes, as process interleaving was provided.

Each PE ran an identical, small (10-15 KB) operating system (OS) kernel which supported process creation and extinction, memory management, message passing (preserving the order of message communications between pairs of processes), and process recovery when dealing with error conditions. The machine design was very effective in realizing the critical software issues involving data allocation and process coordination, and introduced static process creation to simplify the task of message routing and allow programs to be written in extended versions of ordinary sequential programming languages. Most of the programs run on the Cosmic Cube have been simulations of physical systems and scientific computations, since these domains often include naturally parallel computations.

Several hardware enhancements were later developed on the original 64-node Cosmic Cube, increasing the number of PEs from 64 to 128, increasing RAM, adding FIFO master-slave buffers for each physical communication channel, and implementing global synchronization in hardware using wired-ors. Finally, user support for direct message exchanges was provided [538, 670].

The NCube2 implemented a 16-cube, with a total aggregate memory of 512 GBytes. A major design step was the departure from the store-forward routing protocol to wormhole routing, which effectively reduced communication latency by three orders of magnitude, for nonmultihop DMA-based node transfers.

The CM-2 architecture was essentially a large SIMD vector unit. It consisted of an array of up to 65536 PEs connected to a scalar host (front-end). Each "computing node" consisted of a memory unit, a floating point unit, and 32 bit-slice PEs organized into 2 groups of 16 PEs each. Each of the 16 PE groups was connected to a router. The router interface consisted of 4096 router nodes interconnected as a 12-cube. A hardware routing algorithm that could introduce misrouting was used. When the router was "delivering" messages, the PE array was idle.

The control unit at the CM-2 host issued parallel instructions, which broadcasted a sequence of micro-instructions to all PEs to execute that instruction. Data communication between PEs was done using one of three method, the NEWS grids, the global router, and the scan operations. Data could also be communicated between the front-end and the PE array in one of three ways: broadcasting, global combining, and scalar memory bus. In broadcasting, the front-end could broadcast a scalar value to all PEs, e.g., casting a scalar variable (or constant) to a parallel variable (or constant). In global combining, the front-end could obtain the scalar sum, max, min, OR, AND, of a parallel variable. Using the scalar memory bus, the front-end could read or write the memory of a particular PE.

Finally, we note the SGI Origin 2000 system which used a cache coherent distributed shared memory system. The Origin 2000 system [9] was built from a number of nodes, each containing two IBM R10000 processors and their associated second level caches. The interconnection network was a hypercube for small configurations, up to 32 nodes. A directory-based cache coherence protocol was implemented over this fabric in order to provide a system wide cache coherent shared memory. The memory model provided on the Origin 2000 was sequentially consistent. Load-linked and store-conditional instructions were provided for supporting PE synchronization.

# Chapter 5

## Mesh Networks

### 5.1 Topological Properties

In this chapter we consider many aspects of mesh-like architectures. After the definition and analysis of some basic properties, we consider various communication problems, including routing and sorting. We consider several special topics that have been studied for meshes, such as fault tolerance, routing with locality, and PRAM simulation. Finally we consider various modifications of meshes, such as meshes with buses, sparse meshes, and reconfigurable meshes, and discuss how the above problems can be solved on these network topologies.

The main topic of this chapter is *$l$-$k$ routing*. Here $l$ indicates the maximum number of packets any PE has to send, and $k$ the maximum number of packets a PE is going to receive. Most attention goes to the case $l = k$. The special case $l = k = 1$ corresponds to permutation routing. In the analogously defined *$l$-$k$ sorting* problems, the packets are equipped with a key from a totally ordered set, with respect to which the packets must be sorted. Sorting means that the packets must be rearranged so that the packet with rank $r$ ends up in memory location $r$, with respect to some numbering of the memory locations.

We consider the underlying ideas even more important than the algorithms. Almost each of presented algorithms illustrates a new idea, and most of the important ideas will be illustrated. Over the past ten years a rich repertoire of standard tricks has been developed, and our main goal is to make this repertoire accessible. At this point we provide an overview of fundamental ideas, which will be explicitly addressed.

*One-Dimensional Phase.* The majority of algorithms on two- and higher-dimensional meshes are composed of one-dimensional phases. Routing on one-dimensional arrays can be analyzed completely.

*Coallescing.* If the sum of the maxima of the one-dimensional phases is larger than the time needed by any individually routed packet, then it may be profitable to let a packet immediately start a later phase as soon as it has completed an earlier phase.

*Overlaying Uni-axial Algorithms.* On a full port (MIMD) mesh, two uni-axial algorithms can be overlaid, provided that they are "out-of-phase" at all times.

*Spreading.* If the expected queue size of an algorithm is much smaller than the maximum queue size, then it may be advantageous to route packets not exactly to their destination PE, but to a PE close by, so that there is some storage capacity saved.

*Critical Packets.* If the time consumption is highest for an easily distinguishable subset of the packets, then it may be worthwhile to design a second algorithm for these special packets, and to run it in parallel with the main algorithm.

*Copying and Compression.* If the distance bound determines the time consumption, then time can often be saved by sending copies of the packets to different parts of the mesh, or by compressing all packets in a part of the mesh. By the time some missing information becomes known, the copies are already close by.

*2-2 Routing.* On one-dimensional arrays, 2-2 routing costs the same as 1-1 routing: $n$ steps. This means that with no additional cost a second problem can be overlaid.

*Information Gathering.* Sometimes the PEs in which packets reside cannot make the correct decisions on basis of local information only. In that case, they must first gather some information.

*Low Density Information Spread.* If the number of information packets sent over each row and column is $o(n)$, then they can be routed with priority, without causing much delay to the other packets.

*Randomization.* Useful for smoothing phenomena arising due to worst case packet distribution. In many cases, whenever the bisection bound is sharp, first randomizing all packets is for free (as far as the worst case asymptotic performance is concerned). Since randomization does not require global information, it can be performed in order to keep the packets busy in a useful way, while information is gathered that is required for further handling.

*Coloring.* If there are several packets in each PE, one may concentrate on uni-axial algorithms and then color half of the packets white and the other half black, and route these by two orthogonal versions of the same algorithm.

*Derandomization.* Most effects of randomization can also be obtained by first sorting packets in $o(n) \times o(n)$ submeshes, and then handing them out in round robin fashion. This allows us to concentrate on the conceptually simpler randomized algorithms.

*Turning Offline into Online.* Sometimes, one knows how to perform a routing, provided that PEs had global knowledge and sufficient time to perform some precomputation. In many cases, almost the same good performance can also be achieved by an online algorithm.

*Superpackets.* If there is some greater amount of cost connected to handling a packet, e.g., when turning an offline algorithm into an online algorithm, or because of startup delays, it is useful to bundle packets with a common destination into superpackets.

*Submeshes.* In general, operations on submeshes of size $o(n) \times o(n)$ cost only a lower order term of the total time consumption. Thus, all kinds of operations that are too expensive to be performed on the whole mesh can be performed in submeshes. Frequently, packets are routed between submeshes after or before a rearrangement in the submeshes. In this way, we can focus on algorithms with a polynomial (in the number of PEs) number of packets per PE.

*Indexing Schemes.* In sorting problems it is often desirable that PEs with almost identical indices also lie close by (snake-like indexings). However, the "desnakification" technique overcomes this, and allows to sort in essentially the same time, with respect to almost any indexing.

### 5.1.1 Meshes and Tori

A *two-dimensional* $n \times n$ *mesh* consists of $N = n^2$ PEs, arranged in a two-dimensional $n \times n$ grid. Each PE is connected to its (at most) four neighbors. PEs can be identified by their coordinates in the grid: the node at position $(i, j)$, $0 \leq i, j < n$, is denoted by $P_{i,j}$, where position $(0, 0)$ lies in the upper left corner (as in a matrix). All definitions carry on to nonsquare $m \times n$ meshes. A *torus* is a mesh with additional *wraparound links*. These links run between $P_{i,n-1}$ and $P_{i,0}$ for all $0 \leq i < n$, and between $P_{n-1,j}$ and $P_{0,j}$, for all $0 \leq j < n$. Tori have the advantage that they are uniform, i.e., there are no corners that have to be treated in a special way. These definitions are illustrated in Figure 5.1.

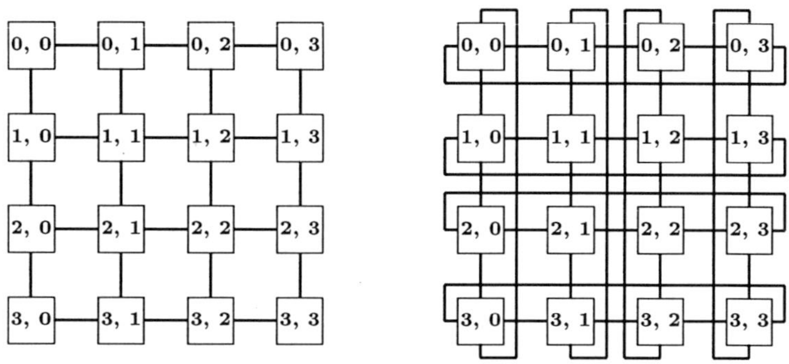


**FIGURE 5.1**
**A $4 \times 4$ mesh on the left, and a $4 \times 4$ torus on the right**

The bisection width of an $n \times n$ mesh equals $n$, and that of a torus $2n$. Their diameters are $2n - 2$ and $2\lfloor n/2 \rfloor$, respectively. More generally, a $n \times s \times n$ $d$-dimensional mesh consists of $n^d$ PEs arranged in a $d$-dimensional grid; $d$-dimensional tori are defined analogously. Next, we summarize the basic properties of these networks.

Meshes and their direct generalizations are attractive because of regularity, scalability, and conceptual simplicity. They are easy to program and are suitable for VLSI implementation. The major drawback is large diameter and small bisection width ($2n - 2$ and $n$, respectively, for two-dimensional meshes). As a consequence, the maximal speedup for problems without much "locality" is $n$, with $n^2$ nodes.

As discussed before, in addition to the SIMD and MIMD models, we distinguish the *uni-axial model*; in every routing step there is one axis along which communication can be performed (either horizontal, or vertical). Along this axis each node can send information to both of its neighbors, and receive from both of them.

**LEMMA 5.1**

*The PEs of a d-dimensional $n \times s \times n$ mesh or torus have degree at most $2d$. The diameter of such a mesh is $d(n-1)$, and that of the torus equals $d\lfloor n/2 \rfloor$. Their bisection widths are $n^{d-1}$ and $2n^{d-1}$, respectively.*

One-dimensional meshes (and tori) are called linear arrays (and circular arrays, respectively).

Two- and three-dimensional meshes are attractive because of their regularity, scalability, and conceptual simplicity. They are easy to program and have crossing-free two- and three-dimensional VLSI layouts. Their major drawback is their large diameter and small bisection width. This implies that except for problems with much locality such as matrix multiplication, the maximum achievable speedup is $\mathcal{O}(n)$ ($\mathcal{O}(n^2)$), with $n^2$ ($n^3$) PEs. On the other hand, networks such as hypercubes impose more constraints on the interconnection modules as the network size increases. Cube connected cycles do not have this problem, but are sometimes harder to program due to their irregularity. Anyway, because of a basic theorem from VLSI layout [663], all planar architectures have an area that is quadratic in their bisection width, while all spatial architectures occupy a volume that increases with the bisection width as the power 3/2. As a consequence, low-dimensional meshes are not bad at all.

### 5.1.2 Indexings

For the two-dimensional sorting problems discussed in Section 5.3, the PEs must be indexed with consecutive numbers from 0 to $N-1$. There are many possible *indexings*. The most natural one is the *row-major* indexing, under which $P_{i,j}$ has index $in + j$. A possible disadvantage is that PEs with consecutive numbers may lie at distance $n$ from each other. This problem is remedied by the *snake-like row-major* indexing, under which the indexing in the rows with odd $i$ is reversed. That is, $P_{i,j}$, with $i$ odd, has index $in + n - j$. There are many more sophisticated indexings. Under the *shuffled-row-major* indexing, $P_{i,j}$ (where $i = \sum_{l=0}^{\log n-1} i_l \cdot 2^l$ and $j = \sum_{l=0}^{\log n-1} j_l$ are the binary expansions of $i$ and $j$) has index $\sum_{l=0}^{\log n-1} (2i_l + j_l)4^l$. These indexings are illustrated in Figure 5.2.

For sorting, *blocked* indexings are even more useful. This means that the mesh is regularly divided in submeshes of size $m \times m$. The PEs in the submeshes are indexed in some convenient way. The submeshes

| | | | |
|---|---|---|---|
| 0 | 1 | 2 | 3 |
| 4 | 5 | 6 | 7 |
| 8 | 9 | 10 | 11 |
| 12 | 13 | 14 | 15 |

| | | | |
|---|---|---|---|
| 0 | 1 | 2 | 3 |
| 7 | 6 | 5 | 4 |
| 8 | 9 | 10 | 11 |
| 15 | 14 | 13 | 12 |

| | | | |
|---|---|---|---|
| 0 | 1 | 4 | 5 |
| 2 | 3 | 6 | 7 |
| 8 | 9 | 12 | 13 |
| 10 | 11 | 14 | 15 |

**FIGURE 5.2**
**(a) Row-major, (b) snake-like row-major, and (c) shuffled row-major indexing**

themselves are also indexed, for example with the snake-like row-major indexing. Then the PE with index $a$ $(0 \leq a < m^2)$ in the submesh with index $b$ $(0 \leq b < (n/m)^2)$ is given index $a + m^2b$. Such an indexing guarantees that all PEs with indices in a certain range stand close together. An example is given in Figure 5.3. Based on space filling curves, one can even derive indexings that guarantee that any pair of PEs, with indexes $a$ and $b$, are at most $\mathcal{O}(\sqrt{|a-b|})$ positions apart [115, 508].

| | | |
|---|---|---|
| $0$ ... $m^2-1$ | $m^2$ ... $2m^2-1$ | $2m^2$ ... $3m^2-1$ |
| $5m^2$ ... $6m^2-1$ | $4m^2$ ... $5m^2-1$ | $3m^2$ ... $4m^2-1$ |
| $6m^2$ ... $7m^2-1$ | $7m^2$ ... $8m^2-1$ | $8m^2$ ... $9m^2-1$ |

**FIGURE 5.3**
**A blocked snake-like row-major indexing with $m = n/3$**

### 5.1.3 Communication Models

Throughout this chapter, we assume the store-forward communication model, unless specified differently. Several mesh variants are distinguished. In the *unique port model* (SIMD), the PEs can use only one globally specified link at each step. In the *full port model* (MIMD), the PEs may communicate with any of their neighbors at every step. In the *uni-axial model* there is one axis along which a PE can communicate with both neighbors at every step. Full duplex communication, i.e., simultaneous send and receive capability, is usually assumed.

| | | | | | |
|---|---|---|---|---|---|
| 0 | 1 | 2 | 3 | 4 | 5 |
| 6 | 7 | 8 | 9 | 10 | 11 |
| 12 | 13 | 14 | 15 | 16 | 17 |
| 18 | 19 | 20 | 21 | 22 | 23 |
| 24 | 25 | 26 | 27 | 28 | 29 |
| 30 | 31 | 32 | 33 | 34 | 35 |

$\longmapsto$

| | | | | | |
|---|---|---|---|---|---|
| 0 | 5 | 1 | 4 | 2 | 3 |
| 30 | 35 | 31 | 34 | 32 | 33 |
| 6 | 11 | 7 | 10 | 8 | 9 |
| 24 | 29 | 25 | 28 | 26 | 27 |
| 12 | 17 | 13 | 16 | 14 | 15 |
| 18 | 23 | 19 | 22 | 20 | 21 |

**FIGURE 5.4**
**The embedding of a $6 \times 6$ torus onto a $6 \times 6$ mesh**

As shown in Figure 5.4, an $n \times n$ torus can be embedded onto an $n \times n$ mesh with dilation 2, since the PEs that are adjacent in the torus on the left are at distance two in the mesh on the right. Hence, any full port (MIMD) torus algorithm can be simulated on a mesh with delay factor 2. Also, any optimal torus algorithm (an algorithm matching the distance or bisection bound [see Section 5.3]), automatically yields an optimal mesh algorithm.

However, the opposite is not true. There are problems that cannot be solved on a torus twice as fast as on a mesh. The simplest example is an instance of the 1-1 sorting problem on one-dimensional arrays which we encounter in Section 5.2 [455].

Since the number of packets to be sent on the mesh is usually large, and the bisection width of the mesh is small, the packet routing time is determined almost entirely by the link transfer rate. Thus, for all asymptotic analyses, such as those performed for the routing algorithms hereafter, store-forward routing is no limitation.

For mesh routing problems, the throughput bound and the connection availability bound are almost never stronger than the bisection bound (see lower bound definitions in Chapter 2). However, these bounds are useful when the traffic through individual links is hard to analyze, and one can either show that for certain instances a lot of packets must cross the bisection, or that possibly many packets must travel a long distance. Notice that the definition of the bisection bound does not require equality of the two parts: for some routing problems, most time is required for getting the packets into or out off a mesh corner. In sorting problems, sometimes stronger lower bounds can be obtained by exploiting a lack of global knowledge.

### 5.1.4 Connection Conflicts

During the routing, it may happen that several packets have to move over the same link. As only one packet can move at a time, the others are going to be delayed. The selection of the packet that moves first is called a *priority strategy*.

There are several priority strategies. A good general purpose strategy is the *farthest-first* strategy. This means that among the competing packets, the packet selected is the one which still has to go the farthest distance. The default assumption is that only the direction in which the move is made is considered. Sometimes the total distance that the packet still has to trace is taken as the basis for the comparison. This latter strategy we call *farthest-total-first* strategy. There are several other strategies, but they do not appear to have algorithmic importance, though sometimes they facilitate the analysis. These strategies require that newly arriving packets are inserted in some nontrivial data structure, which may be expensive when the number of waiting packets is large. A cheaper strategy is to select a packet randomly, or any of the waiting packets (which may correspond to the worst possible scenario).

### 5.1.5 Remarks

For mesh routing, various optimality notions have been considered. Usually, the number of routing steps taken by an algorithm, called the *routing time*, is one of the most important parameters.

The *queue size*, the maximum number of packets a PE may have to store during the course of the algorithm is an important secondary parameter. There are two reasons why the queue size is important. In the first place, if one assumes that the PEs are small nodes in a massive network, then there may be actual physical storage limitations. Secondly, and possibly more relevant, long queues may imply elaborate queue handling, which may slow down routing.

Another aspect of an algorithm is, whether it is *oblivious* or *adaptive*. An algorithm is oblivious if the path of a packet is determined initially based on information available in the source PE where the packet is initiated; this may include randomness, in which case we refer to *randomized oblivious* routing. An algorithm is adaptive if packets adapt their path depending on collisions or traffic information that becomes available. This is a fundamental distinction, and we have encountered a lower bound for oblivious routing, which can be broken by an adaptive algorithm in Theorem 2.3.

A similar distinction is between *offline* and *online* algorithms. An algorithm is called offline if some outside observer with global knowledge and infinite (or polynomial) time precomputes the schedule, which is then tagged to the packets. Such a routing may make sense for fixed patterns. There are many important examples of this: FFT, Warshall's transitive closure algorithm, transposes, affine permutations, all-to-all routing, shuffles, broadcasting, gossiping, etc. In many cases, however, one has a packet distribution with a pattern that is not known in advance. For such cases one should have an online algorithm, that computes a schedule based on limited knowledge. Knowledge can be possibly gathered and handled before making decisions, but this is not for free as it is for an offline algorithm. We encounter algorithms of both types, and also a useful technique for turning offline into online algorithms.

Finally, algorithms are distinguished as *deterministic* and *randomized*. Normally, deterministic algorithms are considered superior than randomized algorithms with the same performance in all other respects. Randomized algorithms use a random number generator to make certain decisions, such as where or when to route a packet. In many cases randomized algorithms are simpler than deterministic algorithms achieving the same performance, and in some cases they are even better. We encounter examples of both cases.

## 5.2 One-Dimensional Arrays

In routing algorithms for higher dimensional meshes, one-dimensional routing is often used as a subroutine. The problems on one-dimensional arrays are sufficiently easy. Given the initial (source) and final (destination) packet distributions, the routing time can usually be determined precisely. In this section, we consider linear and circular arrays, and greedy routing with the farthest-first protocol. The derived results will later be used to analyze $k$-$k$ routing and the routing of randomizations.

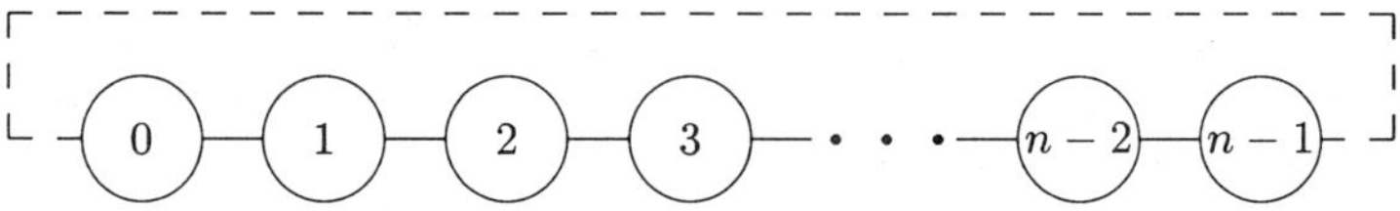


**FIGURE 5.5**
**Linear or circular array with $n$ PEs**

The linear and circular arrays considered consist of $n$ PEs, indexed from 0 to $n-1$ (see Figure 5.5). The directed link from $P_i$ to $P_j$ is denoted by $(i, j)$.

### 5.2.1 Lower Bounds for Routing and Sorting on Arrays

We analyze lower bounds for routing and sorting on one-dimensional arrays in detail.

A basic result is obtained by combining the distance and bisection bound:

***LEMMA 5.2***
*For k-k routing on a linear array with n PEs, there is a packet distribution that requires at least* $\max\{n-1, kn/2\}$ *routing steps. The corresponding lower bound for a circular array is* $\max\{n/2, kn/4\}$.

For $k = 2$ we can combine throughput and distance bound to derive a result that even holds when there is global knowledge:

***THEOREM 5.1***
*For 2-2 routing on the circular array, there is a distribution that requires* $2\lfloor n/3 \rfloor$ *routing steps.*

**PROOF** We assume that $n$ is divisible by 3. Consider the following subdivision of the circular array:

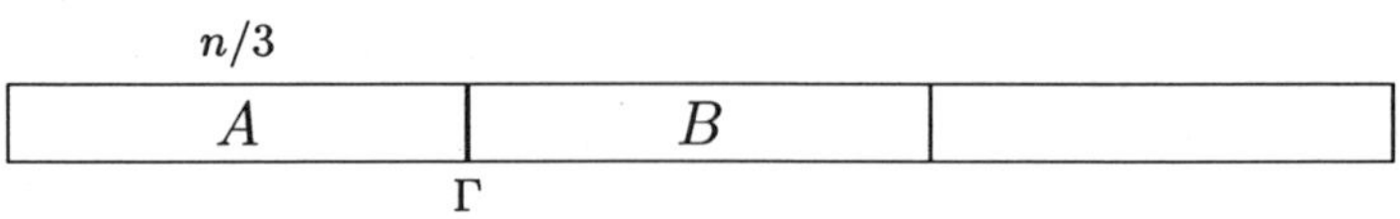

The 2-permutation 2-$S_{n/3}$, which shifts all packets rightwards over $n/3$, is such that all packets residing in a PE in $A$ move to a PE in $B$. If they are all routed via a shortest path, then link $\Gamma$ transfers $2n/3$ packets, requiring at least $2n/3$ routing steps. $\Gamma$ may be relieved by routing a number of packets leftwards. However, then the length of this procedure becomes more than $2n/3$. Hence, this does not decrease the routing time. ∎

For sorting, Mansour and Schulman [459] have shown a trade-off between storage capacity and the number of communication steps. However, the communication model used is slightly different than the ones considered here.

### 5.2.2 Basic Lemmas

We now derive the lemmas that specify precisely the number of routing steps for routing a given distribution of packets to their destination. These lemmas implicitly (or explicitly) underlay most results in the remainder of this chapter. For a packet distribution $R$, $T_{\text{right}}(R)$ denotes the number of rightward routing steps required for routing $R$. $T_{\text{left}}(R)$ is defined analogously. Most lemmas are formulated for $T_{\text{right}}$ only. Obvious analogues hold for $T_{\text{left}}$. Next, many lemmas and theorems are stated for $T_r(R) = \max\{T_{\text{right}}, T_{\text{left}}\}$, and proved only for $T_{\text{right}}$.

We consider routing packets on a linear array using the farthest-first strategy. The following lemma can be derived as a corollary of Lemma 5.4 given below. Because of its rather simple and interesting derivation we prove it independently.

**LEMMA 5.3**

*Given a distribution $R$ on the linear array, such that for any $r$, $0 \leq r \leq n/2$, the subsets consisting of $r$ PEs at the right or left end of the linear array send and receive at most $kr + \alpha$ packets, $k, \alpha \geq 0$. If the farthest-first strategy is used, then $T_{\text{right}}(R) \leq \max\{n+\alpha-1, kn/2+\alpha\}$.*

**PROOF** Consider a packet $p$ that has to be routed from a PE $P_i$ at position $i$, to $P_j$. Without loss of generality, we assume $i < j$. Let $\Gamma$ be the link $(j-1, j)$. Let $\mathcal{S}$ be the set of packets that have to cross $\Gamma$ from left to right, the packets that are routed from PEs $P_{i'}$ with $i' < j$

to $P_{j'}$ with $j' \geq j$. By the farthest-first strategy, $p$ can only be delayed by other packets of $\mathcal{S}$. In the $P_{i'}$ with $i' < j$ there are initially at most $jk + \alpha$ packets. There are at most $(n-j)k + \alpha$ packets with destination $P_{j'}$ with $j' \geq j$. So, $\#\mathcal{S} \leq \min\{j, n-j\}k + \alpha$. The packet $p$ will reach its destination latest, if the number of packets in $\mathcal{S}$ is maximal and all packets of $\mathcal{S}$ are initially located in the left most PEs. From now on we assume this distribution of packets. For $j \leq n/2$, the first packet of $\mathcal{S}$ crosses $\Gamma$ in routing step 1, and the last packet in routing step $jk + \alpha \leq kn/2 + \alpha$. If $j \geq n/2$, the distribution looks like:

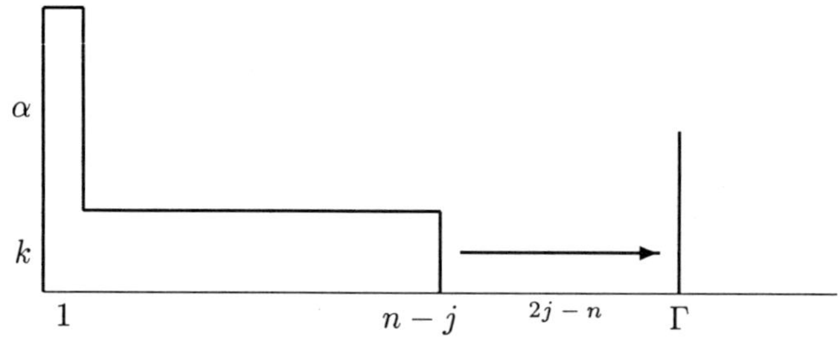


The first packet of $\mathcal{S}$ crosses $\Gamma$ after $2j - n - 1$ steps, and the last packet in step $2j - n + (n-j)k + \alpha$. For $k \geq 2$, it is easy to check that $2j - n + (n-j)k \leq kn/2$, for all $n/2 \leq j \leq n-1$. For $k \leq 1$, $2j - n + (n-j)k \leq n-1$. These maximal values are assumed for $j = n/2$, $j = n-1$, respectively. ▮

In order to prove that randomizing $k$ layers of packets on the linear array can be carried out in about $kn/4$ routing steps, we need a stronger lemma. Let for a given distribution of packets over the PEs and all $i, j$, $a_r(i,j)$ denote the number of packets traveling from left to right through $P_i$ and $P_j$, in this order. For linear arrays the definition implies that $a_r(i,j) = 0$ for all $i \geq j$; for circular arrays it leaves open the possibility that $i > j$ and that the packets go through $(n-1, 0)$ before they reach $j$.

For use in the following lemma we define

$$h_r(i,j) = \begin{cases} a_r(i,j) - 1 & \text{if } a_r(i,j) > 0 \\ -\infty & \text{if } a_r(i,j) = 0 \end{cases} \tag{5.1}$$

***LEMMA 5.4***

*Given an arbitrary distribution $R$ of packets on the linear array, and if the farthest-first strategy is used,* $T_{\text{right}}(R) = \max_{i<j}\{j - i + h_r(i,j)\}$.

**PROOF** Choose $i_{\max}, j_{\max}$ such that $j_{\max} - i_{\max} + h_r(i_{\max}, j_{\max}) = \max_{i<j}\{j - i + h_r(i,j)\}$. A packet starting in a PE with index $\leq i_{\max}$ has to make at least $j_{\max} - i_{\max}$ routing steps before it reaches $P_{j_{\max}}$. At most one packet reaches $P_{j_{\max}}$ from the left in every routing step. This shows that $T_{\text{right}} \geq j_{\max} - i_{\max} + h_r(i_{\max}, j_{\max}) = \max_{i<j}\{j-i+h_r(i,j)\}$.

We prove now the opposite inequality. Consider the subproblem of the routing within the left most $k$ PEs of the array. We consider all packets starting there, assuming that they finish in $P_k$ if they move further to the right in the original problem. These packets are called *k-packets.*

**Claim** All $k$-packets reach $P_k$ within $\max_{\{m \mid m<k\}}\{k - m + h_r(m,k)\}$ steps.

**PROOF** With the farthest-first strategy a $k$-packet can only be delayed by other $k$-packets. Therefore, we can leave the other packets out of consideration. Let $m_{\max}$ be such that $k - m_{\max} + h_r(m_{\max}, k) = \max_{1 \leq m<k}\{k - m + h_r(m,k)\}$. Define $\mathcal{P}_l = \{P_1, \ldots, P_{m_{\max}}\}$, $\mathcal{P}_r = \{P_{m_{\max}+1}, \ldots, P_{k-1}\}$. Consider the following subdivision of the array:

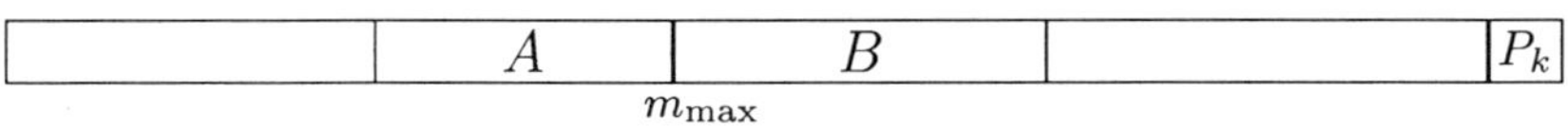


$A$ is a subarray of arbitrary size at the right end of $\mathcal{P}_l$, while $B$ is a subarray of arbitrary size at the left end of $\mathcal{P}_r$. Let $A_k$, $B_k$ be the sets of $k$-packets in $A$, $B$, respectively. The definition of $m_{\max}$ implies that, for all such $A$ and $B$,

$$\#B_k \leq \#B, \tag{5.2}$$

$$\#A_k \geq \#A. \tag{5.3}$$

Equation 5.2 holds because otherwise $m_{\max}$ would have been chosen larger. The Equation 5.3 holds because otherwise $m_{\max}$ would have been chosen smaller. By Equation 5.2, the $k$-packets that start in $\mathcal{P}_r$ move ahead of the $k$-packets coming from $\mathcal{P}_l$. This means that the $k$-packets from $\mathcal{P}_r$ have reached $P_k$ by step $k - m_{\max} - 1$, and that these packets will not delay the $k$-packets from $\mathcal{P}_l$. By Equation 5.3 the $k$-packets from $\mathcal{P}_l$ flow into $P_k$ continuously from step $k - m_{\max}$ on. Thus, the last one of those packets reaches $P_k$ by step $k - m_{\max} + h_r(m_{\max}, k)$. The fact that packets from $\mathcal{P}_l$ really flow into $P_k$ without interruption requires some explanation. These packets will not reach $\mathcal{P}_r$ faster if some of them are moved to the left. Now, we can wipe the $k$-packets to the left in a special way. For all $m$, from $m_{\max}$ down to the smallest index of a PE that sends packets to $P_k$, all $k$-packets in $P_m$ except for one, are moved to $P_{m-1}$. Equation 5.3 assures that there is always a packet left behind. After the wiping, all these packets lie as a long uninterrupted chain in $\mathcal{P}_l$, with possibly a pile at the left most PE that sends a packet to $P_k$. (analogous to the situation depicted in the proof of Lemma 5.3). The packets in the chain are not delayed and thus in every step they move one position rightwards. This proves the Claim. ∎

The claim gives the latest time for packets starting in a PE $P_m$ with $m < k$, to reach their destination. Since the claim holds for arbitrary $k$, $T_{\text{right}}(R) \leq \max_{m<k}\{k - m + h_r(m,k)\}$. This proves Lemma 5.4. ∎

The diameter of a circular array is half the diameter of the linear array of the same length. Does this imply that a distribution of packets can be routed twice as fast? Sometimes it does, sometimes it does not. Most important for us is that randomizations can be routed twice as fast. The following analogue of Lemma 5.4 holds:

***LEMMA 5.5***

*Given an arbitrary distribution R of packets on the circular array, if the farthest-first strategy is used, then* $T_{\text{right}}(R) = \max\{\ \max_{i<j}\{j - i + h_r(i,j)\}, \max_{j<i}\{n + j - i + h_r(i,j)\}\ \}$.

**PROOF** Let $i_{\max}, j_{\max}$ be such that $\max\{h_r(i_{\max}, j_{\max}) + j_{\max} - i_{\max}, h_r(i_{\max}, j_{\max}) + n + j_{\max} - i_{\max}\} = \max\{\ \max_{i<j}\{h_r(i,j) + j - i\}, \max_{j<i}\{h_r(i,j) + n + j - i\}\ \}$. No packet that starts in a PE with index $i_{\max}$ or smaller will reach $P_{j_{\max}}$ before step $j_{\max} - i_{\max}$ if $j_{\max} > i_{\max}$ and $n + j_{\max} - i_{\max}$ if $i_{\max} > j_{\max}$. At most one packet reaches a PE from the right in any step. Thus $T_{\text{right}}$ is at least as large as stated.

The proof of the opposite inequality is analogous to the proof of Lemma 5.4. The main step is the proof of the following:

**Claim** All $k$-packets reach $P_k$ within:

$$\max\{\ \max_{i<k}\{k - i + h_r(i,k)\}, \max_{k<i}\{n + k - i + h_r(i,k)\}\ \}$$

steps.

Renumbering the mesh (by adding $n - k$ in a cyclic way to all the PE indices) does not change the time it takes the packets to reach a certain

PE. Thus, we may assume without loss of generality that $k = n$. In that case the claim follows from the case $k = n$ of the claim in Lemma 5.4. ∎

### 5.2.3 Routing Randomizations

In this section, we analyze the important problem of routing $k$ packets from every PE to a randomly selected destination PE, on a MIMD linear and circular arrays.

Combining Lemma 5.4 and its analogue for routing leftwards, and using Chernoff bounds, we get:

***LEMMA 5.6***

*For a distribution $R$ of packets on the linear array such that from each PE $k$ packets are routed to randomly chosen destinations, if the farthest-first strategy is used, then $T_r(R) \leq \max\{n, kn/4\} + \mathcal{O}((kn \log n)^{1/2})$, with very high probability.*

**PROOF** Routing a randomization with every PE sending 4 packets takes more routing steps than when every PE would send only 1, 2, or 3 packets. So, we may assume $k \geq 4$. We consider $h_r(i,j)$ as defined in Equation 5.1. $h_r(i,j)$ is distributed as the binomial random variable $B(ik, (n-j)/n)$. Hence, by Lemma 5.3, $h_r(i,j) \leq ki(n-j)/n + \mathcal{O}((kn \log n)^{1/2})$, with very high probability. Lemma 5.4 gives $T_{\text{right}} \leq \max_{i<j}\{j-i+h_r(i,j)\} \leq \max_{i<j}\{j-i+ki(n-j)/n\}+\mathcal{O}((kn \log n)^{1/2})$. The same bound holds for $T_{\text{left}}$. Let $f_k(i,j) = j - i + ki(n-j)/n$. We have to find the maximum of $f_k$ over all $0 \leq i \leq j \leq n$. Partial differentiation gives an extreme value $f_k(n/k, n - n/k) = n - n/k$. On the boundary there are two more extremal values: $f(0,n) = n$ and $f(n/2, n/2) = kn/4$. For $k \geq 4$ this latter value is at least as large as the others. ∎

Routing a randomization on a circular array is twice as fast as on a linear array:

***LEMMA 5.7***

*Given a distribution $R$ of packets on the circular array such that from each PE $k$ packets are routed to randomly chosen destinations, if the farthest-first strategy is used, then $T_r(R) \leq \max\{n/2, kn/8\} + \mathcal{O}((kn \log n)^{1/2})$, with very high probability.*

**PROOF** As in the proof of Lemma 5.6 we may assume that $k \geq 4$. We are going to use Lemma 5.5, so we have to show that $h_r(i, j)$ is small enough for arbitrary $i$ and $j$. Without loss of generality, we may assume that $i = 0$. It is easy to check that for any $j$ the expected number of packets going from left to right first through $P_0$, then through $P_j$, equals $k \sum_{l=1}^{n/2-j} l/n \simeq k(n/2-j)^2/(2n)$. For $k \geq 4$ this is smaller than $kn/8-j$. The step to "very high probability" can be made using Chernoff bounds. ∎

### 5.2.4 $k$-$k$ Routing

The *greedy algorithm* routes all packets via the shortest way to their destination. Combining Lemma 5.4 and its analogue for leftward routing gives:

***LEMMA 5.8***

*When the farthest-first strategy is used, the greedy algorithm is optimal for routing $k$-permutations on a linear array.*

This shows that $k$-$k$ routing on a linear array is trivial. The situation on circular arrays is very different, and it is not clear at all how to achieve an optimal routing time. The remainder of this section is devoted to the $k$-$k$ routing problem on a circular array.

We cannot imagine any advantage of sending packets to intermediate destinations and reversing them. Thus, the routing problem is reduced to *directing* the packets: determining for every packet the direction in which it moves. On a linear array there is only one possible direction but on a circular array there is a choice to make. We say that a packet is directed *towards* its destination if the direction gives the shortest way to its destination. Otherwise, the packet is said to be directed *away from* its destination.

Suppose that from a PE of a circular array some packet $p_1$ is directed rightwards and some packet $p_2$ leftwards, traveling $d_1$ and $d_2$ steps, respectively. If $d_1 + d_2 > n$ (meaning that the paths of $p_1$ and $p_2$ overlap), then it is better to reverse the direction of $p_1$ and $p_2$. This certainly does not increase the required number of routing steps and possibly reduces it. This shows that rather than directing the individual packets, we should determine the number of packets to route in either direction, and direct the packets "overlap-free." Thus, an optimal deterministic oblivious local knowledge $k$-$k$ routing algorithm can be represented by $n$ functions $f_i : \{0, 1, \ldots, n-1\}^k \rightarrow \{0, 1, \ldots k\}$, $0 \leq i < n-1$, giving the number of packets in $P_i$ that are directed rightwards, depending on the distances the packets have to travel rightwards. These $f_i$ are referred to as *direction functions.*

At first it seems difficult to perform $k$-$k$ routing on the circular array with less than $kn/2$ routing steps. The crucial idea is that not all packets have to be directed towards their destination. Directing some packets away from their destination helps to bound the number of packets that have to move through a link. The packets can be directed by a PE based on information it initially has. There are several possible approaches. A first idea is to let every PE always direct at least $\alpha k$ packets in both directions. Alternatively, all packets that are within $\beta n$ steps from their destination can be directed towards it, while the other packets are distributed over the two directions as best as possible. For $k = 2$, this leads to the best oblivious algorithm.

There is a simple randomized strategy which leads to almost optimal performance for $k$-$k$ routing on circular arrays. The randomized algorithm requires $kn/4 + o(kn)$ steps. We now describe two versions of a deterministic routing algorithm: the first uses for its decisions only information that is initially available in each PE. In this way, one may lose $n/4$ steps; the second version first gathers some information from the neighboring PEs, which reduces the routing time to $kn/4 + \sqrt{n}$ steps.

With oblivious routing algorithms routing is reduced to the problem of directing the packets suitably. We provide a simple fully local knowledge algorithm which has near-optimal performance. In this algorithm the packets are directed in a weighted way: the farther the packets in a PE have to travel rightwards, the more packets are directed leftwards. This approach can be viewed as a deterministic version of a simple randomized algorithm in which a packet that has to travel $d$ steps rightwards is directed leftwards with probability $d/n$.

For computing the direction all PEs perform the following algorithm:

Algorithm FLKDIRECT

(1) Sort the packets based on the distance $d_i$, $1 \leq i \leq k$, they have to move rightwards. Store them in $a_1, \ldots, a_k$.

(2) Let $j = k - round(\sum_{i=1}^{k} d_i/n)$.

(3) Direct $a_1, \ldots, a_j$ rightwards; direct $a_{j+1}, \ldots, a_k$ leftwards.

We call this algorithm FLKDIRECT, an acronym for *f*ully *l*ocal *k*nowledge *direct*. Notice that FLKDIRECT can be applied for all $k$. For $k = 1$, it directs all packets towards their destinations, for $k \geq 2$ the algorithm is nontrivial.

By its generality FLKDIRECT is a very attractive algorithm, by its simplicity it can be implemented easily. The analysis is a bit tricky though. Because the following lemma is of central importance, implying the near-optimality of FLKDIRECT, its long proof is rendered completely. The reader may want to skip at least part of it.

***LEMMA 5.9***

*When a k-k distribution is directed with* FLKDIRECT, *no link has to transfer more than $kn/4 + n/4$ packets.*

**PROOF** Consider a $k$-$k$ distribution. Let $a_{i,j}$ be the number of packets that must go from $P_i$ to $P_j$. By definition of a $k$-$k$ distribution, $u_i = \sum_{j=0}^{n-1} a_{i,j} = k$, and $v_j = \sum_{j=0}^{n-1} a_{i,j} = k$, for all $0 \leq i, j \leq n-1$.

We analyze the number $H$ of packets that have to go over $(n-1, 0)$. First we consider $\overline{H}$, the (not necessarily integral) number of packets that would go over $(0, n-1)$ if in step 2 the numbers would not be rounded. Moreover, $\overline{H}$ also equals the expected number of packets that would go over $(n-1, 0)$ in the randomized algorithm described above. The packets coming from $P_{n-1}$ contribute $(n-1)a_{n-1,0} + (n-2)a_{n-1,1} + s + 2a_{n-1,n-3} + a_{n-1,n-2}$ to $n\overline{H}$. Summing over all origins we find

$$n\overline{H} = \sum_{i=1}^{n-1} \sum_{j=0}^{i-1} (i-j)a_{i,j}/n. \qquad (5.4)$$

Let $-1/2 < \delta_i \leq 1/2$ be the effect of the rounding for $H_i$, the contribution of $P_i$ to $H$. Then,

$$H = \overline{H} + \sum_{i=1}^{n-1} \delta_i. \qquad (5.5)$$

Combining Equations 5.4 and 5.5, we prove that $H \leq kn/4 + n/2$. Then we refine the proof to obtain the stated result.

From the sum in Equation 5.4 we can split off terms $u_i$ and $v_j$ such that $\sum_{i=1}^{n-1} \sum_{j=0}^{i-1} (i-j)a_{i,j} \leq \sum_{l=1}^{n/2} l u_{n/2-1+l} + \sum_{l=1}^{n/2-1} l v_{n/2+l}$. Once this is shown, it follows that $n\overline{H} \leq k(\sum_{l=1}^{n/2} l + \sum_{l=0}^{n/2-1} l) = kn^2/4$. By Equation 5.5, $H \leq kn/4 + (n-1)1/2$.

The estimate for Equation 5.4 in terms of $u_i$ and $v_j$ can be proven in a standard way, using induction on $n$. For a more instructive proof we consider the case $n = 10$ with help of the following matrix $A$:

| 9 | 9 | 8 | 7 | 6 | 5 | 4 | 3 | 2 | 1 | 0 |
|---|---|---|---|---|---|---|---|---|---|---|
| 8 | 8 | 7 | 6 | 5 | 4 | 3 | 2 | 1 | 0 | 0 |
| 7 | 7 | 6 | 5 | 4 | 3 | 2 | 1 | 0 | 0 | 0 |
| 6 | 6 | 5 | 4 | 3 | 2 | 1 | 0 | 0 | 0 | 0 |
| 5 | 5 | 4 | 3 | 2 | 1 | 0 | 0 | 0 | 0 | 0 |
| 4 | 4 | 3 | 2 | 1 | 0 | 0 | 0 | 0 | 0 | 0 |
| 3 | 3 | 2 | 1 | 0 | 0 | 0 | 0 | 0 | 0 | 0 |
| 2 | 2 | 1 | 0 | 0 | 0 | 0 | 0 | 0 | 0 | 0 |
| 1 | 1 | 0 | 0 | 0 | 0 | 0 | 0 | 0 | 0 | 0 |
| 0 | 0 | 0 | 0 | 0 | 0 | 0 | 0 | 0 | 0 | 0 |
| | 0 | 1 | 2 | 3 | 4 | 5 | 6 | 7 | 8 | 9 |

$A_{i,j}$ is the coefficient of $a_{i,j}$ in Equation 5.4. Subtracting one from all entries in a row or column corresponds to splitting of a term $u_i$ or $v_j$, respectively. $A_{9,0}$ can be eliminated by taking 5 times $u_9$ and 4 times $v_0$. Then $A_{6,0}$ and $A_{7,1}$ are eliminated, and so on. For $n = 10$ this gives $n\overline{H} \leq (5u_9 + 4v_0) + (4u_8 + 3v_1) + (3u_7 + 2v_2) + (2u_6 + v_5) + u_5$. For higher $n$ the process of elimination goes analogously.

In order to show that $H \leq kn/4 + n/4$, we refine our estimates. Consider the matrix $A'$ obtained after subtracting 5 from the entries in row 9 of $A$, 4 from the entries in column 0 of $A$, etc.:

| 9 | 0 | 0 | 0 | 0 | 0 | −1 | −2 | −3 | −4 | −5 |
|---|---|---|---|---|---|---|---|---|---|---|
| 8 | 0 | 0 | 0 | 0 | 0 | −1 | −2 | −3 | −4 | −4 |
| 7 | 0 | 0 | 0 | 0 | 0 | −1 | −2 | −3 | −3 | −3 |
| 6 | 0 | 0 | 0 | 0 | 0 | −1 | −2 | −2 | −2 | −2 |
| 5 | 0 | 0 | 0 | 0 | 0 | −1 | −1 | −1 | −1 | −1 |
| 4 | 0 | 0 | 0 | 0 | 0 | 0 | 0 | 0 | 0 | 0 |
| 3 | −1 | −1 | −1 | −1 | 0 | 0 | 0 | 0 | 0 | 0 |
| 2 | −2 | −2 | −2 | −1 | 0 | 0 | 0 | 0 | 0 | 0 |
| 1 | −3 | −3 | −2 | −1 | 0 | 0 | 0 | 0 | 0 | 0 |
| 0 | −4 | −3 | −2 | −1 | 0 | 0 | 0 | 0 | 0 | 0 |
| | 0 | 1 | 2 | 3 | 4 | 5 | 6 | 7 | 8 | 9 |

The generalization of $A'$ for larger $n$ is clear from the example. An accurate expression for $H$ can be given in terms of the entries $A'_{i,j}$:

$$H = kn/4 - \sum_{i=0}^{n-1} \sum_{j=0}^{n-1} A'_{i,j} a_{i,j}/n + \sum_{i=1}^{n-1} \delta_i.$$

A $k$-$k$ distribution is given by "placing" $k$ packets in every row and column. Now suppose that we have a placement such that

- $\delta_i = 0$, for all $i < n/2$; $\delta_i = 1/2$, for all $i \geq n/2$.
- $a_{i,j} = 0$, for all $i, j < n/2$; $a_{i,j} = 0$, for all $i, j \geq n/2$.

For such a placement $H = kn/4 + n/4$. A placement with larger $H$ value might be obtained only by increasing $\sum_{i=1}^{n-1} \delta_i$ beyond $n/4$. The packets can be rearranged such that indeed $\sum_{i=1}^{n-1} \delta_i > n/4$, but the induced increase of $\sum_{i=0}^{n-1} \sum_{j=0}^{n-1} A'_{i,j} a_{i,j}/n$ is larger. ∎

***THEOREM 5.2***

*A $k$-$k$ distribution directed by* FLKDIRECT *can be routed in $kn/4 + n/4$ steps, for all $k \geq 2$.*

**PROOF** From Lemma 5.9 we know that no link has to transfer more than $kn/4 + n/4$ packets. As no packet has to travel more than $n$ steps, this proves the theorem for $k \geq 3$. For $k = 2$ consider a PE $P$ holding packets $p_1$ and $p_2$, going $d_1$ and $d_2$ steps rightwards. Suppose that $p_1$ is directed rightwards. Then $d_1 \leq d_2$ and $2 - (d_1 + d_2)/n \geq 1/2$. Hence, $2 - 2d_1/n \geq 1/2$, and thus $d_1 \leq 3n/4$. ∎

***LEMMA 5.10***

*The bound given in Theorem 5.2 is sharp.*

**PROOF** For odd $k$, an example is provided by the $k$-$k$ distribution $k$-$L_{n/2}$, under which all packets must be routed $n/2$ positions to the right. Every PE wants to direct $k/2$ packets rightwards. This may be rounded to $k/2 + 1/2$.

For even $k$ we first consider the case $k = 4$, and use the following subdivision of the ring:

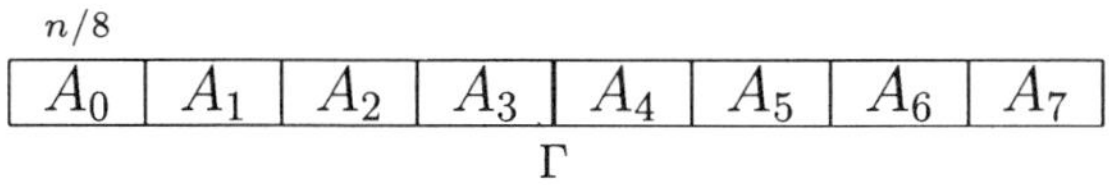


Suppose that all packets from $A_0$ are shifted to $A_7$, all packets from $A_1$ to $A_6$, all packets from $A_2$ to $A_5$, and all packets from $A_3$ to $A_4$. Let $\overline{H}_i$ be the nonrounded number of packets that $P_i$ directs rightwards. Then $\overline{H}_i = 1/2$, for $0 \leq i < n/8$, ..., $\overline{H}_i = 7/2$ for $3/8n \leq i < n/2$. After rounding it may happen that from $A_i$, $i = 0, 1, 2, 3$, $(i+1)n/8$ packets are routed rightwards through $\Gamma$. Thus, in total, $5/4n$ packets may pass through $\Gamma$. The given example can be generalized easily for larger $k$. ∎

The term $n/4$ in Theorem 5.2 is a result of the rounding. It can be reduced by dividing the ring in subchains consisting of $\sqrt{n}$ adjacent PEs each. Within these subchains the rounding can be performed such that the sum of the numbers of packets that are directed rightwards is at most $1/2$ away from the sum of the numbers of packets the PEs in this subchain want to directed rightwards. We call the modified algorithm LKDIRECT:

***THEOREM 5.3***

*A k-k distribution directed by* LKDIRECT *can be routed in* $kn/4 + \sqrt{n}$ *steps, for all* $k \geq 4$.

**PROOF** The rounding can be performed in $\sqrt{n}/2$ steps. After this the term $n/4$ is reduced to $\sqrt{n}/4$. ∎

For $k = 2$ and $k = 3$ the above approach does not give useful results: packets may be sent almost around the ring. Let $S_i$ be the (not necessarily integral) number of packets that $P_i$ directs rightwards.

#### 5.2.4.1 Algorithms for $k = 2$

The above algorithms can also be applied with slight modifications for small $k$. Here we show that for $k = 2$, an algorithm that takes its decisions only on basis of information that is initially available in the PEs runs in $3n/4$ steps. An algorithm that initially gathers some information requires only $2n/3 + o(n)$ steps.

For $k = 2$, $S_i$ equals the number of packets that have to travel less than $n/3$ steps rightwards + half the number of packets that have to travel between $n/3$ and $2n/3$ steps rightwards. This is an alternative formulation of the rule of DYNAMIC DIRECT. No packet has to travel more than $2n/3$ steps. The rounding of the $S_i$ can be performed again in subchains of $\sqrt{n}$ PEs. We get:

***THEOREM 5.4***

*2-2 routing can be performed in $2n/3 + \sqrt{n}$ steps, by a local knowledge algorithm.*

**PROOF** As in the proof of Lemma 5.9, we can show that $H \leq 2n/3$. By the rounding in subchains the additional term is only $\sqrt{n}$. ∎

### 5.2.5 Sorting

There are many possible ideas for sorting. Most of the underlying ideas are not limited in their validity to one-dimensional arrays, but carry over with minimal changes to meshes of arbitrary dimensions. However, there is one very special point with one-dimensional arrays: there are simple randomized and deterministic routing algorithms, which are not based on sorting. This means that on one-dimensional arrays it may make sense to separate the task of finding the destinations of the packets from the task of routing them. As the bisection bound always dominate the time consumption, such details have impact on the lower order terms only, but theoretically and for instances of limited size, it may be nevertheless interesting to study how much can be achieved.

Like routing, sorting on linear arrays is trivial. The simple odd-even transposition idea leads to optimal times for $k$-$k$ sorting, for arbitrary $k$. One must distinguish the case $k = 1$ from the case $k > 1$.

An essential observation, which is used in some variant or another in all mesh sorting algorithms, is that on a linear array, 2-2 routing and sorting does not take more time than 1-1 routing and sorting:

***LEMMA 5.11***

*On a linear array with $n$ PEs, 2-2 routing and sorting can be performed in $n$ steps.*

**PROOF** The routing can be performed in a greedy way, applying the farthest-first strategy. In that case Lemma 5.4 gives that at most

$n$ steps are required. Transposition sort can be performed in the same number of steps. ∎

For randomized $k$-$k$ sorting the following algorithm is useful. After choosing a sufficiently large sample, all packets can guess a good approximation of their rank and thus their destination position within the array. After routing them there, the sorting can be completed by local comparisons. In the following we work out the details.

The central ideas of the algorithm are the following:

Algorithm RANDSORT

(1) Route all packets to a randomly selected intermediate destination.

(2) Divide the ring in subchains of appropriate size $m = o(n)$. Sort the packets within the subchains.

(3) Let $\lfloor r_p n/(km) \rfloor$ be the preliminary destination for a packet $p$ with rank $r_p$ within its subchain.

(4) Route the packets to their preliminary destinations.

(5) Bring the packets to their final destination using transposition sort.

Alternative algorithms use a construction with presplitters and splitters from which the preliminary destination could be estimated. However, after a complete randomization of the packets there is no need for all this: we can just as well use the packets themselves.

***THEOREM 5.5***

RANDSORT *performs $k$-$k$ sorting on a ring in* $\max\{n, kn/4\} + o(kn)$ *steps [632].*

**PROOF** We indicate the main points. With Lemma 5.7 it can be shown that step 1 and step 4 each can be performed in $kn/8 + o(kn)$ steps. This is achieved by sending the packets with probability 1/2 in either direction along the ring. How far a packet can be away from its destination after step 4? Consider a packet $p$ with rank $r + 1$. There are $r$ packets with key smaller than $p$. These packets are routed with probability $m/n$ to the subchain to which $p$ is randomized independently of each other. Hence, we may use Chernoff bounds (see [327]) to conclude that at most $rm/n + \mathcal{O}((rm \log n/n)^{1/2})$ of them are routed to the subchain of $p$. Hence, the maximum for the estimate of the rank of $p$ is $r + \mathcal{O}((rn \log n/m)^{1/2})$. For $r = \Omega(kn)$ this is worst. Then we find an "error" $\mathcal{O}((kn^2 \log n/m)^{1/2})$. If $m = k^{1/3} \log^{1/3} nn^{2/3}$, then the preliminary destination of a packet lies at most a constant number of subchains away from its final destination. ∎

Notice that a deterministic algorithm which is based on sampling techniques with similar performance as RANDSORT exists [632].

## 5.3 Two-Dimensional Arrays

### 5.3.1 Online Routing

In this section, we consider variants of routing on meshes. We will not attempt to list all results for all variants and all models. Most mesh results carry over to tori, and MIMD results carry over to SIMD with the due time factors between them.

#### 5.3.1.1 Permutation Routing

Permutation routing on two-dimensional meshes has attracted a considerable amout of attention.

**Greedy routing**

In the SIMD model, $4n - 4$ steps is a lower bound for almost any problem. For permutation routing, this lower bound is matched by the simple greedy routing strategy, in which all packets are first sent along the rows to their destination columns and then along the columns to their destinations. The routing time on two-dimensional mesh is optimal, but when all packets from a certain row have destination in the same column, they are all wiped into a single node, yielding queues of size $\Omega(n)$. Similarly, assuming the MIMD model, we have:

***LEMMA 5.12***

*On an MIMD $n \times n$ mesh, greedy permutation routing requires at most $2n - 2$ steps.*

**PROOF** Consider the routing in a row of the mesh during the first phase. Initially all packets stand evenly spread out. These packets can be routed to arbitrary destinations without delay, because they move in "lock-step-mode", like geese. Thus, Phase 1 takes at most $n - 1$ steps. The distribution of the packets that must be routed during Phase 2 over the PEs of a column may be uneven, thus we cannot apply the same argument. However, if a packet that has to travel a distance $d$ waits until Step $n - d$ before it sets off, then again no two packets want to use the same link at the same time. Instead of letting a packet wait even if it is not necessary, we can also apply the farthest-first strategy. ∎

Unfortunately, the greedy algorithm may have a catastrophic queue behavior: if all packets in Row 0 want to travel to Column 0, then at the end of Phase 1 they are all crowding together in $P_{0,0}$. Large queues are undesirable for several reasons. If one views the PEs as tiny nodes on a chip, their storage capacity may be very limited. More seriously, the queues must be maintained.

One may believe that the problem is caused by the fact that we perform two separate routing phases: Phase 2 only starts after Phase 1 has been completed. In many cases improvements can be achieved by *coalescing* the routing phases. This means that in the MIMD mesh model Phase 2 starts as soon as there are packets to which it can be applied. In this case this can be done without any additional rules, but if there are packets of several phases using the same links, then we must specify a further strategy to resolve conflicts. The most logical, and the only applied strategy, is to give packets that are still performing an earlier phase absolute priority over those performing a latter phase.

Unfortunately, even when the phases are coalesced the queues are not substantially smaller. For example, consider a distribution of packets with all packets going to Column $n/2$ standing distributed in Row 0 and Row 1: $n/3$ packets in Row 0 and $2/3n$ packets in Row 1, symmetrically distributed around the Column $n/2$, with no such packet in $P_{1,n/2}$. In each of the first $n/3$ steps, $P_{1,n/2}$ is reached by 3 packets. In Step 1 it only receives packets, thereafter, it can get rid off only one. Thus, after $n/3$ steps, $P_{1,n/2}$ holds $2/3n + 1$ packets. Better than before, but still very bad.

If the packets have randomly chosen destinations, then the bad examples are highly unlikely to arise. In fact, Leighton has shown that the queue sizes are bounded by $\mathcal{O}(1)$ with high probability [400]. This is why, in practice, the greedy algorithm mostly works fine, and why most hardware routers apply this strategy, accepting the small probability of traffic jams [607].

Notice that on higher-dimensional meshes the situation is worse. Not only do queues grow large, but also the routing time becomes very bad. For example, consider a three-dimensional mesh with the permutation given by $(x_0, x_1, x_2) \mapsto (x_2, n-1-x_1, x_0)$. Then, in the first routing, all packets in the $n^2$ PEs at positions $(x_0, x_1, 0)$ are wiped together in the one-dimensional subarray $(0, x_1, 0)$. During the second routing, all these packets must travel through the links between $P_{0,n/2-1,0}$ and $P_{0,n/2,0}$. Thus, the routing requires $\Omega(n^2)$ time, instead of $\mathcal{O}(n)$ as it should do.

Notwithstanding the simplicity and the good average case behavior of the greedy algorithm, it is still desirable to design algorithms with a better worst-case performance. This is the topic of the next discussion, focusing on the MIMD (full port) model. In the MIMD model, the mesh diameter $2n-2$ is a lower bound for permutation routing.

The first algorithms achieving $\mathcal{O}(n)$ time on an $n \times n$ mesh were already considered breakthroughs. Then gradually it became clear that in the full port model one can actually match the distance bound of $2n-2$.

Leighton showed that for random destinations the greedy algorithm performs very well, and even the queue size remains small [400]. Thus, the problem with the greedy algorithm only occurs in rare cases, namely, when a large number of packets with destination in the same row all have destination in the same column. A random perturbation may easily destroy such unlucky distributions. Based on this fact, Rajasekaran and Tsantilas introduced a modification of the greedy routing algorithm, which starts by randomizing the packets in stripes. Then, one can see that by using a special treatment of packets, the time lost during randomization can be recovered. The result was a near-optimal randomized permutation routing algorithm [575].

Later, Kunde provided an algorithm similar in spirit to the previous one, but deterministic [367]. In this algorithm, greedy routing was preceded by a sorting phase in column-major order. This phase introduced a similar effect as the randomization in the mesh stripes, and resulted in a near-optimal deterministic algorithm.

Finally, Leighton, Makedon, and Tollis proved "the impossible;" a deterministic algorithm running in $2n-2$ steps with constant size queues (about 1000) [412]. In later papers the queue size was reduced to more practical values [573, 630] and the algorithm has been polished and simplified. The essential idea in all papers since [575] is that "critical packets" (packets moving between opposite corner regions of the mesh) are given priority, so that they do not incur any delay. Queue size estimates essentially depend on sorting algorithms on submeshes of decreasing size.

Next, we present a deterministic routing algorithm which achieves number of steps $T = 2n - 2$ and maximum required queue size $Q = 63$ [630]; this algorithm can be refined further to achieve $Q = 32$.

The presented algorithm depends on two parameters, $m$ and $s$, which are fixed fractions of $n$. The mesh is divided into squares of size $m \times m$. If we need to refer to specific squares, then Figure 5.6 (b) is used. Packets that need to move between opposite corner regions of size $m \times m$ are called *critical*, the other packets are called *noncritical* [575, 412]. The noncritical packets are first *scattered* within squares of size $s \times s$, and then most of them are routed greedily first along rows, and then along columns, though they may be queued up for a certain number of steps. By scattering we mean a redistribution of packets such that the queues remain short during the greedy routing that follows.

In our case, $m = n/6$, and we assume that $n = c \cdot 6^l$, for a constant $c$. Noncritical packets are first scattered in their $s \times s$ submeshes in $T_s$ steps, where $s$ must be chosen such that $T_s \leq m$. After the scattering has been completed, most noncritical packets wait until step $m$, then they move to their destination columns (these are the normal packets, the remaining noncritical packets are treated differently). Here they wait a number of steps, depending on the distance they still have to go: a packet which has to travel $d$ steps along the column waits $2n - 2 - d$ steps before it moves on.

During the scattering, the critical packets move perpendicular to the noncritical packets. This is possible because the scattering is uni-axial. From $T_s + 1$ until $m$ they move along the rows. After this, they move along the columns until the $m \times m$ submeshes reach the opposite side, that is, after $n - m$ steps along the column. Next, they move along the rows until the submeshes have moved a total of $m$ steps along the rows. Now the critical packets that originated in the upper left and lower right corners have reached the regions in which they are routed recursively during the next $2m - 2$ steps. The routing is with respect to their final destinations in the $m \times m$ corner submesh. When the recursive routing is completed, by step $n + 2m - 2$, they move out of the recursive routing region along the rows towards their final destinations. In the meantime, the other critical packets travel along the rows towards the recursive routing region, which they enter after $2n - 4m$ steps, and which they cover after step $2n - 3m$. During the next $2m - 2$ steps, they are routed recursively, and subsequently move along the rows to their final destinations in the corner regions. Notice that $n + 2m - 2 < 2n - 4m$, when $m \leq n/6$. This means that the first recursive routing is over before the traveling critical packets enter the recursive routing regions.

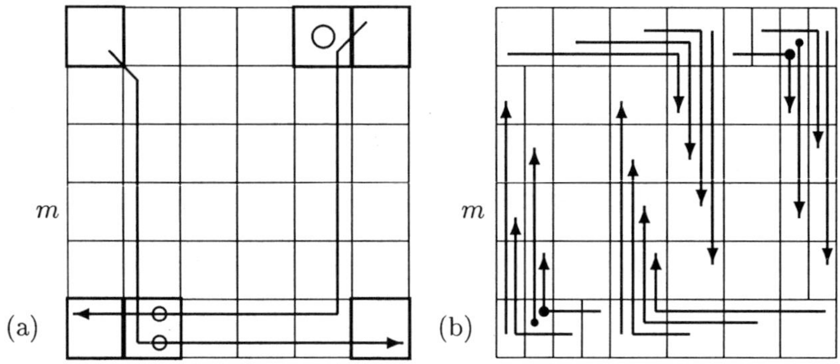


**FIGURE 5.6**
**Routing scheme** $6 \times 6$

In Figure 5.6 (a), the tracks of the critical packets starting in the two upper corners are shown. Their recursive routing region is indicated by small circles. The submesh marked with a bigger circle is the recursive routing region of the critical packets from the lower corners.

In Figure 5.6 (b), the move of the critical packets depends on the scattering algorithm. In the initial parts of the tracks of the special packets, small dots indicate that the packets wait there for $m$ steps, bigger dots indicate that they wait $2m$ steps before they proceed. All other packets move along the given lines without delay.

Because of the recursion, some noncritical packets have to be routed differently. Without precautions, they could be locked into the recursive routing region, and the queues could become $\Omega(\log n)$ long. The noncritical packets are distinguished into *normal* and *special* packets. The special packets are those that originate in some row $i$, for $0 \leq i < m$, and have destination in some column $j$, for $n/2 - 2m \leq j < n/2 - m$, and those that originate in some row $i$, for $n - m \leq i < n$, and have destination in some column $j$, for $n/2 + m \leq j < n/2 + 2m$. All other noncritical packets are normal. The normal packets are routed as described before. The special packets make a detour. The initial parts of their tracks are depicted in Figure 5.6(b). After the special packets have traversed the indicated tracks, they are treated like normal packets;

they move immediately to their destination column, where they wait a number of steps before moving on. The following can be proven easily, see [113] for a detailed proof.

***LEMMA 5.13***

*If routing on submeshes of size $m \times m$, is performed in $2m - 2 + \zeta$ steps, then all packets reach their destination in $2n - 2 + \zeta$ steps.*

**PROOF** The case of critical packets is proved by induction, based on the observation that they need $10m$ steps plus the time required for the recursion to reach their destination.

The analysis of the noncritical packets is facilitated by the conflict freeness. The only uncertainty is the position within its $s \times s$ square to which a packet is scattered. Because we know that packets are not delayed in an unpredictable way, it suffices to determine the maximal length of a path that a noncritical packet may have to traverse. After the scattering, this path is obviously at most $2n - 2 - m$ for normal packets. Also, for special packets, the path length together with possible waiting time is at most $2n - 2 - m$. These packets start to traverse this path after $m$ steps, and, before they start to move along the column, they may be delayed by additional $\zeta$ steps. ∎

We now construct an optimal and a near-optimal algorithm, based on Scheme $6 \times 6$. It turns out that in our algorithms $Q$ is not only a function of $n/s$, but also of a parameter $E$.

***DEFINITION 5.1*** *For a PE $P$, let $e_P$ denote the number of $s \times s$ submeshes from which noncritical packets may be routed in their horizontal routing phase to $P$. The effective width $E(n/s)$ of a routing algorithm is the maximum over all $P$ of $e_P$.*

We also use $e$ without subscript, to denote the maximum of $e_P$ within a subset of the PEs (typically a Section of a row). The name "effective width" is inspired by the fact that for a queue size analysis, the routing of the special packets effectively makes the submeshes of size $s \times s$ wide.

***LEMMA 5.14***

*For Scheme $6 \times 6$, $E(13) = 18$, and $E(28) = 36$.*

**PROOF** $E(n/s) \simeq 1.25n/s$. The precise value is determined by considering how the $m \times m$ and $m \times m/2$ submeshes intersect the $s \times s$ submeshes. ∎

For the first near-optimal algorithm QOPTROUTE1, we assume that $n = 13 \cdot 6^l$, for some $l \geq 0$. Now $s = n/13$, and use SCATTER1 for the scattering: recursive calls are made as long as the scattering time does not exceed $2^1\!/_6 s$ steps. The routing is finished with sorting.

***THEOREM 5.6***
*Algorithm* QOPTROUTE1 *routes permutations with* $T = 2n + \mathcal{O}(1)$ *and* $Q = 31$.

**PROOF** $Q$ is determined by the formula $Q = E(n/s) + n/s$. This can be shown using Theorem 5.5. In a worst case distribution $E(n/s) - 1$ of the submeshes contributing to the queue of a PE $P$ hold one packet with destination in the column of $P$, and one submesh holds $n - E(n/s)$ of these packets. ∎

The first optimal algorithm is called TOPTROUTE1. Let us assume that $n = 28 \cdot 6^l$, for some $l \geq 0$. Take $s = n/28 = m/4^2\!/_3$, and use SCATTER2 to scatter: the recursion is continued until $n = 28$. The routing is finished with the greedy algorithm.

***THEOREM 5.7***
*Algorithm* TOPTROUTE1 *routes permutations with* $T = 2n - 2$ *and* $Q = 63$.

**PROOF** $Q$ is determined by the formula $Q = E(n/s) + n/s - 1$. We need to determine the maximum queue size when the $n$ packets with destination in a certain column are distributed over the $E(n/s)$ submeshes. The maximum is obtained if $E(n/s) - 1$ of these submeshes hold one packet, and one submesh holds all the remaining packets. ∎

Another randomized $2n + \mathcal{O}(\log N)$ MIMD algorithm is given in [320]. This algorithm does not use critical packets, which makes it much simpler. It served as a source of inspiration for several later algorithms. In [325] it was shown how this algorithm can be made deterministic with queue size 2. Applying critical packets again the loss of time can be reduced from $\mathcal{O}(n^{4/5})$ to $\mathcal{O}(1)$:

***LEMMA 5.15***
*Permutation routing on an MIMD mesh can be performed in $2n+\mathcal{O}(1)$ steps using queues of size two [325, 630].*

Although this is an interesting result, the constant hidden in the $\mathcal{O}(1)$ is large. More practical algorithms have routing time below $3n$ and queue size five or six [325].

Unfortunately, the SIMD model appears incompatible with the idea of critical packets, which are supposed to move orthogonally to the noncritical packets. Simulating such a move would double the time consumption. Therefore, to achieve bounded queue lengths, the best results for SIMD permutation routing are obtained by simulating the MIMD algorithms of Kunde [367] (deterministic) and Rajasekaran and Tsantilas [575] (randomized). This gives the following:

***THEOREM 5.8***
*Permutation routing on an SIMD $n \times n$ mesh can be performed in $T = 4n+\mathcal{O}(n/Q)$ deterministically or randomizedly, with constant queue size $Q$ [367, 575].*

### 5.3.1.2 *k-k* Routing

Practically, more relevant than permutation routing (and sorting) are the $k$-$k$ routing (and sorting) problems. Because of the mesh bisection bound, $k$-$k$ routing requires at least $kn/2$ steps. The first nontrivial algorithm for $k$-$k$ routing was presented by Kunde and Tensi [374]. It requires $5kn/4 + o(kn)$ steps. Several improvements gradually reduced the routing time to almost optimal [327, 369, 574].

One of the most important ideas is *packet coloring* [369]; the packets are colored deterministically or at random, white or black. At all times the white packets are routed orthogonally to the black packets. In this way, the routing capacity of the MIMD mesh can be fully exploited. Another idea is to route the packets first to random positions, and from there to their destinations. This idea goes back to Valliant and Brebner [685]. Near-optimal results were first achieved with randomized algorithms [327, 574]. It was then discovered that randomization is superfluous, and can be replaced by sorting packets in sub-meshes and performing unshuffling [332, 370].

***THEOREM 5.9***

*$k$-$k$ routing on MIMD meshes can be performed deterministically in $kn/2 + \mathcal{O}(k^{5/6}n^{2/3})$ steps. The maximum queue size is $k$ [332, 370].*

It is not hard to perform $k$-$k$ routing twice as fast for average case inputs; just omit the randomization phase. An analysis is given in [371]. In [333], several ideas are combined to obtain an algorithm with routing time close to $kn/4$ for average case inputs while never exceeding $kn/2$.

We conclude this subsection by illustrating the main ideas involved in $k$-$k$ routing algorithms.

**Randomized Routing and Sorting**

At first, we present a very simple randomized algorithm, which also has small lower order terms, and which might be preferable over greedy routing for nonrandom distributions. It is easy to turn this algorithm into a sorting algorithm with almost the same performance.

The randomized $k$-$k$ routing algorithm is similar to the algorithms of Rajasekaran and Tsantillas [575] (for permutation routing), and even more similar to the algorithm by Valiant and Brebner for routing $k$-relations on hypercubes [685]. There are two variants, each composed of phases of one-dimensional routing. The first variant consists of one randomization phase followed by greedy routing, while the second performs two randomization phases. The first is preferred if there are some additional costs related to the number of phases and for $k < 8$. The second algorithm is optimal for tori.

*Four-Phase Randomized Routing*

The four-phase algorithm is a variant of the earlier routing algorithm for $k$-relations on the hypercube of Valiant and Brebner [685] for the two-dimensional mesh, completed by an application of the overlapping technique described in [373]. In Valiant and Brebner's algorithm, during Phase 1 the packets are distributed randomly within the hypercube. In Phase 2, they are routed dimension by dimension to their destination. The overlapping technique partitions the problem in two roughly symmetric subproblems which can be solved simultaneously using the same algorithm (where one is rotated by 90°). Our algorithm consists of four phases. The first two randomize the packets along the row and column, while the last two send the packets to their destination along the row and column. The algorithm can also be viewed as a variant of the algorithm of [355] with a much more intensive randomization of the packets. We show that this algorithm performs optimally (apart from a small additional term). Consider a permutation under which all packets have to move over exactly $n$ steps, e.g., the permutation under which the packets from $P_{i,j}$ have to go to $P_{(i+n/2) \bmod n, (j+n/2) \bmod n}$. In total, the links must be used $kn^3$ times. As every PE has 4 links, such a permutation requires at least $n/4$ routing steps and our algorithm solves the $k$-$k$ routing problem with $kn/4 + \mathcal{O}((kn \log n)^{1/2})$ time. At first sight, that seems to be impossible because a lot of packets are sent in a wrong direction during the randomization, thereby wasting connection availability. However, packets are only randomized in a wrong direction when the permutation itself makes sub-maximal use of the available connectivity.

In this and subsequent sections, we assume that the mesh is indexed by column and row position, starting with $(0,0)$ at the lower left position. The PE with index $(i,j)$ is denoted $P_{i,j}$.

Initially the packets are colored randomly green or blue with equal probability. Thereafter, the green packets are routed by

Algorithm GreenRoute

(1) Randomize the packets along the row.

(2) Randomize the packets along the column.

(3) Send the packets along the row to the correct column.

(4) Send the packets along the column to the correct row.

The blue packets are routed 90° out of phase of the green packets by the analogous procedure BlueRoute. The algorithm for routing all packets at the same time is denoted Green/BlueRoute. The effect of the randomization phases is explained below.

***LEMMA 5.16***

*After phase (1) and (2) a packet has equal probability to be in any of the $n^2$ PEs.*

**PROOF** Consider a packet residing in $P_{i,j}$. This packet is after phase (1) and (2) in $P_{i',j'}$ if and only if it was randomized in phase (1) to $P_{i,j'}$ and in phase (2) from there to $P_{i',j'}$. The probability that this happens is $1/n^2$, and is independent of $i', j'$. ∎

Once we know that the packets are really randomly distributed, phase (3) is just an inverse of a randomization and can be performed with the same number of routing steps. Likewise for phase (4). Now, we can use Lemma 5.16 for all four phases. For $k \geq 8$, this gives:

**THEOREM 5.10**

*Routing k-k permutations with Green/BlueRoute on a mesh with wrap-around links takes at most $kn/4+\mathcal{O}((kn\log n)^{1/2})$ routing steps with very high probability.*

In the proof of Theorem 5.10 we neglected the fact that at the start of phase (2), (3), and (4), the packets are not distributed in an entirely regular way over the mesh. This is not a serious problem: going back to the proof of Lemma 5.16 we see that we only need to make sure that any subset of PEs of size $m$ sends and receives at most $km+\mathcal{O}((kn\log n)^{1/2})$ packets.

The expected number of packets that reside in a PE is bounded by $k + \mathcal{O}(1)$. For $k \geq n$, the Chernoff bounds give that the maximal number of packets in any PE is bounded by $k + o(k)$. For small $k$ we cannot bound the number of packets in a single PE. But the Chernoff bounds give that the number of packets in any $m$ PEs is bounded by $km + \mathcal{O}((km\log n)^{1/2})$. Employing a technique of [575], the packets in $m$ adjacent PEs can be smeared out. This takes $\mathcal{O}(m)$ steps. With an additional $\mathcal{O}(m)$ steps, the queue sizes can be bounded by $k + \mathcal{O}((k\log n/m)^{1/2})$. Taking $m = 1$ gives the above result. If $k \leq n^{\alpha}$, for some $\alpha < 1$, we can take $m = k\log n$.

Next, we have:

**THEOREM 5.11**

*If $k \leq n^{\alpha}$, for some $\alpha < 1$, the queue size can be bounded to $k + \mathcal{O}(1)$, with very high probability, at the expense of $\mathcal{O}(n^{\alpha}\log n)$ extra routing steps. For larger $k$ the queue size is bounded by $k + o(k)$.*

Note that $\mathcal{O}(n^{\alpha}\log n) = o(n)$, under the given condition on $k$. This theorem is general and in the same way the queue size of many mesh routing algorithms can be bounded.

*Three-Phase Randomized Routing*

The previously presented four-phase algorithm is simple and general. On a mesh, it can be used to route $k$-$k$ permutations in $kn/2 + o(kn)$

steps. Also, it can easily be generalized for routing on $d$-dimensional meshes with $2d$ phases.

However, the four routing phases are too many. It is easy to see that we can also coalesce phase (2) and phase (3) of GREEN/BLUEROUTE. Effectively, this means that the packets are routed directly to their destination row. But on tori, phase (2) is important for determining the correct routing direction: phase (2) and phase (3) together constitute an optimal routing on circular arrays.

A three-phase mesh routing algorithm has several advantages. The total distance the packets may have to travel is only $3n$, which gives a smaller routing time for $k < 8$. On real networks with startup costs related to every routing operation, we have only three instead of four of these contributions.

**Deterministic Routing**

We now present a very general technique for turning many randomized algorithms for meshes into deterministic algorithms with the same time complexity (except for the lower order terms). In retrospect the underlying idea is already present in the sorting algorithms of Schnorr and Shamir [603], and in the routing algorithm of Kunde [367]. The presented sorting algorithm can also be viewed as an implementation on meshes of Leighton's "column-sort" algorithm [399]. The crucial idea is that whenever randomization is only used for nicely redistributing the packets, the same can also be achieved by sorting the packets in submeshes and shuffling them over the submeshes according to their ranks.

The derandomization technique is based on a combination of local sorting and "unshuffling." More precisely, we divide the mesh into blocks of size $n^\alpha \times n^\alpha$, with $0 < \alpha < 1$. The blocks are indexed. In sorting problems it is required that blocks with consecutive indices are adjacent. The packets in each block are sorted. In a routing problem the packets are sorted based on the index of their destination block, while in a sorting problem based on their key. The packets of each block are then distributed over the blocks. For a $k$-$k$ problem, $k \geq 1$, this is done most frequently by routing the packet of rank $i$ ($0 \leq i < kn^{2\alpha}$) in block $j$ ($0 < j \leq n^{2-2\alpha} = N$) to position $j + \lfloor i/N \rfloor N$ in block $i \bmod N$. Notice that every PE is the destination of exactly $k$ packets.

The advantage of the above *sort-and-unshuffle* operation for sorting on meshes was previously observed by Schnorr and Shamir, who used it in the design of their $3n + o(n)$ sorting algorithm in the one packet model (see Section 5.3.6) [603]. In the following, we demonstrate that this operation can in many cases be employed as a "substitute" for randomization. Following a scheme originally proposed by Valiant and Brebner [685], many randomized algorithms for routing on meshes start by sending the packets to random intermediate destinations. This has the effect of distributing packets with destinations close to each other evenly over the mesh. The sort-and-unshuffle operation simulates this effect in a deterministic manner.

#### 5.3.1.3 $l$-$k$ Routing

The general $l$-$k$ routing problem has attracted less attention [251, 454, 473, 631, 458]. In this interesting class of problems, the PEs send at most $l$ packets, and receive at most $k$ packets. If $k$ and $l$ are almost the same, the best idea is to perform a $\max\{k, l\}$-$\max\{k, l\}$ routing. However, this would be a waste if $k$ and $l$ differ substantially. Then, special algorithms are considered. For $l > k$, the central idea is to color packets such that in the final routing phase, in which packets are routed within their rows or columns towards their destinations, the maximum number of packets in a row or column remains bounded. Thus, there is a strong correspondence with the *token distribution problem*, in which initially each PE holds at most $k$ tokens that must be redistributed such that afterwards each PE holds the same number of packets. For $l < k$, the same steps are performed in reverse order. We first consider some lower bounds for this problem, and then we describe the algorithm in more detail.

**Lower Bounds**

Considering the bisection bound and the number of packets that may have to move over a limited number of links into or out of a corner, we find the following lower bounds for routing $l$-$k$ relations:

***LEMMA 5.17***

*The $l$-$k$ routing problem on an MIMD mesh requires at least:*

$$\max\left\{\frac{k}{2}\cdot\left(\frac{l}{l+k}\right)^{1/2}, \frac{lk}{l+k}, \frac{l}{2}\cdot\left(\frac{k}{l+k}\right)^{1/2}\right\} n$$

*steps.*

**PROOF** For the first bound we consider the following subdivision of the mesh:

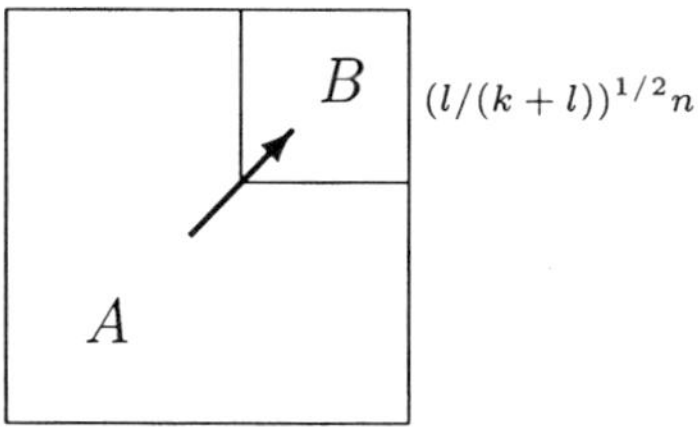


Suppose that every PE in $A$ holds $l$ packets with destination in $B$. It is easy to check that these packets fit into $B$ and that routing them across the boundary of $B$ takes $k/2 \cdot (l/(l+k))^{1/2} \cdot n$ steps.

The second bound is a generalization of the bisection bound. Consider the following subdivision of the mesh:

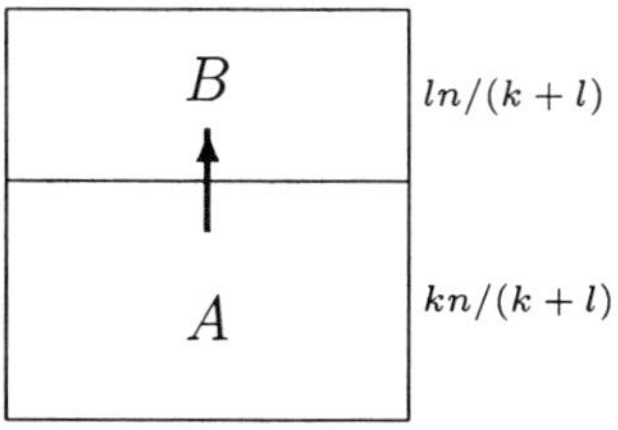


If all PEs in $A$ hold $l$ packets with destination in $B$, then it takes for these $lk/(k+l)n^2$ packets $lkn/(k+l)$ steps to pass across the boundary between $A$ and $B$.

The third bound is obtained analogously to the first, considering the time that is required when all PEs of a $(k/(k+l))^{1/2}n \times (k/(k+l))^{1/2}n$ corner send $l$ packets to the PEs outside this corner. ∎

Which of the bounds in Lemma 5.17 is the strongest depends on the relation between $l$ and $k$. The first bound is strongest when $l \leq k/3$, the second when $k/3 \leq l \leq 3k$, and the third when $l \geq 3k$. When $l = k$, the second bound gives $kn/2$, the bisection bound for the $k$-$k$ routing problem. In the extreme cases the bound becomes very simple:

***COROLLARY 5.1***

*When $l = o(k)$, or $k = o(l)$, then routing $l$-$k$ relations on an MIMD mesh requires at least $(1 - o(1))\sqrt{lk}n/2$ steps.*

### General Algorithms, $l > k$

We consider how to perform $l$-$k$ routing, for $l > k$. For $k > l$ the procedure is analogous. This problem is interesting, because it has a very different character compared to the $k$-$k$ routing. In $k$-$k$ routing, PEs can decide locally (that is, within submeshes of size $o(n) \times o(n)$) how to route their packets. In this way we could establish one-optimality. For $l$-$k$ routing, there are packet distributions which are not distinguishable locally, but which nevertheless must be dealt with in a completely different order to obtain good performance. On the other hand, for each of these relations it is trivial to find a good routing with some global information. We do not even need to know the whole packet distribution; it is enough if a PE knows the number of packets standing in its row, and in its column.

The algorithm we provide is general, and works for all $l$ and $k$. The description is randomized, but the algorithm can be made deterministic using the techniques of Section 5.3.1.2.

(1) For each $0 \leq i < n$, the numbers $r_i$ and $r'_i$ of packets with source and destination, respectively, in Row $i$ are determined and broadcast. For each $0 \leq j < n$, the numbers $c_j$ and $c'_j$ of packets with source and destination, respectively, in Column $j$ are determined and broadcast.

(2) The packets are colored white or black randomly and uniformly. They are colored white with probability $c_j/(r_i + c_j)$.

(3) All white packets use greedy XY routing to randomly chosen intermediate destinations, all black packets use greedy YX routing.

(4) The packets are colored white or black randomly and uniformly. They are colored white with probability $c'_j/(r'_i + c'_j)$.

(5) All white packets use greedy XY routing to their destinations, all black packets use greedy YX routing.

This algorithm is attractive because (a) it is simple, (b) the values of $k$ and $l$ do not need to be known (they are only needed for the analysis), and (c) when $k = l$, it becomes identical to the algorithm of Section 5.3.1.2. The $r_i, r'_i, c_j$ and $c'_j$ can be computed and broadcast in $\mathcal{O}(n)$ steps. However, for relatively small $k$ and $l$ however, it may be undesirable to store the $r_i, r'_i, c_j$ and $c'_j$ at each PE. Alternatively, this information may be stored in a distributed fashion at submeshes of size $2\sqrt{n} \times 2\sqrt{n}$, and then looked up locally. This reduces the number of data that must be stored at each PE to one. One further advantage is that if these numbers are first computed in the submeshes, then they can be added together in less than $2n$ steps (first adding along the rows, then along the columns). All these details are more or less trivial. Nontrivial is the proof that the algorithm actually works for $k \neq l$. Optimal results have been established for $k = o(l)$, or $l = o(k)$.

**THEOREM 5.12**

*An arbitrary l-k routing problem can be solved on an MIMD mesh in* $\max\{\frac{k}{2}(\frac{l}{l+k})^{1/2}, \frac{lk}{l+k}, \frac{l}{2}(\frac{k}{l+k})^{1/2}\}n + \mathcal{O}(n)$ *steps [454, 631].*

**Problem 8** *Can l-k routing be performed in the time given by Lemma 5.17?*

Not considering constants, Lemma 5.17 and Theorem 5.12 were already suggested in Ex. 1.246 and 1.247 of [401]. Unlike the algorithm in [454], the algorithm of [631] works without knowing the values of $l$ and $k$. Furthermore, the routing time seems close to optimal for most inputs. In [473, 458], algorithms are presented with performance within a logarithmic factor from optimal for *all* inputs. This property is called *strong adaptivity* in [473]. We say that algorithms with this property are *competitive.*

#### 5.3.1.4 Token Distribution vs. $l$-$k$ Routing

Actually, [473] deals with *token distribution* [535]. Token distribution refers to rearranging "tokens" which are initially distributed inhomogeneously through the network, so that afterwards all nodes hold approximately the same number of them. The main difference compared to routing is that the final position of the tokens is *not* prescribed.

$l$-$k$ routing and token distribution are closely related. After counting the number of tokens and attributing a rank, an $l$-$m$ routing algorithm can be used for smoothing the packet distribution. Here, $m$ is the average number of packets. On the other hand, if one has a token distribution algorithm, then one can first smooth the packets, then perform an $m$-$m$ routing with $m \leq \min\{l, k\}$ as before, and finally perform the inverse of a token distribution.

This approach need not be competitive for arbitrary $l$-$k$ routing problems (which may be much cheaper than the token distribution), but for $l$-1 routing it is. In that case, we need only perform a token-distribution and a $1 - 1$ routing. This gives the following:

**LEMMA 5.18**

*If $1-1$ routing can be performed while preserving the locality, and if there is a competitive token distribution algorithm, then also the $k$-1 routing problem can be solved competitively.*

### 5.3.2 Offline Routing

Optimal results can be obtained for special permutations, e.g., for the class of bit-oriented *affine permutations.*

**THEOREM 5.13**

*Affine permutations can be routed on an SIMD mesh in $4n-4$ steps using queues of size two [498, 532].*

The algorithm is rather simple, and is based on earlier work by Pease [532] and Nassimi and Sahni [498]. By encoding the permutation in only $\mathcal{O}(\log^2 n)$ bits, precomputation (essentially a matrix inversion) takes $\mathcal{O}(\log^3 n)$ time. An extension to a larger class of affine permutations is given in [622].

**LEMMA 5.19**

*Normal Ascend/Descend permutations can be performed in $2n-2$ steps on the SIMD $n \times n$ mesh [486].*

For MIMD meshes, Krizanc proved a $2.5n$ steps general offline permutation routing algorithm with queue size two [351]. Kaklamanis, Krizanc, and Rao presented a $2n-1$ steps MIMD algorithm with queue size four [320]. Smaller queues are obtained by considering the $n \times n$ mesh as a Cartesian product of two $n$ node, one-dimensional arrays. Using Theorem 2.8 we obtain:

**COROLLARY 5.2**

*Any permutation can be offline routed on uni-axial $n \times n$ mesh in $3n-3$ steps using queues of size one [47].*

**COROLLARY 5.3**

*Any Omega or Inverse Omega permutation can be offline routed on the MIMD $n \times n$ mesh in $2n-2$ steps using queues of size one [717].*

Precomputing the schedule for the above algorithm takes $\mathcal{O}(n^2 \log n)$ steps [126].

Another important class of offline routing problems are the transpositions. In a transposition, the node at position $(x, y)$ has to exchange its data with the node at position $(y, x)$. If trivial greedy routing is applied, then many packets have to move through two of the corner nodes, but with some care one can show that:

***THEOREM 5.14***

*If every node of an $n \times n$ mesh holds $k$ packets, then these packets can be transposed in $(1 - 1/\sqrt{2})kn + o(kn)$ steps [326]. This bound is sharp [171].*

#### 5.3.2.1 Locality Preserving Routing

It may happen that all packets which have to be routed actually have to travel only a short distance. For such cases it would be a pity to apply the sophisticated algorithms which inevitably lead to a routing time of at least $\max\{2\,n, kn/2\}$. However, if $d$ is the maximum distance any packet has to go, it is not hard to obtain a routing time of $\mathcal{O}(kd)$. This is called *routing with locality* or *locality preserving routing.* If $d$ is known globally, one can apply a variant of the basic algorithm by Rajasekaran and Tsantilas in [575]. If $d$ is not known, then each packet may use its own distance $d'$ as an estimate. Therefore it performs a randomization in a range of size $d'$ along the columns, followed by greedy routing; first along the rows, then along the columns. A detailed description of such an algorithm is given in Chapter 6 of [624]. Other variants are possible too. To make such an algorithm deterministic requires more effort, but at the cost of extra routing time.

It is not true that every individual packet arrives within a constant factor from the optimal time, since the applied farthest-first strategy may cause a packet which has to travel a short distance to be delayed many steps. However, globally these algorithms achieve a constant competitive ratio.

### 5.3.3 Dynamic Routing Problems

In dynamic routing each node generates packets with a fixed rate $\lambda$. Each generated packet is assumed to have a random destination. There is a trivial upper bound on the generating rate $\lambda$; approximately one half of the packets generated in the left half of the mesh have their destinations in the right half. As at most $n$ packets can cross the bisection in every step (we consider an MIMD mesh), the mesh will get more and more congested, eventually resulting in infinite delays, for $\lambda > 4/n$. Leighton proved that for all $\lambda < 4/n$, greedy routing along row and column to their destinations normally works fine [400]. This results has been further improved by Kahale and Leighton in [316]. If several packets compete for a link, then the one that has to move farthest in this direction is given priority.

***THEOREM 5.15***

*If the generating rate in an $n \times n$ MIMD mesh is less than the network capacity $4/n$, then the maximum delay incurred by any packet in any window of $T$ steps is bounded by $\mathcal{O}(\log T + \log n)$, with high probability. The maximum queue size is $\mathcal{O}(1 + \log T/\log n)$ [400].*

In [478] this result was generalized and an alternative, more powerful analysis technique is introduced. Even more general, not limited to fixed interconnection networks like meshes or hypercubes, are the results by Harchol-Balter and co-workers [253, 254]. In [254], it is shown how to simplify the computation of upper bounds on the expected packet delay in a dynamic situation for the class of "Markovian queuing networks."

## 5.3.4 Fault Tolerant Routing

### 5.3.4.1 *P*-faulty model

For dynamic faults we want to minimize the expected lifetime of a packet and maximize the probability that the packet reaches its destination. To realize these goals, packets are routed as long as possible via shortest paths, and the best shortest path route is the one with the most alternatives. The $Z^2$*-policy* states that precedence should be given to propagation of messages towards the diagonal [36]. The diagonal with respect to node $P_{0,0}$ is defined as the set of nodes $\{P_{i,i} | \, 0 \leq i \leq n-1\}$. Thus, a packet residing in $P_{i,j}$ with destination in $P_{0,0}$, tries to move to $P_{i,j-1}$ if $i < j$, to $P_{i-1,j}$ if $i > j$, and to either of them if $i = j$.

***THEOREM 5.16***

*The $Z^2$-policy is optimal [36].*

A negative result on the effectiveness of shortest path routing is given in:

***LEMMA 5.20***

*For large $n$, the probability of successful completion of shortest path routing on the $p$-faulty $n \times n$ mesh approaches to zero [229].*

Another approach, *sidetracking*, attempts to move a packet along a shortest path from its current position to its destination. If at any instant a nonfaulty node along a shortest path can not be found, the packet is sent to a randomly chosen nonfaulty node from the remaining neighboring nodes. Thus, this approach corresponds to oblivious non-minimal path routing. Although the probability of successful routing still converges to zero, good empirical results were reported (also for the hypercube) in [230].

In an algorithm by Raghavan [565], instead of sending a packet along its master path $P$ (defined by Valiant and Brebner's randomized algorithm), the packet is broadcast within a routing region $R(P)$, defined to contain all nodes within distance $c \log n$ from $P$, where $c$ is a constant. This region is such that if there is a live path between the packet's source and destination, the path will likely be contained in $R(P)$.

***THEOREM 5.17***

*For all $p \leq 0.29$, online permutation routing can be performed on a $p$-faulty $n \times n$ mesh in $\mathcal{O}(n \log n)$ steps, with high probability [565]. The maximum queue size is $\mathcal{O}(\log^2 n)$.*

Mathies improved the above bound on $p$ to about 0.40 [464]. The most surprising result is

***THEOREM 5.18***

*There exists a constant $p > 0$, such that $p$-faulty $n \times n$ mesh can emulate a fault-free $n \times n$ mesh with constant slowdown, with high probability [127].*

#### 5.3.4.2 Worst case model

The above algorithms give online adaptations. For static faults, stronger results can be obtained by embedding a fault-free network. Kaklamanis et al. stated that "in many cases the routing time is the same as if there were no faults in the array, up to constant factors" [317]. Among other things they prove:

***THEOREM 5.19***

*Permutation routing among all live processors on an $n \times n$ mesh with $f \leq n/3$ faulty nodes can be performed in $\mathcal{O}(n+f^2)$ steps using constant size queues, with high probability [317].*

Further improvements are given by Cole, Maggs, and Sitaraman [127]. They prove that, for some constant $\varepsilon > 0$, $n^{1-\varepsilon}$ worst case faults can be handled in such way that the computation is slowed down only by a constant factor. In practice, the above algorithms are insufficient, since constant factors are considerably large. We would like to have algorithms that tolerate a small number of faults with small delays.

**Problem 9** *Can an $n \times n$ mesh with $x$ random faults emulate a fault-free mesh with delay factor $1/(1 - \mathcal{O}(x/n^2))$, for all $x = 1, 2, \ldots$?*

### 5.3.5 Cut-Through - Wormhole - Hot-Potato Routing

The study of *cut-through routing* on meshes was initiated by Makedon and Symvonis in [456]. Subsequent improvements reduced the time complexity close to the lower bound.

***LEMMA 5.21***

*For routing a permutation on an MIMD mesh, with packets consisting of $b$ flits each, $bn/2 + n/b + 3n/2 + o(bn)$ steps are sufficient [327, 456, 574].*

This is obtained by applying the same algorithm as in $k$-$k$ routing. It is interesting to consider whether cut-through routing is essentially harder than $k$-$k$ routing (when $k = b$).

**Problem 10** *Can cut-through permutation routing be solved in $bn/2 + o(bn)$ steps?*

A weaker result than Theorem 5.15 holds for dynamic cut-through routing.

**LEMMA 5.22**
*We consider cut-through routing on an $n \times n$ MIMD mesh with packets of $b$ flits. If the generating rate is less than $1/(bn)$, then the maximum delay incurred by any packet in any window of $T$ steps is bounded by $\mathcal{O}(b(\log T + \log n))$, with high probability. The maximum queue size is $\mathcal{O}(b + b \log T / \log n)$ [400].*

For *wormhole routing*, theoretical analysis of congestion is more difficult, since packets which have partially crossed a link can block other packets for arbitrary time. However, *delayed greedy* wormhole permutation routing is within a logarithmic factor of being optimal [195]. In delayed greedy routing each message waits at its source for a random initial delay before moving greedily towards its destination.

**THEOREM 5.20**
*For worms of length $b$, the delayed greedy wormhole routing can be performed on an $n \times n$ MIMD mesh in $\mathcal{O}(bn \log n + n^2 / \log n)$ steps, with high probability [195].*

The situation remains largely unclear for dynamic wormhole routing:

**Problem 11** *Is there a dynamic wormhole routing algorithm that assures a constant fraction of the network capacity can be used while all packets reach their destination with minimal delay?*

For average case instances, *hot-potato routing* can be performed in $2n + \mathcal{O}(\log n)$ steps (or less), as shown in [623] (incomplete analysis) and [191] (complete analysis). In the same paper Feige and Raghavan presented a randomized algorithm with linear time complexity. Schuster and Newman [504] were the first to give a deterministic linear-time algorithm. This algorithm was improved by Kaufmann, Lauer, and Schröder [324].

**LEMMA 5.23**
*Hot-potato permutation routing on an MIMD mesh can be performed in $7n/2 + o(n)$ steps [324, 504].*

Newman and Schuster combined the requirements of hot-potato and wormhole routing. They achieve a bound of $\mathcal{O}(b^{2.5}n)$ steps for routing worms of length $b$ on an $n \times n$ mesh [505]. In [631] it is shown that under a realistic condition the time delay is only $\mathcal{O}(b^{1.5}n)$. The idea behind [505] is ingenious; temporarily blocked worms are kept cycling around "parking lots" and as soon as the "highway" running along the parking lot is free they take off. Bar-Noy et al. presented a randomized algorithm running in $\mathcal{O}(bn)$ steps [43]. Roberts and Symvonis achieved the same result with an an offline algorithm [588]. These results make it likely that the following question can be answered affirmatively, either by applying derandomization [332], or techniques for converting offline into online algorithms [652].

**Problem 12** *Is there a deterministic online algorithm for hot-potato wormhole permutation routing, with worms of length b, running in* $\mathcal{O}(bn)$ *time?*

In all hot-potato routing algorithms, packets may take a detour. The algorithms are such that packets which have to go further make fewer detours, and therefore the overall completion time is close to optimal. However, individual packets may get delayed considerably. In [110] it is shown that such a flexibility is essential; lower bounds are constructed for adaptive routing algorithms which use routing of packets along shortest paths. Such algorithms are called *minimal adaptive routing algorithms*. For an $n \times n$ mesh, with maximal storage capacity $k$ in each node, it is shown that for a large class of minimal adaptive routing algorithms there are inputs which require at least $\Omega(n^2/k^2)$ steps.

The topics of wormhole and hot-potato routing have been considered for more general networks in [143, 237, 474, 475, 578]. Particular attention is devoted to vertex-symmetric networks (among them tori and butterflies). More practical aspects of wormhole routing, taking into account that not only the network links have finite speed, but that also the rate with which the nodes can supply data to the network is finite, are considered in [248, 628].

### 5.3.6 1-1 and $k$-$k$ Sorting Algorithms

We will consider both 1-1 and $k$-$k$ sorting. When no ambiguity results, 1-1 sorting will be called simply *sorting*.

#### 5.3.6.1 One Packet Model

In the earliest papers on mesh sorting it was assumed that at all times every node can hold only one packet. Under this assumption one can derive interesting lower bounds using the so called "joker zone arguments."

***LEMMA 5.24***

*Sorting on an MIMD mesh in the one packet model requires at least* $3n - \mathcal{O}(n^{1/2})$ *steps for sorting in (snake-like) row-major order [366, 603].*

Deeper methods are required for finding lower bounds for all indexing schemes.

***LEMMA 5.25***

*There is no indexing scheme, with respect to which sorting in the one packet model can be performed in less than* 2.27 $n$ *steps [252].*

Thompson and Kung [665] presented the first $\mathcal{O}(n)$ sorting algorithm. Schnorr and Shamir [603] designed the first optimal algorithm.

***THEOREM 5.21***

*Sorting on an MIMD mesh in one packet model with respect to the snake-like row-major indexing can be performed in* $3n + \mathcal{O}(n^{3/4})$ *steps [449, 603, 665]. On a* $m \times n$ *mesh the algorithm takes* $m+2n+o(n+m)$ *steps.*

#### 5.3.6.2 Limited Instruction Sets

Many attempts have been made to design sorting algorithms on $m \times n$ meshes in which few basic operations are alternated. The simplest algorithm performs rounds of the following type: first the columns are sorted, then the rows (the odd rows one leftwards, the even rows rightwards). This algorithm is called *shearsort.* Applying the 0-1 principle (which states that a comparison-based sorting algorithm that sorts an arbitrary 0-1 distribution is correct (see following section and also [401, p. 141]), it is easy to prove that $\lceil \log m \rceil + 1$ rounds are enough (here $m$ is the number of rows):

***LEMMA 5.26***

*Shearsort sorts* $mn$ *numbers on an* $m \times n$ *mesh in snake-like order in* $(m+n)(\lceil \log m \rceil + 1)$ *steps [597, 602].*

Kutyłowski and Wanka have analyzed the complexity of this algorithm more precisely. They show that the above bound is sharp when the number of columns $n$ is not a power of 2, but that $\min\{\log n, \lceil \log m \rceil + 1\}$ rounds are sufficient when $n$ is a power of 2 [377].

An extension of this algorithm by Schnorr and Shamir includes an extra phase in each iteration, after the two basic sorting phases [603]. This phase corresponds to a shuffle permutation applied to $i^{th}$ row, $1 \leq i \leq m$ for $rev(i) = i \bmod n^{1/2}$ times. The resulting *revsort* algorithm iterates these three phases $\lceil \log \log m \rceil$ times.

**LEMMA 5.27**

*Revsort sorts $mn$ numbers on a $m \times n$ mesh in snake-like order in $(m+n)\lceil \log \log m \rceil$ steps [603].*

The next natural question is whether it is possible to sort in a constant number of phases. Marberg and Gafni achieved this by including more sorting and shuffle steps in each iteration. Their algorithm is called *rotatesort.*

**THEOREM 5.22**

*Rotatesort sorts $mn$ numbers on a $m \times n$ mesh in snake-like or row-major order with 16 row and column phases [460]. This takes $\mathcal{O}(n+m)$ steps in total.*

They also prove a lower bound on the number of required phases:

**THEOREM 5.23**

*Sorting on a $m \times n$ mesh, $m, n \geq 8$, into any sorting order requires at least five phases consisting of column/row sorts and shuffles [460].*

**Problem 13** *Is it possible to bridge the gap between the lower bound, 5, and the upper bound, 16, on the number of phases consisting of row/column sorts and shuffles necessary to sort on a mesh?*

For special shapes of a mesh, there are better upper bounds; Leighton's *columnsort* algorithm consists of alternating column sorting and row rearrangements [399].

***THEOREM 5.24***

*For $m \geq n^2$, columnsort sorts $mn$ numbers on a $m \times n$ mesh into column major order using a sequence of eight phases of row/column sorts and transpose/shifts [399]. In total this takes $\mathcal{O}(n + m)$ steps.*

More abstractly, the basic idea of columnsort is that in order to sort $N$ elements, one must first sort all $N^{1/3}$ subsets of size $N^{2/3}$ and perform an unshuffle. Then all subsets are sorted again, and another rearrangement is performed. Two more sortings of pairs of consecutive subsets complete the sorting. This fundamental idea has appeared (sometimes in disguise) several times. Of course, the algorithm can also be applied recursively: in order to sort the subsets of size $N^{2/3}$, it is sufficient to sort subsets of size $N^{4/9}$ or larger. As on a square $n \times n$ mesh, the number of nodes in a row $n \geq (n^2)^{4/9}$, columnsort can be applied to sort with a constant number of phases on meshes of arbitrary shape.

*Bubblesort* algorithms cycle through a small set of local comparisons in a fixed order. Unfortunately, straightforward approaches result in $\Omega(n^2)$ sorting time, even on average [600]. A positive result has been reported by Ierardi:

***THEOREM 5.25***

*On an $n \times n$ mesh, bubblesort can, on average, be run in $\mathcal{O}(n \log^{1/2} n)$ time [297].*

Related results have been obtained by Schwiegelsohn and Kutyłowski ea. [297, 376, 604]. In [376], it is shown how to obtain periodic sorting networks of depth only 5 out off comparator-based sorting networks. In doing so, a factor of $\mathcal{O}(\log n)$ is lost in time consumption. This is a very strong result, but for meshes, the algorithm of Ierardi, which is away from optimal by only a factor $\mathcal{O}(\log^{1/2} n)$, is still slightly better.

### 5.3.6.3 Multi-Packet Model

If we drop the unnatural restriction that nodes can hold only one packet, better results are achievable. By first concentrating all packets in the central $n/2$ columns and then applying the algorithm of [603], Kunde achieved a sorting time of $2.5\,n + o(n)$ steps [369]. The first algorithm which almost matches the diameter bound with delay $2n + o(n)$, was discovered by Kaklamanis and Krizanc [318]. This probabilistic algorithm was successfully derandomized in [332]. The latter algorithms use blocked snake-like row-major indexing. Krizanc and Narayanan considered more natural indexing schemes when all keys are either zero or one [354]. In [625] it was shown that sorting with respect to row-major indexing can be solved optimally in $2n + o(n)$ steps on an MIMD mesh using queues of size five [318, 319, 332, 353, 354, 369, 625].

We now focus on describing an optimal 1-1 sorting algorithm, starting with some basic results that facilitate the proof of correctness.

**Basic Lemmas**

***LEMMA 5.28***

**0-1 Principle** *A comparison-based sorting algorithm $\mathcal{A}$ that sorts any distribution consisting of zeros and ones, sorts any distribution [343].*

**PROOF** We prove this by contradiction. Assume that there is a packet distribution that is not properly sorted by $\mathcal{A}$. Then there are at least two mixed up packets, $p_1$ and $p_2$, which in the properly sorted order would stand consecutively. Let $k$ be the key of $p$, and let $k_1$ be the key of $p_1$. Now start again with the original distribution, but with all keys of packets $p$ replaced by 0 if $k \leq k_1$, and by 1 if $k > k_1$. $p_1$ and $p_2$ end up in the same positions as before, because in any comparison the same decision is made. So, this $0 - 1$ distribution is not correctly sorted. ∎

The condition that the algorithm is comparison-based is required, because otherwise the algorithm might explicitly exploit the fact that there are only zeros and ones, e.g., by counting the number of ones, giving them a rank, and routing all packets.

In some of our algorithms we need a similar result:

***LEMMA 5.29***

**0-1-2 Principle** *Assume that during a certain stage of a sorting algorithm, applied to an arbitrary distribution with only zeros, ones, and twos, the ones may at worst be spread out over a certain bounded range. If the same sorting algorithm is applied to a general packet distribution, the packets with common value cannot stand spread out over a larger range.*

### Optimal 1-1 Sorting

We start with a uni-axial algorithm for 1-1 sorting in row-major order. It runs in $2^1/2n + o(n)$ steps. Asymptotically this is much faster than the algorithms previously discussed in Section 5.3.1.2. This algorithm is obtained by combining a new insight in merge sorting and the desnakification technique, with knowledge about sorting with splitters. Then, it is turned into a near-optimal bi-axial algorithm. Without loss of generality, we assume that all packets have different keys.

*Uni-Axial Sorting*

The mesh is divided in $s \times s$ submeshes. In the algorithm of this section $s = n^{5/6}$, and $m = n/s = n^{1/6}$. We distinguish packets and *splitters*. The splitters are copies of a small subset of the packets. They are broadcast and the packets estimate their ranks by comparison with the splitters. This widely known idea (going back to work of Reischuck [586], and Reif and Valiant [585]) has been used for randomized [318, 330, 571] and deterministic [332] sorting on meshes. Sometime, as in the $k$-$k$ sorting algorithm of Section 5.3.1.2, we do not need splitters because the packets are fully distributed over the mesh, and thus reliable estimates of the ranks of the packets can be obtained by local comparison among the packets themselves. However, in the case of 1-1 sorting this idea does not lead to efficient algorithms. The splitters allow us to spread the necessary information rapidly, while the packets are involved in more useful operations.

First we give the algorithm for selecting and routing the splitters:

Algorithm SPLITTER-ROUTE

(1) In every submesh, sort the packets. Copy the packets with ranks $im^2$, $0 \leq i \leq s^2/m^2 - 1$: the splitters.

(2) In every submesh $B_{i,j}$, $0 \leq i, j < m$, rearrange the splitters such that they stand in the positions $(i', j')$ of $B_{i,j}$, with $is/m \leq i' < (i+1)s/m$, and $js/m \leq j' < (j+1)s/m$.

(3) Send the splitters along the rows. A splitter starting in position $(i', j')$ of $B_{i,j}$ drops copies in the positions $(i', j')$ of $B_{i,l}$, for all $0 \leq l < m$.

(4) Send the splitters along the columns. A splitter starting in position $(i', j')$ of $B_{i,j}$ drops copies in the positions $(i', j')$ of $B_{l,j}$, for all $0 \leq l < m$.

***LEMMA 5.30***

SPLITTER-ROUTE *takes* $2n + \mathcal{O}(s)$ *steps to complete. No link has to transfer more than* $\mathcal{O}(s)$ *packets. Finally, each PE holds precisely one splitter, and all splitters are available in every* $s \times s$ *submesh.*

**PROOF** Steps (1) and (2) take $\mathcal{O}(s)$ steps, Steps (3) and (4) take less than $n$ steps. The rearrangement is such that the splitters in $B_{i,j}$ stand in "subsubmesh" $(i, j)$. After the broadcast these splitters occupy the subsubmeshes $(i, j)$ in all submeshes: the splitters from different submeshes perfectly fit next to each other. This arrangement also assures that during Steps (3) and (4), a link has to transfer at most $m/2s/m = s/2$ splitters. ∎

When splitters and packets want to use the same link, priority is given to the splitters. By Lemma 5.30, this delays the packets by at most $\mathcal{O}(s)$. For the packets we perform a kind of $m$-way merge algorithm:

Algorithm 11SORT

(1) In every submesh, sort the packets in row-major order.

(2) In every submesh $B_{i,j}$, $0 \le i, j < m$, shift the packets in row $l$, $0 \le l < s$ to row $l$ of $B_{i,(j+l) \bmod (m/2)}$, and copies to row $l$ of $B_{i,(j+l) \bmod (m/2)+m/2}$.

(3) In all columns, sort the packets downwards.

(4) In every submesh, determine for every packet its "rank," the number $r$, $0 \le r \le s^2$ of splitters that are smaller. The preliminary destination of a packet $p$ with rank $r$, lies in section $S_l$, with $l = \lfloor rm^2/s \rfloor$. If $\lfloor (rm^2 - m^4)/s \rfloor = l - 1$, then create a copy $p'$ of $p$ with preliminary destination in $S_{l-1}$. If $\lfloor (rm^2 + m^4)/s \rfloor = l + 1$, then create a copy $p'$ of $p$ with preliminary destination in $S_{l+1}$. Discard the splitters, and the (copies of) packets that have preliminary destination in the other half of the mesh.

(5) In every submesh, sort the packets in column-major order on their preliminary destination column-bundles.

(6) In every row, route the packets to the first PEs in their preliminary destination column-bundles that hold less than two packets.

(7) In each submesh, sort the packets in row-major order on their preliminary destination section.

(8) In every column, route the packets to the sections of their preliminary destinations.

(9) In every section, sort the packets.

(10) In every section $S_l$, $0 < l \leq mn - 1$, throw away the $m^4$ packets with the smallest keys, and in each $S_l$, $0 \leq l < mn - 1$, throw away the $m^4$ packets with the largest keys. Redistribute the remaining $ks$ packets within $S_l$.

After Step (3), there are in a $0 - 1$ distribution at most $m^2$ "dirty" rows. For a general distribution this means that a packet resides at most $m^2 - 1$ rows away from its destination row. These three steps take $2n + \mathcal{O}(s)$ steps, just as SPLITTER-ROUTE. So, we may assume that after Step (3) the splitters are available in the submeshes. Step (4), ..., Step (10), resemble the final steps of KKSORT2 for $k = 1$.

*Analysis*

As the algorithm is given, it is not entirely correct. It is *not* true that, as in KKSORT2, exactly $m^4$ packets must be thrown away on both sides of every section: 11SORT orders the packets, but the sections do not necessarily hold exactly $s$ packets. Fortunately, the numbers of packets that must be thrown away in a section on the low and high side, respectively, can be determined.

Consider some section $S$ and the sections from which it may receive packets after Step (3):

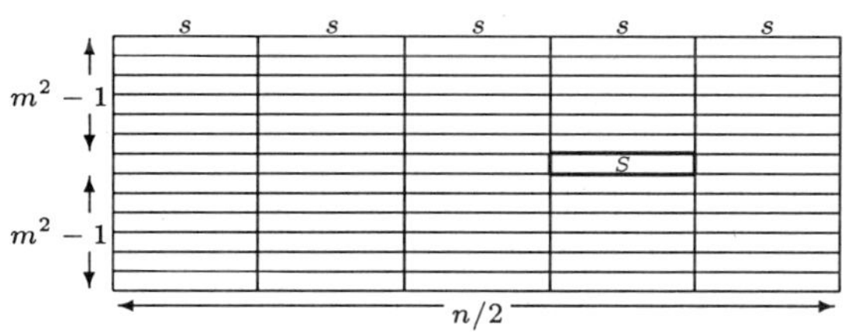


$(2m^2 - 1)n/2$ packets are stored at these sections, among which are the $s$ packets with destination in $S$. After Step (8), these $s$ packets all reside in $S$, but also some packets that do not belong in $S$. How can we figure out which packets to keep, and which packets to throw away? Suppose that $S$ is the $l$-th section, $(m^2 - 1)n/s \leq l < m^2n/s$, in the involved (whole) rows. Then finally $S$ should hold the packets with ranks $r$, $ls \leq r < (l + 1)s$ from among the $(2m^2 - 1)n$ packets. Analogously to the uni-axial model merge algorithms discussed previously, we could copy *all* packets to $S$, sort them, and throw away the smallest $ls$ packets and the largest $(2m^2 - 1)n - (l + 1)s$ packets. This gives a correct but very inefficient algorithm. However, it is not necessary to copy all packets to $S$. It is sufficient, if for each contributing section $i$, the counters, the numbers $under_{S,i}$ and $over_{S,i}$ of packets that are *not* sent to $S$ (because they are definitely too small or definitely too large), respectively, are known in $S$. The counters can easily be determined in Step (4). They can be transferred to $S$ during the subsequent steps, in parallel with the packets. As every section sends and receives only $\mathcal{O}(m^3)$ counters in total, they can be routed without causing substantial delay. The numbers $Under_S = \sum_i under_{S,i}$ and $Over_S = \sum_i over_{S,i}$ can be computed in Step (9). Finally, in Step (10), the smallest $ls - Under_S$ packets and the largest $(2m^2 - 1)n - (l+1)s - Over_S$ packets in $S$ are thrown away, leaving exactly the $s$ packets belonging in $S$. Now we get the following:

**THEOREM 5.26**

*Uni-axial* 1-1 *sorting in row-major order can be performed in* $2^1\!/\!2n + \mathcal{O}(n^{5/6})$ *steps. The queue size is five.*

**PROOF** Let $s = n^{5/6}$. For the routing time and correctness, we only have to prove that Step (6) can be performed in $n/2 + \mathcal{O}(s)$ steps. All other steps can be performed in $\mathcal{O}(s)$ steps.

The estimate of the rank of a packet, $rm^2$, is accurate up to $m^4$. This means that for some section $S_l$, only a packet (or its copy) with actual destination in some PE $P_k$, with $k \in [ls - 2m^4, (l+1)s + 2m^4]$, may get preliminary destination in $S_l$. Hence, at most $s^2 + 4m^4 s$ packets have preliminary destination in any submesh $B_{i,j}$. By the sorting in Step (5), they are distributed almost optimally over the rows of row-bundle $i$; at most $s + \mathcal{O}(m^4)$ packets stand in any row. The $m^2 s$ packets with destination in $B_{i-1,j}$ and $B_{i+1,j}$, that may stand in row-bundle $i$ have no serious influence. This shows that Step (6) can be performed as specified; no PE in $B_{i,j}$ has to receive more than two packets.

We consider the routing time of Step (6). For a rightwards moving packet $p$, residing in some PE $P_{i,j}$ and moving to column $l$, with $j, l < n/2$, we are interested in the number $h_l$ of packets within row $i$ that go to some column $k$, with $k \geq l$. By the above analysis, we know that $h_l \leq n/2 - l + \mathcal{O}((n/2 - l)m^4/s) \leq n/2 - l + \mathcal{O}(s)$. Now, $p$ is delayed at most $h_l$ times, and hence $p$ finishes Step (6) within $n/2 + \mathcal{O}(s)$ steps.

A PE may hold up to four (copies of) packets during Steps (4) and (5). In addition, Step (4) can be organized such that a PE holds at most one splitter or counter. Hence, $Q \leq 5$. ∎

*Near-Optimal Bi-Axial Sorting*

Essentially 11SORT consists of three main routing phases: horizontal, vertical, and horizontal (Steps (2), (3), and (6)). These phases take $n$, $n$ and $n/2$ steps, respectively. The links between the left and right half are not used anymore after step $n/2$. Thus, it may happen that a packet $p_1$ that stands in column 0 after phase (1) is routed to a preliminary destination in column $n/2 - 1$ in phase (3). This is unnecessary since a copy of $p_1$ stands in column $n/2$. In a uni-axial algorithm this observation does not lead to a faster algorithm. There may be a packet $p_2$, after phase (1) in column $n/2 - 1$ and with preliminary destination in column 0, which has to travel $n/2$ steps in phase (3). On the other hand, in a bi-axial algorithm, it is possible to coalesce the phases. Then $p_2$ can start Phase 3 after $3n/2 + \mathcal{O}(s)$ steps, and reach its preliminary destination after $2n + \mathcal{O}(s)$ steps.

We work out these ideas. Only Step (4) is changed. Instead of discarding the packets that have their destinations in the other half, we now perform the following:

> In all columns $j$, $0 \leq j < n/2$, discard the (copies of) packets that have preliminary destination in some column $j'$, with $j' \geq 2j$. For $n/2 \leq j < n$, discard the packets with $j' < 2j - n$.

Notice again that, by this rule, exactly one of the copies of a packet reaches every possible destination section. The steps are coalesced. Most importantly, this means that Step (3) begins in column $j$ after $n/2 + |n/2 - j|$ steps, and Step (6) after $3n/2 + |n/2 - j|$ steps.

***THEOREM 5.27***

*Bi-axial* 1-1 *sorting on an MIMD mesh in row-major order can be performed in* $2n + \mathcal{O}(n^{5/6})$ *steps using queues of size five [318, 319, 332, 353, 354, 369, 625].*

**PROOF** A packet that starts Step (6) after $2n - d + \mathcal{O}(s)$ steps, has to travel at most $d$ steps to reach the column-bundle of its preliminary destination. We check this for a packet $p$ that is routed in Step (2) to some column $j$, with $j < n/2$. $p$ starts Step (6) after $2n - j + \mathcal{O}(s)$ steps. In Step (4) the preliminary destination of $p$ is determined. $p$ survives only when it goes to some column $l$, with $l < 2j$, while $p$ has to travel at most $j$ steps. By a refinement of the analysis in the proof of

Theorem 5.26, it can be shown that $p$ is not delayed more than $2j - l$ times. Hence, Step (7) can start in all submeshes after $2n + \mathcal{O}(s)$ steps. ∎

In fact this algorithm is still *locally* uni-axial; every PE uses only horizontal or vertical links.

Although asymptotically optimal, all algorithms cited in Theorem 5.27 have lower-order terms that tend to dominate the overall sorting time for small meshes. In this perspective, valuable contributions have been given by Thompson and Kung [665], Nassimi and Sahni [497], Kumar and Hirschberg [363], and Lang et al. [387]. Lately several ideas were combined to reduce the dominant term by almost a factor of two:

**LEMMA 5.31**

*For all $n$, sorting in row-major order can be performed in $4.75\,n$ steps using queues of size nine. The routing model is uni-axial. For $n = 2^l$ sorting can be performed in $4.5\,n$ with queue size six, and for $n = 3^l$ in $4\frac{1}{3}\,n$ with queue size five [625].*

**Problem 14** *What is the best sorting algorithm for small values of $n$, say $n \leq 256$?*

The theory of $k$-$k$ sorting has been developed hand in hand with that of $k$-$k$ routing. Kunde [369] gave a deterministic $kn + o(kn)$ sorting algorithm. The first almost optimal algorithm was developed by Rajasekaran in [327], improved to $kn/2 + o(kn)$ in [331, 571]. These algorithms are randomized. In [332, 370] it is shown how to achieve $kn/2 + o(kn)$ deterministically. The proof of this Theorem together with the algorithms is provided next.

*Example of Derandomization*

A simple example illustrates how sort-and-unshuffle can be applied to derandomize a sorting algorithm.

We consider the $k$-$k$ sorting problem for $k = n^4$. We choose $\alpha = 0$, that is, each block consists of a single PE. The mesh is indexed with a snake-like row-major scheme. Consider the following simple randomized algorithm:

Algorithm RANDSORT

(1) All packets are routed to a random destination with probability 1/2 row-first, and with probability 1/2 column-first.

(2) The packets in each PE are sorted. The packets with rank $i$ are routed to PE $\lfloor i/n^2 \rfloor$.

(3) Sort all pairs of PEs $(2i, 2i+1)$, $0 \le i < n^2/2$. Sort all pairs of PEs $(2i-1, 2i)$, $0 < i < n^2/2$ . Repeat these steps $\mathcal{O}(1)$ times.

***LEMMA 5.32***

*After Step (2) of* RANDSORT, *each packet is at most $\mathcal{O}(1)$ PEs away from its destination, with high probability.*

**PROOF** Consider a packet $p$ with destination in PE $j$, $0 \le j < n^2$. There are less than $k(j+1)$ packets with destination before $p$. Each of these packets is routed in Step (1) with probability $n^2$ to PE $j$. Thus, using Chernoff bounds, the number of these packets in $j$ can be bounded by $(j+1)kn^{-2} + \mathcal{O}((jk)^{1/2}/n)$. Hence, the maximum index of the PE to which $p$ is routed in Step (2) is given by $j' = \lfloor ((j+1)kn^{-2} + \mathcal{O}((jk)^{1/2}/n))/(kn^{-2}) \rfloor = j + \mathcal{O}(1)$. ∎

With Lemma 5.4 and using Chernoff bounds it can be easily shown that the algorithm runs in $kn/2 + \mathcal{O}(k + (kn \log n)^{1/2})$ steps. The algorithm can be made deterministic by replacing Step (1) with:

(1′) The packets residing in each PE are sorted. The packet $p$ with rank $i$, $0 \le i \le n^4 - 1$ is routed to $P_{i \bmod n^2}$. Then, $p$ is routed row-first if $i \bmod (2n^2) < n^2$; else $p$ is routed column-first.

We refer to this algorithm as HIGHKSORT. The correctness follows from the following lemma:

***LEMMA 5.33***

*After Step 2 of* HIGHKSORT, *each packet is at most one PE away from its destination.*

**PROOF** Consider a packet $p$ that resides in some PE $P$ after Step (1′). Since there are $n^2$ regularly interspaced packets from each PE in $P$, the rank of $p$ among the packets that originally resided in any PE $P_j$, $0 \le j < n^2$, can be estimated to within $n^2$. Hence, $p$ can determine its global rank to within $n^4$, the number of packets residing in a single PE. ∎

Note that the way in which the packets are routed in Step (1′) and Step (2) ensures that the packets are distributed evenly.

***THEOREM 5.28***

HIGHKSORT *performs $k$-$k$ sorting for $k = n^4$ in $kn/2 + \mathcal{O}(k)$ steps on a mesh.*

**PROOF** Under the routing operations in Step (1′) and Step (2), each PE sends exactly $n^2$ packets to each other PE. Each of these steps can be achieved in a perfectly regular way in $kn/4$ steps. Step (3) takes $\mathcal{O}(k)$ steps. ∎

The routing operations in Step (1′) and Step (2) are so regular that they can be easily implemented in such a way that no PE ever holds more than $k$ packets. We can show the following lemma:

***LEMMA 5.34***

HIGHKSORT *can be implemented to run with maximal queue size $k$.*

**PROOF** We prove that each PE holds at most $k/2$ white packets at all times. Consider a phase in which the white packets move horizontally. The distribution is such that initially the number of packets that leave $P$ over its right link equals the number of packets that enter $P$ over this link. Since $k$ is large and because of the packet distribution, the farthest-first strategy guarantees that $P$ receives a packet if and only if it sends a packet. ∎

Note that the movement of the packets in the deterministic algorithm is more regular than in the randomized one. The running times and queue sizes of the two algorithms are comparable.

*Sorting for General k*

In fact, HIGHKSORT can be easily modified into an algorithm that performs $k$-$k$ sorting optimally for all $k \geq 8$. The mesh is divided into blocks with side length $n^{2/3}/k^{1/6}$. We assume that this number is an integer. Let $m = k^{1/3}n^{2/3}$ denote the number of blocks. The size of the blocks is chosen such that the number of packets in each block equals $m^2$. In principle we can now apply HIGHKSORT with the blocks playing the role of PEs, but we still have to specify precisely where every packet is going to be routed during Step (1′) and Step (2). We assume that the mesh is indexed with a blocked snake-like row-major scheme. We get the following algorithm:

Algorithm KKSORT

(1a) The packets residing in each block are sorted. The intermediate destination of a packet $p$ of rank $i$, $0 \leq i < m^2$, lies in block $i \bmod m$. If $i \bmod (2m) < m$, then $p$ is colored white, else $p$ is colored black.

(1b) The blocks are divided into subblocks with side lengths $n^{1/3}/k^{1/3}$. The packets of each block are rearranged such that the packets with intermediate destination in block $(i, j)$, $0 \leq i, j < k^{1/6}n^{1/3}$, appear in subblock $(i, j)$.

(1c) The white packets are routed along the row to the column of blocks of their intermediate destination. The packets in the $j_2$th column of subblocks in the $j_1$th column of blocks, $0 \leq j_1, j_2 < k^{1/6}n^{1/3}$, are routed as a block to the $j_1$th column of subblocks in the $j_2$th column of blocks. The black packets are routed analogously.

(1d) The white packets are routed along the column to the block of their intermediate destination. The packets in the $i_2$th row of subblocks in the $i_1$th row of blocks, $0 \leq i_1, i_2 < k^{1/6}n^{1/3}$, are routed as a block to the $i_1$th row of subblocks in the $i_2$th row of blocks. The black packets are routed analogously.

(2a) The packets residing in each block are sorted. The preliminary destination of a packet $p$ of rank $i$, $0 \leq i \leq m^2 - 1$, lies in block $\lfloor i/m \rfloor$. If $i$ is even, then $p$ is colored white, else $p$ is colored black.

(2b) Identical to Step 1.b.

(2c) Identical to Step 1.c.

(2d) Identical to Step 1.d.

(3) Sort all pairs of blocks $(2i, 2i+1)$, $0 \leq 1 < m/2$. Sort all pairs of blocks $(2i-1, 2i)$, $0 < i < m/2$.

The correctness of the KKSORT algorithm follows from the same argument as in the proof of Lemma 5.33, replacing "PE" by "block:"

***LEMMA 5.35***

*After Step (2d) of* KKSORT, *each packet is either in its destination block, or in a block that is adjacent in the snake-like ordering of the blocks.*

The colorings in Step (1a) and Step (2a) assure that from each block precisely $m/2$ packets are routed row-first to each other block, and $m/2$ packets column-first. This is essential for the proof of the following:

***LEMMA 5.36***

*(1c), (1d), (2c) and (2d) of* KKSORT *each take at most* $\max\{n, kn/8\}$ *steps on a mesh.*

**PROOF** We analyze the lemma for the routing along the row in Step (1c) of the white packets. The analyses for the black packets and for the other steps are analogous.

The PEs in block $(i, j)$ send $m/2$ packets row-first to any other block. Therefore, during Step 1.c $k^{1/6}n^{1/3}m/2 = k^{1/2}n/2$ packets are routed from block $(i, j)$ to any block $(i, j')$ in the same row of blocks. By the rearranging in Step (1b) these packets are evenly distributed over the rows of block $(i, j)$. Hence, we can concentrate on the routing problem on a chain in which $k^{2/3}n^{1/3}/2$ packets are routed from each section of length $n^{2/3}/k^{1/6}$ to each other section. Within each section the packets are sorted according to the distance they have to go. In order to apply Lemma 5.4, we consider how many packets have to go from the first $i$ sections to the last $j$ sections, $i + j \leq k^{1/6}n^{1/3}$. These are $ijk^{2/3}n^{1/3}/2$ packets. The gap the packets have to bridge is $(k^{1/6}n^{1/3} - i - j)n^{2/3}/k^{1/6}$. Hence, the routing can be performed in $\max_{i,j}\{ijk^{2/3}n^{1/3}/2 + (k^{1/6}n^{1/3} - i - j)n^{2/3}/k^{1/6}\}$ steps. For $k \geq 8$ the maximum is assumed for $i = j = k^{1/6}n^{2/3}/2$, and equals $kn/8$. Otherwise the maximum is assumed for $i = j = 0$, and equals $n$. We do not have to consider "cuts" within the sections because there the packets stand in the correct order. ∎

Combining Lemmas 5.35 and 5.36 gives the following:

***THEOREM 5.29***

KKSORT *performs k-k sorting deterministically in* $\max\{4n, kn/2\} + \mathcal{O}(k^{5/6}n^{2/3})$ *steps on an MIMD* $n \times n$ *mesh [327, 331, 332, 369, 370, 625, 629].*

The constant of the lower order term, $\mathcal{O}(k^{5/6}n^{2/3})$, depends on the applied algorithm for the local sorting operations. It is minimized when KKSORT itself is applied recursively. For the sorting in Step (3), the algorithm has to be modified slightly.

***COROLLARY 5.4***

*For $k \geq 8$, $k$-$k$ sorting on a mesh can be performed in $kn/2 + (4 + o(1))k^{5/6}n^{2/3}$ steps.*

**PROOF** Let $s = n^{2/3}/k^{1/6}$. During KKSORT, $k$-$k$ sorting on meshes of size $s \times s$ must be applied in Steps (1a) and (2a). In Steps (1b) and (2b), the packets are rearranged within meshes of size $s \times s$. Each of these four steps can be performed in $ks/2 + o(ks)$ steps. The sorting operations in Step (3) can be performed in $ks + o(ks)$ steps each. ∎

The small constant of the additional term indicates that our results may be of practical value. The results of this section were stated only for meshes. Generalizations for tori are immediate. As the routing steps on which the algorithms are based, and which determine the leading term of the sorting times, can be performed twice as fast on a torus, the problems on a torus can be solved almost twice as fast as on a mesh. Without further modification we can show with a refinement of Lemma 5.34 that the queue size is at most $k+4$. When the packets are accurately timed we get:

***LEMMA 5.37***

*For $k \geq 2$, KKSORT can be implemented to run with maximal queue size $k$.*

### 5.3.7 Global Communications and Algorithms

Research on intensive communication schemes for wormhole-routed systems has focused on small or large packets [213].

For example, Peters and Syska considered the broadcasting problem on two-dimensional tori and showed that it can be performed in optimal $2\lfloor \log_5 n \rfloor$ steps [537]. Their ideas have been generalized to three-dimensional tori in [161].

Total exchange can be performed trivially in just $N/d$ steps on a $d$-dimensional store-forward mesh. For wormhole-routed meshes it is less obvious how to organize the routing so that the total cost is minimal. Optimal gossiping algorithms in higher-dimensional meshes are based on a definition of diagonals, e.g., see section on three-dimensional sparse meshes [313, 669]. If the packets are large, a store-forward approach yields the best results.

For a nice discussion and proofs of time complexity of general parallel mesh algorithms, such as convex hull, Hough transforms, matrix multiplication, Gaussian elimination, tridiagonal systems, connected components, transitive closure, and shortest path problems involving mainly store-forward communication schemes, we refer the reader to the excellent coverage in [401].

## 5.4 Mesh-Like Networks

We provide some references for popular variants of the mesh architecture, outlining some important known results.

### 5.4.1 Higher Dimensional Arrays

A *d-dimensional* $n \times \cdots \times n$ *mesh* (or *torus*) consists of $N = n^d$ processors arranged in a $d$-dimensional cube. On a $d$-dimensional torus the lower bound for $k$-$k$ routing is $kn/4$. The routing and sorting algorithms of [327, 332, 370, 374, 408, 625, 700] are applicable to higher-dimensional meshes and tori. Optimal time, matching the bisection bound within lower-order terms, is assured as long as there are at least $k \geq 4d$ packets at every node.

In case of permutation routing, optimal randomized results for two-dimensional tori and three-dimensional meshes are established in [320]. Similar results can be obtained by applying a deterministic algorithm to results from from [332] and [625]. For higher-dimensional meshes, Kunde [368] gave some preliminary results, which were improved by Suel. In [651] it is shown how to sort in $5/4dn + o(n)$ on a $d$-dimensional mesh, and in $3/4dn + o(n)$ on a $d$-dimensional torus. Still there remains a considerable gap with the lower bounds.

**Problem 15** *Find lower and upper bounds for permutation routing and sorting on meshes of dimension* $d > 3$ *and tori of dimension* $d > 2$*?*

Greedy routing to independently chosen random destinations is analyzed in [400]. The fault-tolerant $Z^2$-policy can be generalized to higher-dimensional meshes [36]. Theorem 2.6 can be used to offline routing of any permutation in optimal time. Normal Ascend/Descend permutations can also be performed in optimal time [485]. Also, affine permutations can be routed with at most $2d(n-1)$ routing steps.

For $l$-$k$ routing on $d$-dimensional meshes, one can apply algorithms from [376, 454, 631] but considering the constant in the leading term, these are not even close to optimal.

**Problem 16** *Find algorithms, randomized or deterministic, for $l$-$k$ routing which perform well for higher-dimensional meshes?*

## 5.4.2 Bus-Based Meshes

Another class of networks are those that are equipped with some buses, along which communication can be performed in one step. There are many variants: the buses can exist instead of, or in addition to, the ordinary mesh links; the buses can be fixed or reconfigurable; the buses can be partitionable or not; they can allow broadcasting, or provide only point-to-point communication; they may allow for conflicts or not. In this section we are not going to discuss algorithms for all possible combinations of these models. Instead, we present a few models, for which there exist algorithms that exploit the specific features in a nice way.

### 5.4.2.1 Meshes of Buses

A *mesh of buses*, consists of $n^2$ PEs, arranged in an $n \times n$ grid, interconnected by one bus per row and one per column. The buses are such that in every step, only one connected PE may write a number to it, which may be received by all connected PEs. The maximum of the numbers that are distributed over the PEs can be computed in $n^{2/3}$ steps, and this is optimal.

### 5.4.2.2 Meshes With Buses

A *mesh with buses*, has one bus per row and column of the grid in addition to the normal mesh links. The advantage is that now local operations can be performed in parallel, whereas on a mesh of buses, there is no such thing as locality. Notice that on a mesh with buses, $\Theta(n^2)$ packets can be exchanged which is much more than the $\Theta(n)$ on a mesh of buses. On the other hand, the bisection width of both types of networks is comparable. So, the difference between them is most noticeable on problems that do not route a large amout of data over the bisection, but benefit from the possibility of performing local operations. On a mesh with buses, the maximum of a set of numbers that is distributed over the PEs can be computed in $\mathcal{O}(n^{1/4})$ steps. For meshes with buses, sorting is only slightly more complex than routing, since the buses are excellent for rapid broadcasting and gathering. Routing has been investigated in [328, 421, 572, 652].

#### 5.4.2.3 Sparse Meshes

Due to the small bisection width of ordinary two-dimensional meshes, the maximum achievable speedup with $n^2$ PEs is only $\Theta(n)$ for most problems. A simple architecture which tries to reduce the number of PEs to $n$, by placing PEs on the diagonal, is the *sparse mesh*.

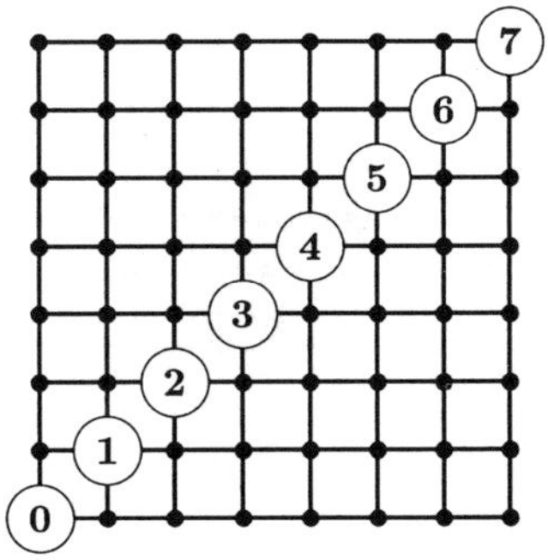


**FIGURE 5.7**
**A sparse mesh with** 8 **PEs**

A sparse mesh with $n$ PEs consists of a two-dimensional $n \times n$ grid of buses, with PEs connected to buses at the diagonal. The horizontal buses are called row-buses, the vertical buses are called column-buses. $P_i$, $0 \leq i < n$, can send data along the $i$-th row-bus and receive data from the $i$-th column-bus. In one *step*, $P_i$ can send one packet of standard length to an arbitrary $P_j$. The packet first travels along the $i$-th row-bus to position $(i, j)$, and then turns into the $j$-th column-bus. Bus conflicts are forbidden; the algorithm must guarantee that in every step a bus is used for the transfer of only one packet. The sparse mesh network for $N = 8$, with an $8 \times 8$ network of buses, is illustrated in Figure 5.7; the large circles indicate the PEs with their indices, the smaller circles the links between the buses. Notice that the described system is that of an XY-wormhole-routed two-dimensional grid (such as the Intel Paragon), out of which we remove all nondiagonal PEs.

The standard length of the packets should be chosen so that the *startup time*, i.e., the time required to establish a connection, does *not* dominate the *transfer time*, i.e., the time needed to actually transfer the bytes over the network. For large systems, this implies that the standard length increases with $n$, because of a distance-dependent delay. Typical startup times are currently 10 $\mu$s or more. Even if this was reduced by a factor of $10^3$, which would mean tremendous progress, then a signal would still travel $10 \cdot 10^{-9} \cdot 3 \cdot 10^8 = 3$ m in one startup time. So, in comparison to the startup time, node distance is not very important yet, and for some time to come distance may even decrease. One may easily prove that $h$-relations, for $h = \Omega(n^\epsilon \log n)$ can be can be routed on an $n \times n$ sparse mesh in $\lceil 1/\epsilon \rceil (h + o(h))$ steps, and thus permutations are routed in $\mathcal{O}(1)$ steps.

#### 5.4.2.4 Higher-Dimensional Sparse-Meshes

A sparse mesh is nothing more than a crossbar with the same PE at the end of each row and column. However, the topology with PEs at the diagonal is simple, but more importantly, it allows for an elegant generalization to higher dimensions.

***DEFINITION 5.2*** *A subset of a d-dimensional $n \times s \times n$ grid is a diagonal, if the projection of all its points on any of the $(d-1)$-dimensional coordinate planes is a bijection.*

For this we need the following definition, which is a simplification of the definition in [312].

$$\begin{aligned} Diag_d = \{(x_0, x_1, \ldots, x_{d-1}) \,|\, 0 \le x_0, x_1, \ldots, x_{d-1} < n, \text{ and} \\ (x_0 + x_1 + s + x_{d-1}) \bmod n = 0\}\,. \end{aligned}$$

Then, we have:

***LEMMA 5.38***

*In a d-dimensional $n \times s \times n$ grid, $Diag_d$ is a diagonal [312].*

The $d$-dimensional sparse mesh, consists of the $n^{d-1}$ PEs on $Diag_d$, interconnected by $dn^{d-1}$ buses of length $n$. The PE at $(x_0, \ldots, x_{d-1})$ is denoted $P_{x_0, \ldots, x_{d-1}}$.

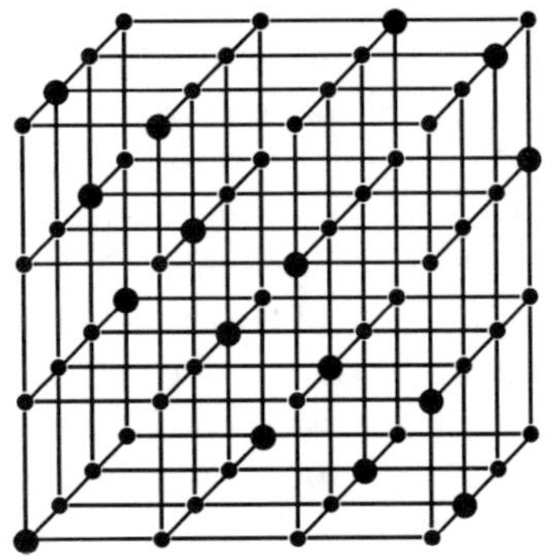

**FIGURE 5.8**
**A three-dimensional $4 \times 4 \times 4$ sparse mesh with 16 PEs**

The routing is *dimension-ordered*. Thus, after $t$ substeps, $0 \leq t < d$, a packet traveling from $P_{x_0,\ldots,x_{t-1},x_t,\ldots,x_{d-1}}$ to $P_{x'_0,\ldots,x'_{t-1},x'_t,\ldots,x'_{d-1}}$ has reached position $(x'_0,\ldots,x'_{t-1},x_t,\ldots,x_{d-1})$. The scheduling should exclude bus conflicts. $Diag_3$ is illustrated in Figure 5.8; PEs are placed so that if they were occupied by towers in a three-dimensional chess game, none of them could capture another.

Now, it is even harder to find a conflict-free allocation of the buses. In contrast to the two-dimensional sparse mesh, it is no longer automatically true that any permutation can be routed offline in one step. For example, in a four-dimensional sparse mesh, under the permutation $(x_0, x_1, x_2, x_3) \mapsto (x_2, x_3, x_0, x_1)$, all packets at the positions $(0, 0, x_2, n - x_2)$, $0 \leq x_2 < n$, have to pass through position $(0, 0, 0, 0)$.

Three-dimensional sparse meshes are practically the most interesting generalization. Notice that on a three-dimensional $n \times n \times n$ sparse mesh, $h$-relations for $h = \Omega(n^{\epsilon} \log n)$ can be routed in $\lceil 2/\epsilon \rceil (h + o(h))$ steps.

#### 5.4.2.5 Reconfigurable Meshes

In a *reconfigurable mesh* all nodes are connected to one big bus, which can be decomposed into many smaller buses. It is surprising that the following holds:

***THEOREM* 5.30**

*On a reconfigurable $n \times n$ mesh $n$ numbers can be sorted in a constant number of steps [51, 699].*

The algorithms are based on rotate- and columnsort. Pushing a columnsort-based algorithm to the limit, the number of steps can be decreased below 40. An extension to sorting $n^{d-1}$ numbers on a $d$-dimensional reconfigurable mesh is simple [509]. Theoretically interesting is the following:

**Problem 17** *How many steps are required for sorting $n$ numbers on a reconfigurable $n \times n$ mesh?*

Another feature of reconfigurable meshes is that they allow routing and sorting extremely sparse packet distributions in time given by the bisection bound [329]. Also [50] deals with this problem, and in [652] sparse dynamic problems are treated. The theory of reconfigurable meshes is rich and we could only mention a few results. Nakano gives an extensive and up-to-date list of publications [494].

### 5.4.3 Meshes with Diagonals

Hardly any theoretical papers deal with *meshes with diagonals*, in which each node has up to eight neighbors. Surprisingly, most routing and sorting times can be reduced by at least a factor of two. We provide (without proof) the following:

***LEMMA 5.39***

*On an $n \times n$ mesh with diagonals, $k$-$k$ routing and sorting can be performed in $2kn/9 + o(kn)$ steps [372]. On a torus with diagonals, this problem can be solved in $kn/12 + o(kn)$ steps [627].*

### 5.4.4 Coated Mesh

Another mesh variant places PEs on the outside periphery of the network, while the internal mesh nodes provide the communication network. This topology is also common with interconnecting ATM switches. The motivation to consider such a network is that on a traditional mesh with $n^2$ nodes the bisection width is only $n$. This implies that for most problems the best speedup which can be achieved is much smaller than the number of nodes. By placing nodes on the outside one obtains the *coated mesh*, which saves substantially on the hardware costs, without losing much of the communication power. Such architectures are particularly suited for PRAM simulations [111, 420]. Various routing problems are considered in [112].

## 5.5 Allocation and Scheduling Strategies

We focus on mesh allocation techniques, since job scheduling algorithms which have been discussed in detail for the hypercube topology in Section 4.4 are also applicable to mesh systems.

*Contiguous allocation* schemes on the mesh limit communication paths between PEs and avoid interprocess interference. However, this restriction also reduces the chance of successfully allocating a job, thus increasing fragmentation. The following proposed contiguous mesh allocation schemes are listed in terms of increasing performance.

*Two-dimensional buddy system* (TDB). With TDB, the allocated subsystem is always a square submesh with $2^n \times 2^n$ nodes, i.e., the size of a requested submesh is rounded up to the the nearest power of two [426]. Thus, TDB suffers from internal fragmentation and all types of external fragmentation. The implementation is based on a single bit vector, which is used to keep track of allocation and deallocation processes.

*Frame sliding* (FS). FS reduces fragmentation by assigning submeshes of arbitrary size [117]. FS slides the requested submesh (called frame) over the available mesh to locate any free submesh on which to execute the job. Thus, FS does not cause any internal fragmentation. Two bit vectors are used; one keeps track of previously allocated nodes, while the other one scans the available submesh.

*Adaptive scan* rotates the submesh requested, thus increasing the probability of job allocation [168]. Variations of adaptive scan include: (a) the *busy list* which replaces bit vectors with linked lists [617], and (b) the *free list* which uses a linked list of available submeshes for allocation [438]. The free list provides faster allocation time at the expense of extra deallocation time.

*First-* and *best-fit* allocation provide full submesh recognition, thus preventing external fragmentation due to imperfect subsystem recognition [337, 721]. Both algorithms are implemented using bit vectors. Previous studies indicate that only slight performance improvement can be obtained by using a full recognition strategy.

*Limit* allocates a job to a smaller size submesh if the required submesh can not be allocated. Since the probability of finding a smaller size submesh is higher, average job turnaround time is reduced.

Contiguous allocation suffers from physical fragmentation due to insufficient resource and dynamic job scheduling. To alleviate this problem, various *noncontiguous allocation* schemes have been proposed. Although such schemes described below are usually proposed for a mesh multicomputer, their use can be extended to any particular architecture [99].

*Random* allocation implements a random PE selection.

*Naive* algorithm allocates a submesh obtained using a first-fit algorithm during a row scan.

*Multiple buddy system* (MBS) is similar to TDB in that it subdivides the system into square submeshes with side lengths equal to a power of two. If a request for a particular submesh size can not be satisfied, MBS searches recursively for smaller submeshes by repeatedly halving the initially requested mesh size. Thus, MBS provides only partial contiguity, e.g., for a $32 \times 32$ mesh MBS allocates contiguously only square jobs with side lengths equal to 1, 2, 4, 8, 16, and 32. Observe that, assuming a uniform job distribution, the probability of such a submesh request is just 6/1024; thus, many jobs are still allocated noncontiguously.

*Adaptive noncontiguous allocation* (ANCA) allocates jobs on noncontiguous PEs, if fragmentation prevents a job from being allocated to a contiguous region [99]. Despite the small increase in communication time due to noncontiguous allocation, internal fragmentation due to insufficient resource and dynamic job scheduling can be completely removed if all jobs can be allocated completely noncontiguously. Thus, variants of the ANCA algorithm are obtained by using an application dependent adaptability metric, which refers to the number of times that a submesh request can be split into smaller subsystems, i.e., zero adaptability corresponds to strictly contiguous allocation.

*Restricted size reduction* (RSR) allocates jobs on a reduced size submesh (obtained by repeatedly folding the longer mesh side), if a submesh of required size cannot be located [99]. Thus, queuing delay is reduced at the expense of longer execution time. Internal fragmentation due to insufficient resource and dynamic job scheduling may be completely removed if size reduction allows all jobs to be executed on a single PE. Notice that restrictions on size reduction depend both on the individual architecture, and also on the application, especially since application memory requirements increase with smaller systems.

For two-dimensional mesh systems, both RSR and ANCA schemes reduce internal fragmentation resulting from insufficient resource and dynamic job scheduling by trading increased queuing time for communication time. However, although both RSR and ANCA improve average turnaround time, it is not certain which variant is optimal for an actual implementation [99]. In [443], NAS benchmarks have been used for an experimental evaluation of noncontiguous allocation.

A reliability analysis of TDB and FS allocation on two-dimensional mesh is based on examining the maximum recognition ability of the mesh allocation algorithms [480]. For full recognition allocation schemes the analysis turns out to be much harder, and results in an open enumeration problem relating to the alignment of working nodes in different rows and columns. In the same study, the availability of mesh-connected computers using allocation schemes is computed using Markov Chain analysis and relevant approximations.

## 5.6 VLSI Layout and Customization

While for a two-dimensional mesh topology, the VLSI layout is straightforward, for a two-dimensional torus topology the layout should be arranged to minimize the wire length between any two adjacent node. For high-dimensional topologies, both the wire length and area should be minimized.

For customization issues we refer the reader to the discussion in the corresponding section of Chapter 4, which provides interesting comparisons with the hypercube topology.

## 5.7 Commercial Mesh Systems

### The Intel Paragon

Unlike previous Intel hypercube systems (the iPSC/1, iPSC/2, and iPSC/860), the Intel Paragon employs a low-dimensionality communication network (a mesh) and relies on a heterogeneous design [276] consisting of I/O (disk, frame buffers) nodes, a mesh of computing nodes (for batch and interactive execution), and service nodes with editors and compilers. Like the Caltech Cosmic Cube, each node of the Paragon is implemented on one board. It consists of a bus connected to an i860 XP application processor, a FP unit, one i860 XP message processor, an external I/O unit (for nodes at the left and right columns of the mesh), performance monitoring hardware, a network interface, and up to 128

MB of memory. The router at each mesh intersection is a $5 \times 5$ crossbar switch. It has 10 buses, 8 for communication with NEWS (north, east, west, south) neighbors, and 2 for data transfer with the board PE attached to the mesh at that intersection. Routing is pipelined using wormhole routing with hardware support for virtual channeling (to resolve deadlocks).

**The Cray-T3E**

According to the current TOP500 list of supercomputers, 13 sites of the first 20 fastest supercomputers in the world are actually Cray-T3E systems [175]. The Cray-T3E is a *self-hosted* massively parallel (MPP) machine. It does not require front-end or host machines [607].

The Cray-T3E is configured as 3D torus, and like the Paragon, it comes with three kinds of processors:

- *Command processors* are organized as in conventional multiprocessor Unix machines. All single processor programs, like editors, compilers, and login shells run on command processors.
- *I/O processors* provide access to user file systems via HIPPI interfaces.
- *Application processors* are reserved to run parallel programs. Spare processors are provided to dynamically replace faulty nodes using fault-tolerant routing.

The machine is running *Unicos/mk*, a Unix-like operating system for MPP machines based on the Chorus microkernel.

The Cray-T3E implements a logically shared address space over physically distributed memories (with up to 2 GB per processor). Several atomic, synchronization, and collective operations are supported on all Cray MPP and PVP systems by calling Cray ShMem library functions.

The synchronization operations are based on a pair of atomic instructions that the DEC Alpha processor executes in sequence: a *load-linked* loads a link register with the value stored at a memory location, while a *store-conditional* stores a register value at that location if and only if no other processor has performed a store (or store-conditional) at that location since the load-linked read it. Finally, the register is set to one if the store succeeded, or zero if the store failed. ShMem library calls can be used within Fortran and C/C++ programs.

The Cray-T3E provides a plethora of atomic operations on arbitrary memory locations, allowing an unlimited number of synchronization variables. The atomic operations provided by the Cray-T3E are Fetch& Inc, Fetch&Add, Compare&Swap, and (masked) Swap. Although Test& Set is not provided, it can easily be implemented using Fetch&Store or Compare&Swap.

Distributed shared memory (ShMem) operations on the Cray-T3E (and other Cray systems), which are available to users and operating system programmers, include atomic and remote read/write as shown in Table 5.1.

**Table 5.1** Read-modify-write and remote read/write in ShMem

| Generic Name | ShMem_ Operation | Description |
|---|---|---|
| Fetch&Inc | `short_finc(&a,n)` | return $M_n(a)$; $\langle m_n(a) = m_n(a) + 1\rangle$ |
| Fetch&Add | `short_fadd(&a,b,n)` | return $m_n(a)$; $\langle m_n(a) = m_n(a) + b\rangle$ |
| Fetch&Store | `swap(&a,-1,b,n)` | return $m_n(a)$; $\langle m_n(a) = b\rangle$ |
| Compare&Swap | `short_cswap(&a,b,c,n)` | if $m_n(a) \neq b$ : return $m_n(a)$ else : return $m_n(a)$; $\langle m_n(a) = c\rangle$ |
| (Masked) Swap | `short_mswap(&a,b,c,n)` | return $m_n(a)$; $\langle m_n^i(a) = c^i\rangle$, $\forall i \in \{0, \ldots, 31\} : b^i = 1$ |
| Remote Read | $PE_p$ : `get(&a,&b,c,n)` | $\langle m_p(a) = m_n(a); \ldots; m_p(a + c - 1) = m_n(b + c - 1)\rangle$ |
| Remote Write | $PE_p$ : `put(&a,&b,c,n)` | $\langle m_n(a) = m_p(b); \ldots; m_n(a + c - 1) = m_p(b + c - 1)\rangle$ |

The operands are generally short (32-bit) integers; some operations allow also for other data types, such as int, long, float, and double. A definition of each operation listed above is provided below:

- *Fetch_Φ(a, b)* operations return the memory contents $M(a)$, while storing at location $a$ the function $\Phi(M(a), b)$.

- *Compare&Swap(a, b, c)* primitives atomically compare the content of memory location $M(a)$ with a replacement value $b$, and store a third value $c$ if they match; the operation also returns a condition flag indicating either success or failure.

- *(masked) Swap(a, b, c)* operations store selected bits from $c$ in $M(a)$. Selection is done by the mask $b$. It returns the previous content of $M(a)$.

Shared memory (ShMem) locks provide atomicity during parallel update operations, while barriers help in avoiding certain race conditions.

Synchronization barriers allow a set of participating processors to determine when all processors have signaled some event (typically reached a certain point in their execution of a program). Barrier routines, e.g., (`shmem_barrier_all()`) or simply (`barrier()`), are implemented in hardware on the Cray-T3E using combining trees. Eureka "find" operations are also provided. Also, *shmem_set__lock(a)* and *shmem_clear_lock (a)* routines together provide mutual exclusion. The first call to *shmem_set__lock(a)* (acquire lock) returns immediately. A subsequent call returns after a *shmem_clear_lock(a)* (release lock) corresponding to the first *shmem_set__lock(a)* is executed. This ensures that only one process is holding the lock at a time.

Cache coherence is implemented on the Cray-T3E using cache-invalidate protocols. However, the T3E is not stream coherent; prefetching must be turned off to avoid data race conditions which may cause inconsistent results, program aborts, or hangs. Due to adaptive packet routing, which is turned on in some models, the Cray-T3E may also reorder instructions from a given PE. Since successive calls to `shmem_put` are not guaranteed to arrive in order, special calls to `shmem_quiet` (or `shmem_fence`) are always needed to verify that all previous puts from a particular PE to all other PEs (respectively, to a given PE) have executed in order. The design and use of efficient ShMem synchronization functions has been studied in [239].

# Chapter 6

# Other Parallel Networks

We have previously considered various static and dynamic networks. More specifically, we have covered multistage interconnection networks in Chapter 3, hypercube networks in Chapter 4, and mesh networks in Chapter 5. In this chapter, we will examine other interconnection networks such as de Bruijn networks, shuffle-exchange network, and circulant network. We will also examine the class of Cayley graphs which are proposed and used more recently in the construction of interconnection networks. In fact, we will see in Section 6.4 that the two very popular networks: butterfly and cube-connected-cycles (CCC) are Cayley graphs. De Bruijn networks and shuffle-exchange networks are not Cayley graphs. However, they can be shown (also in Section 6.4) to be coset graphs of Cayley graphs.

Hypercubes, mesh networks, circulant networks, butterfly, de Bruijn, CCC, and shuffle-exchange networks are among today's popular interconnection networks for parallel processing systems. Architectures based on these networks have been built in industry, research, laboratories, and academic institutions. In particular, hypercubes and mesh networks are among the most widely used networks. Meshes and their variations are attractive because of regularity, scalability, and conceptual simplicity (see Chapter 5). They are easy to program and are suitable for VLSI implementation. The hypercube is favorable because of its efficiency in implementing a broad class of algorithms and its structural uniformity that can greatly simplify programming [593]. However, the major drawback of a mesh network is its large diameter and small bisection width while the major shortcoming of the hypercube is its high degree. In fact, the degree of a binary $n$-dimensional hypercube is $n$ which increases proportionally with the dimension $n$. These disadvantages have led to the development of several bounded-degree networks such as butterfly and cube-connected-cycles [559]. These two networks have degrees bounded by a constant and can emulate the hypercube with little or no slowdown on a variety of important computational problems and algorithms [267]. Both butterfly and CCC networks, like hypercube, are Cayley graphs which possess some algebraic properties such as symmetry and recursive structure. These properties will be discussed in some details in Section 6.4.

Butterfly and CCC networks are intended to be approximations of the hypercube in the sense that they can run many hypercube algorithms including the ascend-descend algorithms [559]. However, these two networks have to use $N \log_2 N$ nodes to emulate the hypercube with $N$ nodes in the binary case. De Bruijn and shuffle-exchange networks are developed which, in the binary case, have $N$ nodes as the hypercube and can approximate the hypercube on many computational tasks. These two networks have constant degree, as in the case for butterfly and CCC. The $N$-node binary de Bruijn network has diameter exactly $\log_2 N$ and is pancyclic [58, 266, 282]. It also contains an $(N-1)$-node complete binary tree. These properties are useful for computation. We will discuss the binary as well as the $k$-ary de Bruijn networks in Section 6.1 and binary and $k$-ary shuffle-exchange networks in Section 6.2. In graph theoretic terms, these two networks have been generalized to include $N$ nodes for any arbitrary positive integer $N$ [89, 180, 299, 440, 649].

In Section 6.3, we include the class of circulant networks which are special Cayley graphs. Circulant networks are graphs and digraphs with circulant adjacent matrices in which each row after the first is a cyclic shift of its previous row. Circulant matrices arise in the study of periodic or multiply symmetric dynamical systems. In particular, they have applications in the theory of crystal structure [593].

## 6.1 De Bruijn Networks

The $n$-dimensional $k$-ary de Bruijn network, denoted as $D(k,n)$, is modeled as a directed graph (or digraph) with $k^n$ nodes. Each node is labeled as an $n$-bit $k$-ary number $a_1a_2\cdots a_n$ where $a_i \in [0,k-1]$. Node $a_1a_2\cdots a_n$ is adjacent to $a_2a_3\cdots a_{n-1}\alpha$, where $\alpha \in [0,k-1]$. Each node in $D(k,n)$ has indegree and outdegree equal to $k$. Self-loops do exist at certain nodes. For example, in $D(2,3)$, node 000 is adjacent to itself. So is node 111. The undirected version of $D(k,n)$, where each link

$(a, b)$ is changed to an undirected edge $\{a, b\}$ and self-loops and multiple edges are deleted, is denoted as $UD(k, n)$. The two graphs $D(2, 3)$ and $UD(2, 3)$ are shown in Figure 6.1 and Figure 6.2, respectively.

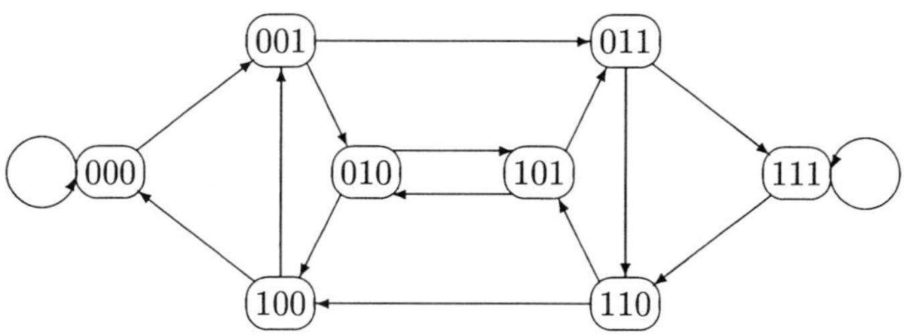


**FIGURE 6.1**
**De Bruijn graph** $D(2, 3)$

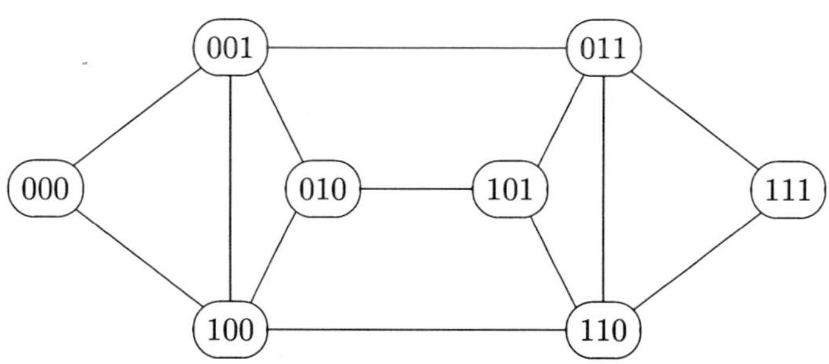


**FIGURE 6.2**
**De Bruijn graph** $UD(2, 3)$

The concept of a de Bruijn digraph was originally defined in 1946 in [160]. Since then, there are at least two equivalent ways to define these digraphs. The following is defined by Imase and Ito [298], and by Reddy, Pradhan, and Kuhl [584] independently.

**DEFINITION 6.1** *The de Bruijn digraph $D(k,n)$ is the digraph with nodes labeled as integers in $[0, N-1]$, where $N = k^n$. There is an arc $(x,y)$ from $x$ to $y$ if and only if $y \equiv dx + a \bmod k^n$, where $0 \leq a \leq k-1$.*

The second equivalent definition deals with iterated line digraphs (see [160]). Let $G$ be a digraph. The line digraph $L(G)$ has $V(L(G)) = E(G)$ and $E(L(G)) = \{((u,v),(v,w)) \mid v \in V(G)\}$. In other words, there is an arc from node $(u,v)$ to node $(v,w)$ in $L(G)$ because arc $(u,v)$ is incident to the node $v$ and arc $(v,w)$ is incident from the node $v$ in $G$. Indeed, we have the following definition for $D(k,n)$.

**DEFINITION 6.2** *Let $K_t^*$ be the complete symmetric directed graph on $t$ nodes with each node having a self directed loop. Let $L(G)$ be the line digraph of $G$. Then $D(k, n+1) = L^n(K_k^*)$.*

In Figure 6.3, we exhibit the digraphs $D(2,1) = L^0(k_2^*)$, $D(2,2) = L^1(k_2^*) = L^1(D(2,1))$, and $D(2,3) = L^2(k_2^*) = L^1(D(2,2))$.

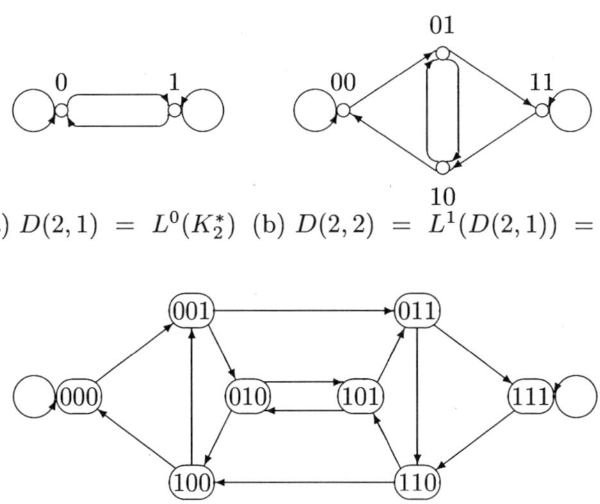


(a) $D(2,1) = L^0(K_2^*)$ (b) $D(2,2) = L^1(D(2,1)) = L^1(K_2^*)$

(c) $D(2,3) = L^1(D(2,2))$

**FIGURE 6.3**
**De Bruijn graphs $D(2,1)$, $D(2,2)$, and $D(2,3)$**

The de Bruijn digraphs arose from the design of shift registers and digital computers. It is also related to a kind of combinatorial object called *de Bruijn sequence.* Let $k, n$ be positive integers where $k \geq 2$ and $n \geq 1$. The topic of a periodic binary de Bruijn sequence $s(2, n)$ existed in the mathematical literature since 1894, but continued to be rediscovered from time to time in several different contexts [298]; the sequence $s(k, n)$ is defined as follows:

***DEFINITION 6.3*** *The $k$-ary de Bruijn sequence $s(k, n)$ is a circular $k$-ary sequence of length $k^n$ in which each of the $k^n$ different $k$-ary $n$-tuples appears exactly once as a subsequence of length $n$.*

It is clear that in $s(2,2)$, $s(2,3)$, and $s(3,2)$ we have the following sequences and all the corresponding $k$-ary subsequences of length $n$,

$$\begin{array}{ll} s(2,2): 0110; & 01, 11, 10, 00 \\ s(2,3): 10111000; & 101, 011, 111, 110, 100, 000, 001, 010 \\ s(3,2): 122001102; & 12, 22, 20, 00, 01, 11, 10, 02, 21. \end{array}$$

We note that the $k^n$ different $k$-ary $n$-tuples in the $k$-ary de Bruijn sequence $s(k, n)$ in fact represent the nodes in the de Bruijn digraph $D(k, n)$. Moreover, a sequence in $s(k, n)$ would induce an order of the $k$-ary $n$-tuple which are nodes in $D(k, n)$. This order gives rise to a Hamiltonian cycle in $D(k, n)$ as $01 \to 11 \to 10 \to 00 \to 01$ in $s(2, 2)$ above. Hence, we have:

***THEOREM 6.1***
*The de Bruijn digraph $D(k, n)$ consists of a Hamiltonian cycle.*

In 1971, Lempel [417] showed a result stronger than that in Theorem 6.1 by proving the following theorem, the proof of which we omit here.

***THEOREM 6.2***
*Let $k, n, t$ be any three positive integers. Then there exists a closed sequence (ordered cycle) of $t$ $k$-ary digits in $[0, n-1]$ such that the $t$ subsequences of $n$ consecutive digits are all different for every $t$, $2 \leq t \leq k^n$.*

***COROLLARY 6.1***
*For every $t$, $2 \leq t \leq k^n$, $D(k, n)$ contains a cycle of length $t$.*

We have seen that the de Bruijn digraph $D(k,n)$ contains many cycles. From Theorem 6.2 and Corollary 6.1, we know that $D(k,n)$ contains a cycle of length $t$, $2 \leq t \leq k^n$ and that each of these cycles comes from the $t$ subsequences of $n$ consecutive digits. In fact, $D(k,n)$ has many good properties other than that in Corollary 6.1. We now examine the distance between any two nodes in $D(k,n)$. Let $a = a_1a_2\cdots a_n$ and $b = b_1b_2\cdots b_n$ be any two nodes in $D(k,n)$, where $a_i, b_i \in [0, k-1]$. We easily see that the distance $d(a,b)$ between $a$ and $b$ is $n$ by the following steps:

$$\begin{aligned} a = a_1a_2\cdots a_n &\longrightarrow a_2a_3\cdots a_nb_1 &&\longrightarrow a_3a_4\cdots a_nb_1b_2 \longrightarrow \cdots \\ &\longrightarrow a_nb_1b_2\cdots b_{n-1} &&\longrightarrow b_1b_2\cdots b_n = b\,. \end{aligned}$$

Hence diameter $d(D(k,n))$ of $D(k,n)$ is bounded above by $n$. On the other hand, we can show that $d(\mathbf{0},\mathbf{1}) = n$ where $\mathbf{0} = \underbrace{000\cdots 0}_{n}$ and $\mathbf{1} = \underbrace{111\cdots 1}_{n}$. Hence $d(D(k,n)) \geq n$. Therefore we have the following:

***THEOREM 6.3***

*The diameter of the de Bruijn digraph $D(k,n)$ is $n$.*

It can be shown that the diameter of the de Bruijn graph $UD(k,n)$ is also $n$, but the proof is more involved [204]. Shortest path routing on the de Bruijn graph $UD(k,n)$ is related to string matching and quite elaborative. Two of these algorithms using a match matrix and prefix tree can be found in [204].

The de Bruijn digraph $D(k,n)$ is a regular digraph with indegree and outdegree equal to $k$. Since nodes such as $\mathbf{0} = 000\cdots 0$ and $\mathbf{1} = 111\cdots 1$ do have self loops, the connectivity of $D(k,n)$ can not exceed $k-1$. On the other hand, it was shown in [130] that the connectivity of $D(k,n)$ is at least $k-1$. Hence we have the following:

***THEOREM 6.4***

*The connectivity of $D(k,n)$ is $k-1$.*

We now turn to the problem of broadcasting on de Bruijn graph $UB(k,n)$. Some definitions and terminologies are needed. Broadcasting is a process where message dissemination is made to all members of the network by initiating the message by one member and by placing a series of calls over the communication links of the network. In our model, each call involves only two nodes, each call requires one unit of time, a node can participate in only one call per unit of time, and a node can only call an adjacent node. We have the following definitions.

***DEFINITION 6.4*** *Let $G$ be a connected graph and $x$ be a node in $G$. The broadcast time $b(x)$ of the node $x$ is the minimum of time units required for broadcasting a message from $x$ to all other nodes in $G$. The broadcast time of the graph $G$ $b(G)$ is the maximum broadcast time of any node $x$ in $G$. That is, $b(G) = \max\{b(x) \mid x \in V(G)\}$.*

The following theorem gives the broadcast time of $UD(k, n)$ [590]:

***THEOREM 6.5***

*Let $UD(k, n)$ be the de Bruijn graph of maximum degree $2k$ and diameter $n$. Then $b(UD(k, n)) = \frac{k+1}{2} \cdot n + \frac{k}{2}$.*

**PROOF** Let the node $a_1a_2\cdots a_n$ be the node where we would like to broadcast. Let us broadcast from this node successively to nodes $a_2a_3\cdots a_n0$, $a_2a_3\cdots a_n1$, $\cdots$, $a_2a_3\cdots a_n(k-1)$, and then to the nodes $(k-1)a_1a_2\cdots a_{n-1}$, $(k-1)a_1a_2\cdots a_{n-1}$, $\cdots$, $0a_1a_2\cdots a_{n-1}$. The node $a_2a_3\cdots a_ni$ is called at time $i+1$ and node $ja_1a_2\cdots a_{n-1}$ is informed at time $2k-j$. For any other node $b_1b_2\cdots b_n$ that first receives the call at time $t$, broadcast using the following two rules:

(i) If the call arrives from a left neighbor $\alpha b_1b_2\cdots b_{n-1}$, broadcast it to the right neighbor in the order $b_2b_3\cdots b_n0$, $b_2b_3\cdots b_n1$, $\cdots$, $b_2b_3\cdots b_n(k-1)$. It follows that the node $b_2b_3\cdots b_ni$ receives the call at time $t+i+1$.

(ii) If the call arrives from a right neighbor $b_2b_3\cdots b_n\alpha$, broadcast it to the left neighbor in the order $(k-1)b_1b_2\cdots b_{n-1}$, $(k-2)b_1b_2\cdots b_{n-1}$, $\cdots$, $0b_1b_2\cdots b_{n-1}$. It follows that the node $jb_1b_2\cdots b_{n-1}$ will be informed at time $t+k-j$.

Utilizing these rules, we see that node $c_1c_2\cdots c_n$ is called at time at most $\sum_i(c_i+1)$, and at time at most $k+\sum_i(k-c_i)$. Hence it is called at time:

$$t = \min\left\{\sum_i(c_i+1), k+\sum_i(k-c_i)\right\}.$$

Then
$$\begin{aligned} 2t &\le k + \sum_i ((c_i + 1) + (k - c_i)) \\ &= k + \sum_i (k+1) \\ &= k + (k+1)n \end{aligned}$$

and $b(UD(k,n)) \le \frac{k+1}{2}n + \frac{k}{2}$. ∎

The proof of Theorem 6.5 can be modified to compute the broadcast time for $D(k,n)$.

As we have discussed in the beginning of this chapter, de Bruijn networks have plenty of good properties. We now turn to the embedding problems. An *embedding of a graph $G$ into a graph $H$* is a mapping of nodes of $G$ (called guest graph) to nodes of $H$ (called host graph) and of edges of $G$ to paths in $H$ between the images of their end nodes. The *dilation of the embedding $f$* is the maximum length in $H$ of the image under $f$ of any edge of $G$. The *congestion of $f$* is the maximum number of times an edge of $H$ is used by the images of edges of $G$. The *load of $f$* is the maximum number of nodes of $G$ mapped to any node of $H$ under $f$. If an embedding of $G$ into $H$ has constant dilation, congestion, and load, then the network based on the graph $H$ will be able to simulate an arbitrary computation on the guest graph $G$ with only a constant factor slowdown. In this direction, it has been shown that there is an embedding of binary $n$-dimensional hypercube into binary $n$-dimensional de Bruijn graph $UD(2,n)$ of the same order and diameter with dilation less than or equal to $n/2$ [267].

Similar result also holds for meshes. It was also shown that an $N \times N$ mesh of trees graph can be embedded into a $4N^2$-node binary de Bruijn graph with constant dilation, load, and congestion where $N = 2^n$ [280]. The *$N \times N$ mesh of trees graph* is constructed by having an $N \times N$ array of nodes, where $N = 2^n$, and then establishing an $N$-leaf complete binary tree on the $N$ nodes of each row and column in the $N \times N$ array.

Permutation routing and parallel sorting problems have also been studied. In particular, there exist efficient permutation routing and sorting schemes for de Bruijn networks. It was shown in [59] that any deterministic oblivious routing scheme for permutation routing on $D(k,n)$ nodes, in the worst case, should take $\Omega(\sqrt{N})$ steps under the one-port model (where $N = k^n$). This lower bound improves the lower bound in [31] for $D(k,n)$ (the lower bound in [31] is $\Omega(\sqrt{N/d})$ for any graph of $N$ nodes with degree $d$) and was shown indeed to be a tight one. Efficient deterministic nonoblivious permutation routing which runs in $O(k \cdot n^2)$ and parallel routing which runs in $O((\log k) \cdot k \cdot n^2)$ have been obtained for the de Bruijn digraph $D(k,n)$ [59].

When considering de Bruijn graphs or digraphs as parallel architecture, we have so far discussed a host of structures and properties which are certainly advantage for de Bruijn graphs and digraphs. We have seen that de Bruijn networks have either the best or almost the best diameter, connectivity, regularity, routing, sorting, and symmetry. The de Bruijn digraph $D(k,n)$ has a fixed degree $k$ which is much better than the hypercube networks. It has an optimal diameter which is much better than the mesh network. Although $D(2,n)$ can simulate the hypercube and mesh networks with dilation $n/2$, it can simulate the mesh of trees network with a constant slowdown [228, 280]. One important factor in the VLSI design of an interconnection network is its scalability. From the definition, we know that $D(k,n+1)$ is the line digraph of $D(k,n)$. However, $D(k,n)$ is not a subgraph of $D(k,n+1)$. This is probably the only perceived drawback of de Bruijn networks when compared to the hypercube or mesh networks. Fortunately, when we examine the structure of de Bruijn networks a step deeper and more closely, some subtle relationships do exist between $D(k,n)$ and its subgraphs [194, 417].

***DEFINITION 6.5*** *Let $H$ be a subgraph of the graph $G$. The density of $H$ in $G$, denoted by $den(H : G)$, is defined to be*

$$den(H : G) = |E(H)|/|E(G)| .$$

***DEFINITION 6.6*** *Let $H$ be a subgraph of the graph $G$. $H$ is said to be a building block for $G$ if there exists a subgraph $\bar{H}$ in $G$ that is the disjoint union of several copies of $H$. The efficiency of $H$ as a building block for $G$, denoted as $eff(H,G)$, is defined by $den(\bar{H} : G)$.*

We note that a subgraph of $D(2,n)$ of density 0 is a building block of efficiency 0 for any $D(2,N)$ with $N \geq n$. A subgraph of $D(2,n)$ of density 1 (that is, $D(2,n)$ itself) can not be a building block for any larger de Bruijn digraph since the disjoint union of $2^{N-n}$ copies of $D(2,n)$ has $2 \cdot 2^n \cdot 2^{N-n} = 2^{N+1}$ edges and $D(2,N)$ has the same number of edges that is connected. Figure 6.4 shows a subgraph $H$ of $D(2,3)$ with density $6/16 = 3/8$ while Figure 6.5 shows two copies of the subgraph $H$ which can be wired together to form $D(2,4)$. Each edge in $H$ is labeled as an $(n+1)$-bit binary string $x$ which connects the nodes with $n$-bit labels $x^R$ and $x^L$.

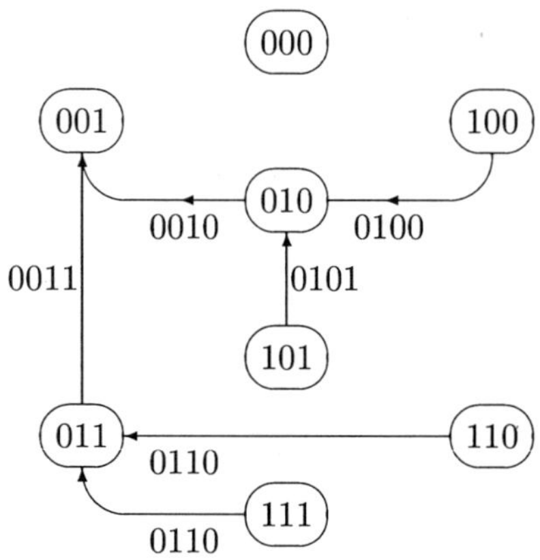


**FIGURE 6.4**
**Subgraph H of $D(2,3)$ with density $= 3/8$**

In Figure 6.5, we see two copies of $H$ from Figure 6.4 to form $D(2,4)$. Since $D(3,4)$ has 32 edges and $H$ has 6 edges, we have $eff(H : D(2,4)) = 12/32 = 3/8$ which is the same as the density of $H$ in $D(2,3)$. In fact, the digraph $H$ in Figure 6.4 is a building block of efficiency 3/8 for any $D(2,N)$ with $N \geq 3$ as we will see later. This leads to the following definition.

***DEFINITION 6.7*** *A universal de Bruijn building block of order $n$ is a subgraph of $D(2,n)$ that is a building block for any de Bruijn digraph $D(2,N)$ with $N \geq n$.*

The following theorem shows that it is quite easy to compute the efficiency of any universal de Bruijn building block (also see [417]).

***THEOREM 6.6***

*Let $H$ be a universal de Bruijn building block of order $n$. Then for all $N \geq n$, we have $eff(H : D(2,N)) = den(H : D(2,n))$.*

**PROOF** By Definition 6.6, we have $eff(H : D(2,N)) = den(\bar{H} : D(2,N)))$ which in turn is equal to $|E(\bar{H})|/|E(D(2,N))|$. But since

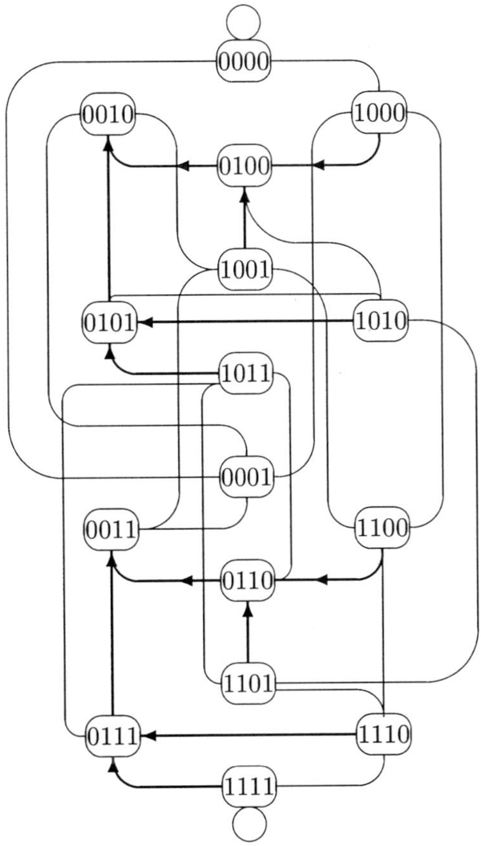


**FIGURE 6.5**
**Two copies of $H$ to form $D(2,4)$**

$|E(\bar{H})| = |E(H)| * (|V(D(2,N))|/|V(H)|)$, we have

$$\begin{aligned} eff(H : D(2,N)) &= \frac{|V(D(2,N))| * |E(H)|}{|V(H)| * |E(D(2,N))|} \\ &= \frac{2^N * 2^{n+1} \cdot den(H : D(2,n))}{2^n * 2^{N+1}} \\ &= den(H : D(2,n)). \end{aligned}$$

∎

If $S$ and $T$ are sets of binary strings, we say that $S$ *covers* $T$ if every string in $T$ has a substring in $S$. Similar, we say that $S$ *prefixes* $T$ if every string in $T$ has a prefix in $S$. We say that a set of strings $S$ is *irreducible* if no string in $S$ is a substring of any other. We define $D_n(S)$ to be the subgraph of $D(2,n)$ obtained by deleting from $D(2,n)$ all edges whose labels have a prefix in $S$. The following theorem is proved in [6]. We omit the proof here.

***THEOREM 6.7***

*If $S$ is irreducible and covers the set of binary strings of length $n$, then $D_n(S)$ is a universal de Bruijn building block of order $n$.*

The set of $S = \{1, 000\}$ is irreducible and covers the set of binary strings of length 3. $D_3(S)$ is exactly the same as the subgraph $H$ in $D(2,3)$ in Figure 6.4. Therefore Theorem 6.7 asserts that the digraph $D_3(\{1,000\})$ is a universal de Bruijn building block with efficiency 3/8 as we stated before. From Theorem 6.7, we see that in order to search for universal de Bruijn building block of order $n$, it suffices to find irreducible set $S$ which can cover the set of binary strings of length $n$. Moreover, the derived $S$ with smaller $cost(S)$ would give larger efficiency (the efficiency of $D_n(S)$ equals $1 - cost(S)$) where $cost(S)$ is defined as $\sum_{s \in S} 2^{-|s|}$. In fact, universal de Bruijn building blocks of order $n$ whose efficiency approaches 1 when $n$ approaches infinity exist.

The above discussion is related to the VLSI design of de Bruijn networks. Theorem 6.7 shows that the union of $2^{N-n}$ copies of $D_n(S)$ forms a subgraph (i.e., $D_N(S)$) of the big de Bruijn digraph $D(2,N)$. Therefore, the de Bruijn digraph $D(2,N)$ can be constructed by adding the edges missing from $D(2,N)$ to the union of $2^{N-n}$ copies of $D_n(S)$. For any $n$ and $N$, $N > n$, this can always be done provided we can construct $S$. Now we examine properties of de Bruijn digraph similar to the building block approach but from graph isomorphism viewpoint. We will see that the de Bruijn digraph $D(k,n)$ can be decomposed into an appropriate number of node-disjoint copies of de Bruijn digraphs with smaller number of nodes.

Let $G$ and $H$ be two digraphs (or graphs). The *Kronecker product* $G \otimes H$ of $G$ and $H$ is the digraph with $V(G \otimes H) = V(G) \times V(H)$ and $E(G \otimes H) = \{((x_1, y_1), (x_2, y_2)) \mid (x_1, x_2) \in E(G) \text{ and } (y_1, y_2) \in E(H)\}$.

For a digraph $G$, a *factor of* $G$ is a spanning subdigraph of $G$. If $F_1, F_2, \cdots, F_t$ are arc-disjoint factors of $G$ such that $\bigcup_{1 \le i \le t} E(F_i) = E(G)$, then we write

$$G = F_1 \oplus F_2 \oplus \cdots \oplus F_t \, ,$$

and say that $G$ is the *arc sum* of the factors $F_i$, $i = 1, 2, \cdots, t$. This arc sum is said to be a *factorization of* $G$ into $F_i$'s, $i \in [1, t]$. If $F_i$'s are all isomorphic to a factor, say $F_i \cong H$ for all $i$, $i \in [1, t]$, then we say that $G$ has an *isomorphic factorization into* $H$ and that $H$ *divides* $G$. In this case, we write $H|G$. The following two lemmas are easily proved.

**LEMMA 6.1**

*Let $G$, $H$, and $F$ be digraphs such that $F|H$, then $G \otimes F \mid G \otimes H$.*

**LEMMA 6.2**

*Let $G, H$ be digraphs, then $L(G \otimes H) \cong L(G) \otimes L(H)$.*

A digraph $G$ with $|V(G)| = n$ is said to be *cyclic transitive* if the nodes of $G$ can be labeled by numbers in $[0, n-1]$ such that the cycle permutation $(0, 1, 2, \cdots, n-1)$ is an isomorphism of $G$. For a cyclic transitive $t$-regular digraph $G$ of order $n$ we would like to see if $L^m(G)$ is a factor of $L^p(G) \otimes L^q(G)$ where $m = p + q + 1$. Since $L^m(G)$ has to have the same number of nodes as $L^p(G) \otimes L^q(G)$, we count the number of their nodes of each digraph. We have

$$\begin{aligned} |V(L^p(G) \otimes L^q(G))| - |V(L^m(G))| &= t^p \cdot n * t^q \cdot n - t^m \cdot n \\ &= t^{p+q} \cdot n^2 - t^m \cdot n \\ &= t^{m-1} \cdot n^2 - t^m \cdot n \\ &= t^{m-1} \cdot n^2 - t^{m-1} \cdot n \cdot t \\ &= t^{m-1}(n^2 - nt) \\ &= t^{m-1} \cdot n \cdot (n - t). \end{aligned}$$

Let $w = t^{m-1} \cdot n \cdot (n-t)$ and we see that by adding $w$ isolated nodes to $L^m(G)$, the resulting digraph has the same number of nodes as $L^p(G) \otimes L^q(G)$ where $m = p + q + 1$. The following lemma was proved in [194].

**LEMMA 6.3**

*Let $G$ be a $t$-regular cyclic transitive digraph of order $n$. Let $p, q, m$ be positive integers such that $m = p + q + 1$. Then*

$$(L^m(G) \cup wK_1) \mid L^p(G) \otimes L^q(G), \text{ where } w = t^{m-1} \cdot n \cdot (n-t).$$

We now return to de Bruijn digraphs. It is simple to show that $D(2, pq)$ can be embedded into $D(2^p, q)$ with dilation 1. It is shown in [217] that $D(k, pq)$ is a subdigraph of $D(k^p, q)$. In fact, the following stronger result holds surprisingly (see [194]).

**THEOREM 6.8**

*Let $D(k, n)$ be the $k$-ary $n$-dimensional de Bruijn digraph with order $k^n$ and $n = pq$. Then $D(k, pq) \mid D(k^p, q)$.*

**PROOF** By Definition 6.2, $D(k, pq) = L^{pq-1}(K_k^*)$. By Lemma 6.3, we have

$$L^{pq-1}(K_k^*) \mid L^{q-1}(K_k^*) \otimes L^{(p-1)q-1}(K_k^*)$$

and

$$L^{pq-1}(K_k^*) \mid L^{q-1}(K_k^*) \otimes L^{q-1}(K_k^*) \otimes L^{(p-2)q-1}(K_k^*)$$

by Lemma 6.1 and 6.3. Continuing to expand on right-hand side of the vertical bar, we have

$$D(k, pq) \mid \underbrace{L^{q-1}(K_k^*) \otimes L^{q-1}(K_k^*) \otimes \cdots \otimes L^{q-1}(K_k^*)}_{p \text{ times}}.$$

By Lemma 6.2, the RHS is isomorphic to $L^{q-1}(K_k^* \otimes K_k^* \otimes \cdots \otimes K_k^*)$ which is isomorphic to $L^{q-1}(K_{k^p}^*)$. But this is equal to $D(k^p, q)$ by Definition 6.2. Hence we conclude that $D(k, pq) \mid D(k^p, q)$. ∎

Theorem 6.8 has many applications in studying combinatorial and computational properties of de Bruijn digraphs. We only list two corollaries which are immediate from the theorem.

**COROLLARY 6.2**

*Any de Bruijn digraph $D(k^p, q)$ with $t$ faulty arcs where $t < k^{p-1}$ contains a de Bruijn digraph $D(k, pq)$ with no faulty arc.*

**COROLLARY 6.3**

*The de Bruijn digraph $D(k^p, q+1)$ can be decomposed into $k^{p-1}$ node-disjoint copies of $D(k, pq+1)$.*

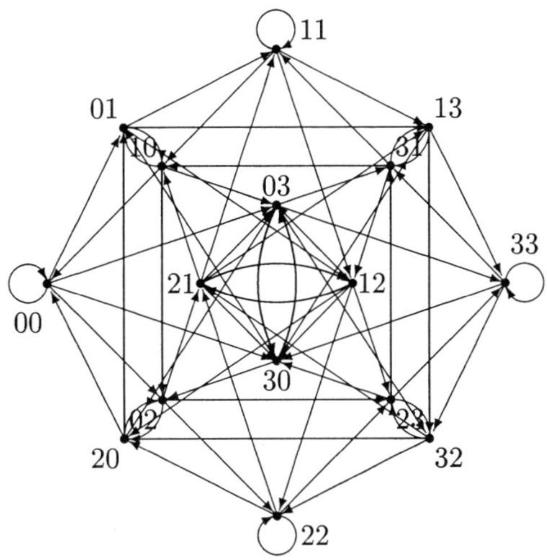


**FIGURE 6.6**
**An isomorphic factorization of $D(4,2)$ into $D(2,4)$**

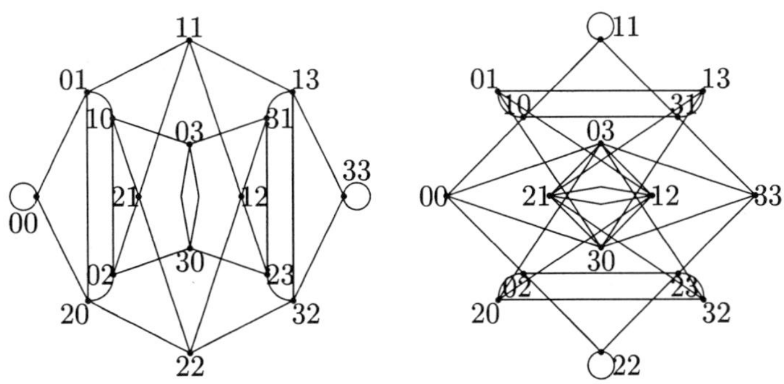


**FIGURE 6.7**
**Factors of $D(4,2)$ isomorphic to $D(2,4)$**

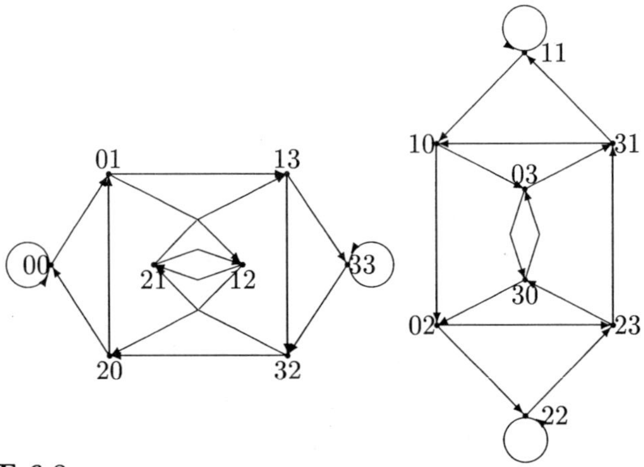


**FIGURE 6.8**
**Two copies of node-disjoint subdigraphs of $D(4,2)$ isomorphic to $D(2,3)$**

Figure 6.6 shows an isomorphic factorization of $D(4,2)$ into $D(2,4)$ while Figure 6.7 exhibit these two factors of $D(4,2)$ which are isomorphic to $D(2,4)$ (with retained edge-directions). Figure 6.8 shows two node-disjoint subdigraphs of $D(4,2)$ which are isomorphic to $D(2,3)$.

When $k = 2, p = 1$, and $q = 2$, Corollary 6.3 states that taking 2 copies of $D(2,3)$ and adding appropriate 32 arcs would result in the digraph $D(4,2)$. Recalling Theorem 6.7, we see that 2 copies of $H$ ($D(2,3) - S = D_3(S)$) together with other 20 arcs would form $D(2,4)$. The thrust of using the building block in Theorem 6.7 is that it is universal as long as the set $S$ can be found. It also builds up binary de Bruijn digraph $D(2,n)$ vertically. Although the building block (the factor subdigraph) in Corollary 6.3 is not universal, one advantage is that any de Bruijn digraph $D(k,n)$ with $n = pq + 1$ can serve as a building block (factor to form a larger de Bruijn digraph $D(k',n')$ where $k'$ is always greater than $k$). In fact, if $k' = k$, then $p = 1$ and $n = q + 1$. Hence $n' = n$ and $D(k',n') = D(k,n)$.

The building block construction in Theorem 6.7 and other work have been used to construct a very large Viterbi decoder (whose floorplan is the digraph $D(2,13)$) which will be used on NASA's Galileo mission. The decomposition and construction in Corollary 6.3 has yet, as far as we know, to be designed and implemented.

## 6.2 Shuffle-Exchange Digraphs

In this section, we study $k$-ary $n$-dimensional shuffle-exchange digraph which was defined by Stone [649] in the binary case. The binary shuffle-exchange digraphs have been studied extensively. See, for example, the excellent treatment contained in Leighton's book [401].

Let $a_1 a_2 \cdots a_n$ be a sequence of $n$ numbers from $[0, k-1]$. We have:

**DEFINITION 6.8** *The $k$-ary $n$-dimensional shuffle-exchange digraph $S(k,n)$ is defined to be the digraph where each node $a_1 a_2 \cdots a_n$ corresponds to a unique $k$-ary $n$-bit number. There is an arc from node $a = a_1 a_2 \cdots a_n$ to node $b = b_1 b_2 \cdots b_n$ if either:*

*(i) $b$ is a left cyclic shift of $a$ (shuffle edge); or*

*(ii) $a$ and $b$ differ in exactly the last position (exchange edge).*

It is clear that the shuffle-exchange digraph $S(k,n)$ has $k^n$ nodes and $k^{n+1}$ arcs. Condition (i) states that the node $a = a_1 a_2 \cdots a_n$ is adjacent to $a_2 a_3 \cdots a_n a_1$ by a shuffle-edge and that it is adjacent to $k$ other nodes $a_1 a_2 \cdots a_{n-1} x$, where $x \in [0, k-1]$ and $x \neq a_n$ by exchange of edges in Condition (ii). $S(k,n)$ is a $k$-regular digraph with self-loop existing at any node which has the same digit at all $n$-bit positions. There is another way to define $S(k,n)$ as node labeled with numbers in $Z_N$ where $N = k^n$ (Hsu and Liu [280]). Let $k$ and $n$ be positive integers with $k \leq N$ where $N = k^n$. Let $p = \lceil N/k \rceil = \lceil k^{n-1} \rceil$. We have:

**DEFINITION 6.9** *The $k$-ary $n$-dimensional shuffle-exchange digraph $S(k,n)$ is the digraph with nodes labeled as elements in $[0, N-1]$ where $N = k^n$. Each node $x \in [0, N-1]$ with $ip \leq x < (i+1)p$ for some $i$ and $x \equiv b \bmod k$ is adjacent to:*

*(i) $kx + i \bmod N$ (the shuffle edge); and*

*(ii) $x + j$, where $j \in [-b, -1] \cup [1, k-1-b]$ (the exchange edges).*

We illustrate Definitions 6.8 and 6.9 using some examples. Figure 6.9 is $S(2,3)$ using Definition 6.8 while Figure 6.10 depicts $S(2,3)$ using Definition 6.9. Figure 6.10(b) is a redrawing of Figure 6.10(a) $i$th node placed on a cyclic position. Each of these two definitions has its advantages. For example, using alphabets in $[0, k-1]$ as we stated in Definition 6.8 enables us to see the shifting process of alphabets across positions while the use of numbers in $[0, N-1]$ where $N = k^n$ in Definition 6.9 allows us to understand more on the adjacency of nodes as addition of numbers. In general, it can be said that Definition 6.8 is more combinatorial while Definition 6.9 is more algebraic.

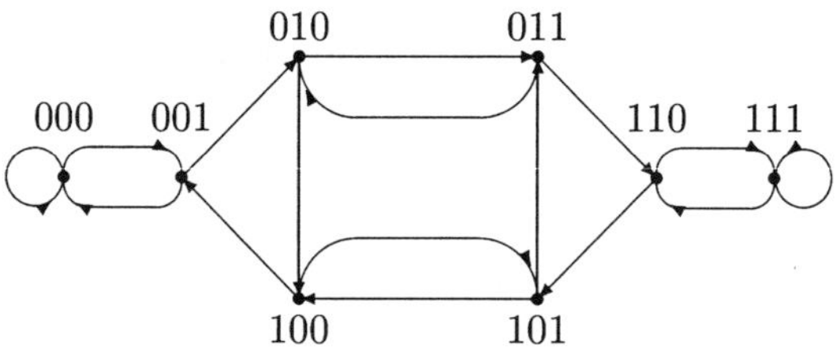


**FIGURE 6.9**
**$S(2,3)$ using Definition 6.8**

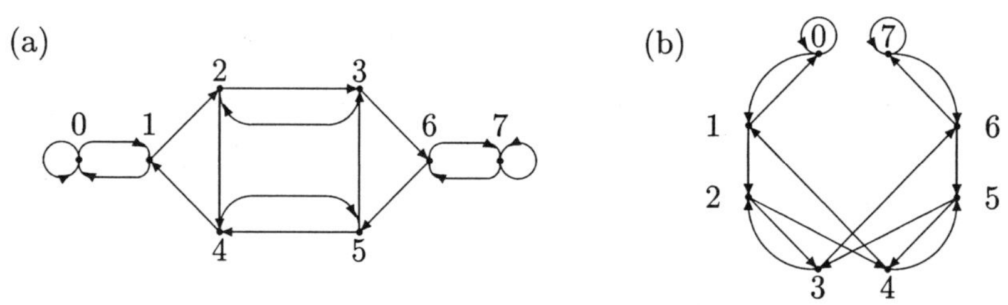


**FIGURE 6.10**
**$S(2,3)$ using Definition 6.9**

It is interesting to note that in a shuffle-exchange digraph $S(k,n)$, the nodes are grouped into $k^{n-1}$ $k$-complete symmetric digraphs (as building blocks). In fact, as we will see soon, a shuffle-exchange digraph $S(k,n)$ will lead to a de Bruijn digraph $D(k,n-1)$ if we identify (or shrink) each of the $k^{n-1}$ $k$-complete symmetric digraphs as a node. This transformation can be understood in a clearer way by examining conditions (i) and (ii) in Definition 6.9. When shrinking each of the $k$-complete symmetric digraphs into one supernode, condition (ii) is no longer needed and should be combined with condition (i). In this case, Definition 6.9 is the same as Definition 6.1 which defines a de Bruijn digraph. This process can be reversed to produce a shuffle-exchange digraph when one is given a de Bruijn digraph. We summarize this two-way transformation as follows.

**REMARK 6.1** Given a shuffle-exchange digraph $S(k,n)$ with nodes labeled as $a = a_1a_2\cdots a_n, a_i \in [0,k-1]$, there exists a de Bruijn digraph $D(k,n-1)$ with nodes labeled as $a_1a_2\cdots a_{n-1}*$ (where $*$ is the wildcard notation for any number in $[0,k-1]$) which we denote as $\bar{a} = a_1a_2\cdots a_{n-1}$. Two nodes $\bar{a} = a_1a_2\cdots a_{n-1}$ and $\bar{b} = b_1b_2\cdots b_{n-1}$ are adjacent if there exist two nodes $a = a_1a_2\cdots a_{n-1}\alpha$ and $b = b_1b_2\cdots b_{n-1}\beta$, where $\alpha, \beta \in [0,k-1]$, in $S(k,n)$ and $a$ is adjacent to b. Conversely, given a de Bruijn digraph $D(k,n)$ with nodes labeled as $\bar{a} = a_1a_2\cdots a_n, a_i \in [0,k-1]$, there exists a shuffle-exchange digraph $S(k,n+1)$ with nodes labeled as $a = a_1\cdots a_na_{n+1}$ where $a_{n+1} \in [0,k_1]$. Node $a = a_1\cdots a_na_{n+1}$ is adjacent to node $b = a_2a_3\cdots a_{n+1}a_1$ in $S(k,n+1)$ if $\bar{a} = a_1a_2\cdots a_n$ is adjacent to $\bar{b} = a_2a_3\cdots a_{n+1}$ and $a = a_1a_2\cdots a_{n-1}\alpha$ is adjacent to node $b = b_1b_2\cdots b_{n-1}\beta$, for all $\beta \in [0,k-1]$ and $\beta \neq \alpha$. ∎

Figure 6.11 exhibits the shuffle-exchange digraph $S(3,3)$ where each cluster of nodes $a_1a_2\alpha, \alpha \in [0,2]$ is a complete symmetric digraph of 3 nodes, while its transformed de Bruijn digraph $D(3,2)$ is shown in Figure 6.12.

The close relationship between shuffle-exchange digraph and de Bruijn digraph as described in Remark 6.1 demonstrates that these two networks are very similar to each other. This is (one of) the reason(s) that these two networks are referred to as shuffle-oriented networks (see [180, 282] and [361]) because they all involve the shuffle operation. This close correspondence may help us study their combinatorial property such as diameter. However, it is not straightforward, and in fact not true as we will see later, that the existence of a Hamiltonian cycle in the de Bruijn digraph $D(k,n)$ would lead to the existence of a Hamiltonian cycle in the corresponding shuffle-exchange digraph $S(k,n+1)$.

Now we turn to computing the diameter of the shuffle-exchange digraph $S(k,n)$. Consider two nodes $a = a_1a_2\cdots a_n$ and $b = b_1b_2\cdots b_n$, where $a_i, b_i \in [0,k-1]$, in $S(k,n)$. We construct a path from $a$ to $b$ by first traversing an exchange edge (if necessary) to get the edge $a_1a_2\cdots a_{n-1}a_n \to a_1a_2\cdots a_{n-1}b_1$. Then we continue to convert $a_i$ to $b_{i+1}$ for each $i$ during the subsequent shuffle-exchange operations:

$$a_i\cdots a_{n-1}b_1\cdots b_i \to a_{i+1}\cdots a_{n-1}b_1\cdots b_ia_i \to a_{i+1}\cdots a_{n-1}b_1\cdots b_{i+1}.$$

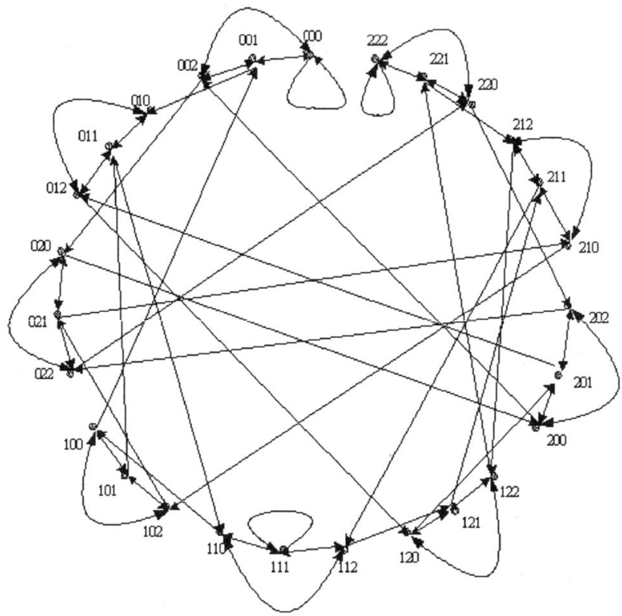


**FIGURE 6.11**
**Shuffle-exchange digraph $S(3,3)$**

The resulting path has just $n-1$ shuffle edges and at most $n$ exchange edges, for a total of at most $2n-1$ edges. Hence the diameter of $S(k,n)$ is less than or equal to $2n-1$. It is easy to show that the distance $d(\mathbf{0},\mathbf{1})$ between nodes $\mathbf{0} = 000\cdots0$ and $\mathbf{1} = 111\cdots1$ in $S(k,n)$ is $2n-1$. Hence $d(S(k,n)) = 2n-1$.

The problem of finding a Hamiltonian cycle in $S(k,n)$ turns out to be less easy. From Remark 6.1, $S(k,n)$ can be transformed to the de Bruijn digraph $D(k,n-1)$. However, the existence of a Hamiltonian cycle in $D(k,n)$ does not guarantee the existence of a Hamiltonian cycle in $S(k,n)$. By closely examing the relationship between and structures in these two networks, we aim to find conditions imposed on a Hamiltonian cycle in $D(k,n-1)$ so that this cycle can be transformed to a Hamiltonian cycle in $S(k,n)$. Let $C = e_1e_2e_3\cdots e_m$ be a closed walk in $D(k,n)$ that traverses all $k^n$ nodes. By Remark 6.1, we can transform

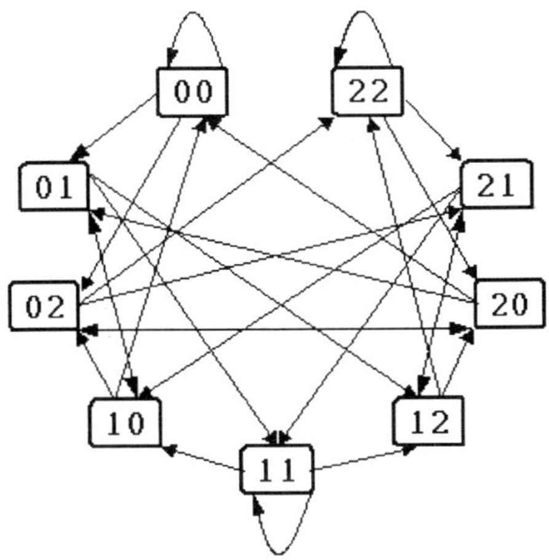


**FIGURE 6.12**
**The de Bruijn digraph for $S(3,3)$**

each $e_i$ to $e'_i$ in $S(k, n+1)$. Let

$$C' = \{e'_1, e'_2, e'_3, \cdots, e'_m\}.$$

The following theorem holds.

***THEOREM 6.9***
*Let $C$ and $C'$ be as defined above. If $C'$ satisfies the following conditions:*

*(1) There do not exist two distinct $i$ and $j$ such that $e'_i$ and $e'_j$ are adjacent to the same node in $S(k, n+1)$,*

*(2) For each $i \in [0, p-1]$, there exists $j$ such that $e_j$ goes to supernode $(i)$ in $D(k, n)$, $e_{j+1}$ leaves $(i)$ and $e'_j = (x, y)$, $e'_{j+1} = (z, w)$, for distinct $y$ and $z$ in $GS(k, kp)$,*

*then we can construct a Hamiltonian cycle in $S(k, p+1)$ based on $C$.*

As examples, we first take $S(5,2)$ which is related to $D(5,1)$. In $D(5,1)$, we have the following closed walk: $(0) \to (3) \to (2) \to (1) \to$

$(4) \rightarrow (0)$. We then extend this closed walk to a Hamiltonian cycle in $S(5,2)$ as follows:

$$\begin{array}{ll} 00 \rightarrow 01 \rightarrow 02 \rightarrow 03 & (0) \\ \rightarrow 30 \rightarrow 31 \rightarrow 33 \rightarrow 34 \rightarrow 32 & (3) \\ \rightarrow 23 \rightarrow 20 \rightarrow 22 \rightarrow 24 \rightarrow 21 & (2) \\ \rightarrow 12 \rightarrow 10 \rightarrow 11 \rightarrow 13 \rightarrow 14 & (1) \\ \rightarrow 41 \rightarrow 42 \rightarrow 43 \rightarrow 44 \rightarrow 40 & (4) \\ \rightarrow 04 \rightarrow 00 & (0). \end{array}$$

In general, the Hamiltonian cycle $(0) \rightarrow (k-2) \rightarrow (k-3) \rightarrow \cdots \rightarrow (1) \rightarrow (k-1) \rightarrow (0)$ in $D(k,1)$ satisfies the two conditions in Theorem 6.9. So we can extend it to a Hamiltonian cycle in $S(k,2)$ for $k > 1$.

***THEOREM 6.10***
*The shuffle-exchange digraph $S(k,2)$ is Hamiltonian for $k > 1$.*

We note that it has been shown recently by Feldmann and Mysliwietz [194] that the undirected shuffle-exchange graph $US(2,n)$ has a Hamiltonian path for all $n > 1$.

## 6.3 Circulant Networks

In this section, we discuss circulant networks. We force on circulant networks defined on cyclic addition groups $Z_N$ where $N$ is a positive integer greater than or equal to 3. The study of circulant networks and adjacency matrices (circulant matrices) dates back to the 1930s and 40s in the investigation of periodic or multiple symmetrical systems with application in the theory of crystal structure (see [2]).

Let $G(N; s_1, s_2, \cdots, s_k)$ be the digraph with $N$ nodes, labeled with elements from $Z_N$, and each node $x$ is adjacent to $k$ other nodes $x + s_i$, $i = 1, 2, \cdots, k$. The network $G(N; \pm s_1, \pm s_2, \cdots, \pm s_k)$ is the undirected version of $G(N; s_1, s_2, \cdots, s_k)$ where each node $x$ is adjacent to $2k$ other nodes $x \pm s_i$, $i = 1, 2, \cdots, k$. The addition or subtraction is taken modulo $N$. In this book, we are mainly concerned with the case $s_1 = 1$.

For a given digraph $G(N; s_1, s_2, \cdots, s_k)$, let $d(G(N; s_1, s_2, \cdots, s_k))$ be its diameter. Let $d(N)$ be the minimum of $d(G(N; 1, s_2, \cdots, s_k))$ among all $s_i's, i = 2, \cdots, k$. We are interested in finding $s_2, s_3, \cdots, s_k$ so that the diameter would achieve the minimum $d(N)$. It is helpful to find the upper and lower bounds for $d(N)$. In particular, if we find some values of $s_2, s_3, \cdots, s_k$ so that the diameter is equal to the lower bound, then we are sure this diameter is optimal. To simplify the matters at issue, we look at the case $k = 2$.

Let $d(N) = \min_s\{d(G(N;1,s))\}$. It is simple to get the upper bound of $2\sqrt{N} - 2$ by taking $N = t^2$. It follows that the node $N-1$ can be reached by $(t-1)$ steps of $s$ and $(t-1)$ steps of one. Hence $d(0, N-1) = (t-1) + (t-1) = 2t - 2 = 2\sqrt{N} - 2$. We claim that any other nodes can be reached in less than $2t - 2$ steps. Hence $d(N) \leq 2\sqrt{N} - 2$.

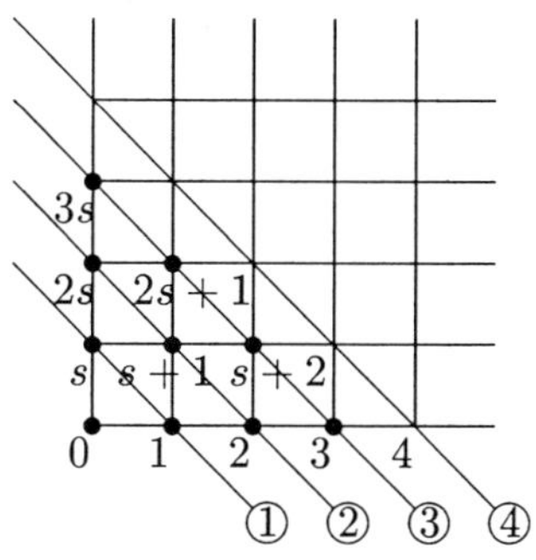


**FIGURE 6.13**
**Circulant network** $G(N;1,s)$

In the network $G(N;1,s)$, starting from the node 0, we can reach $t+1$ nodes at distance $t$ (see the diagonal line in Figure 6.13). Hence with diameter $d$ from the node 0, we can reach totally $1+2+3+\cdots+(d+1)$ nodes. This sums to $\frac{(d+1)(d+2)}{2}$ which should be greater than or equal to the number of nodes in the network $G(N;1,s)$. We have

$$\frac{(d+1)(d+2)}{2} \geq N$$

and

$$d \geq \sqrt{2N} - \frac{3}{2}\,.$$

This lower bound is not tight. In fact, Wong and Coppersmith [706] improved this lower bound to $\sqrt{3N} - 2$, a bit better but it is tight for some $N$.

**THEOREM 6.11**
*Let* $d(N) = \min_s\{d(G(N;1,s))\}$*, where* $s$ *runs over all integers in* $[1, N-1]$*. Then we have*

$$\left\lceil \sqrt{3N} \right\rceil - 2 \leq d(N) \leq 2\sqrt{N} - 2\,.$$

As we stated before, the lower bound in Theorem 6.11 is tight for some $N$. For example, $d(G(16;1,7)) = 5$ and $\lceil\sqrt{3N}\rceil - 2 = 5$ for $N = 16$. Hence it is optimal. This example is typical and rather educational in the sense that it is often mistaken that by taking $s = \sqrt{N}$ we might get an optimal diameter. As the two figures in Figure 6.14 show, we have $d(16;1,\sqrt{16}) = 6$, but $d(16;1,7) = 5$.

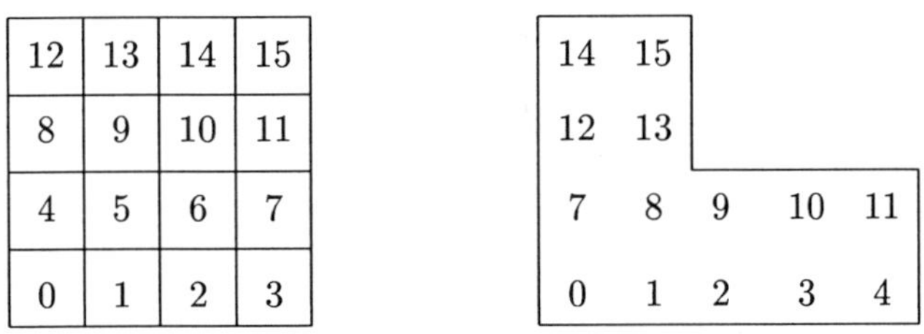


(a) $d = 6$ (b) $d = 5$

**FIGURE 6.14**
$G(16;1,4)$ **and** $G(16;1,7)$

Another example is obtained by Erdos and Hsu [188] in which $N = 3t^2 + 3t$, $s = 3t + 2$, and $d(G(N;1,s)) = \sqrt{3N} - 2$.

In general, the following is obtained by Wong and Coppersmith [706]:

**THEOREM 6.12**
*Let* $d(N)$ *be the minimum of the diameters of the networks* $G(N; 1, s_2, \cdots, s_k)$ *where* $s_i \in (1, N-1]$*. Then we have*

$$(k!N)^{1/k} - \frac{1}{2}(k+1) \leq d(N) \leq kN^{1/k} - k\,.$$

Although the lower bound in Theorem 6.12 was found in the 1990s, the optimal solutions were not obtained until 1987 (see [205, 291]) and 1990s (see [58, 188, 279, 203]). The upper bound $3\sqrt[3]{N}-3$ in Theorem 6.12 when $k=3$ has been improved a bit to $(2.7)\sqrt[3]{N}-3$ (see [188]) and later by Hsu and Jia [279] to $\sqrt[3]{16}\sqrt[3]{N}-3$. The lower bound $\sqrt[3]{6N}-3$ has been improved by Fiduccia, Forcade, and Zito [203] to $\sqrt[3]{25/3}\sqrt[3]{N}-3$ in the Abelian case (recall that $Z_N$ is a cyclic group and, of course, is an Abelian group also).

In the case when $k=2$, let $lb(N)=\lceil\sqrt{3N}\rceil-2$. As we know, the lower bound is achieved for infinitely many values of $N$. On the other hand, there are infinitely many values of $N$ for which $lb(N)$ can never be achieved. For example, $d(3t^2;1,3t+2)=d(3t^2)$, but $lb(3t^2)=\lceil 9t^2\rceil-2=3t-2$ (see [188, 205, 291]). Therefore, it is interesting to study the function $f(N)=d(N)-lb(N)$. Computer search shows that there are several $N$'s, $1\leq N\leq 30000$, for which $d(N)-lb(N)=3$; and there are three $N$'s, $1\leq N\leq 75000$ such that $d(N)-lb(N)=4$. These are $N=53749$ for which $d(N)=d(G(N;1,985))=404$, $N=64729$ for which $d(N)=d(G(N;1,394))=443$, and $N=69283$ for which $d(N)=d(G(N;1,1764))=458$. This gap is confirmed by Coppersmith (private communication to one of us) who showed the existence of an infinite set of values of $N$ for which $d(N)$ is as large as $\sqrt{3N}+c(\log N)^{1/4}$ for some constant $c$.

Infinitely many values of $N$ and families of digraph $G(N;1,s)$ for which $d(N)$ is known and often equal to the lower bound $lb(N)$ have been found following two different approaches. The problem has been visualized as a geometric pattern by Fiol et al. [205]. When the link $(x,x+1)$ is represented by a horizontal segment and the link $(x,x+s)$ by a vertical one, the distance between two nodes is calculated by adding the number of horizontal and vertical segments between them. It is then possible to represent $G(N;1,s)$ by a tessellation of the plane. The network $G(9;1,7)$ is depicted in Figure 6.15. The optimization problem consists of finding a title of $N$ unit squares which periodically tessellates the plane with minimum diameter, then searching for a title with the values of $s$ (or values of $s_1,s_2$ in general) that enable this construction. It was reported in [205] that given $N$, there are different pairs of $s_1,s_2$ which give optimal diameter and one of these pairs is with $s_1=1$ with few exceptions. The first such exception is when $N=450$. In this case, the optimal diameter is achieved with $s_1=2,s_2=185$

and gives $d(G(450;2,185)) = 35$. However, the best diameters with $s_1 = 1$ give $s = 59$, $d(G(450;1,59)) = 36$ and $s = 34$, $d(G(450;1,34)) = 37$. Infinitely many values of $N$ are obtained with $d(N) = lb(N)$. For example, $d(G(3t^2+t;1,3t)) = d(G(3t^2+2t;1,-3t)) = 3t-1$, $d(G(3t^2+2t+1;1,3t+1)) = d(G(3t^2+3t+1;1,3t+2)) = d(G(3t^2+4t+1;1,-3t-2)) = 3t$, and $d(G(3t^2+4t+2;1,3t+3)) = d(G(3t^2+5t+2;1,3t+4)) = d(G(3t^2+6t+2;1,-3t+4)) = 3t+1$. Similar results have been obtained by Hwang and Xu [291].

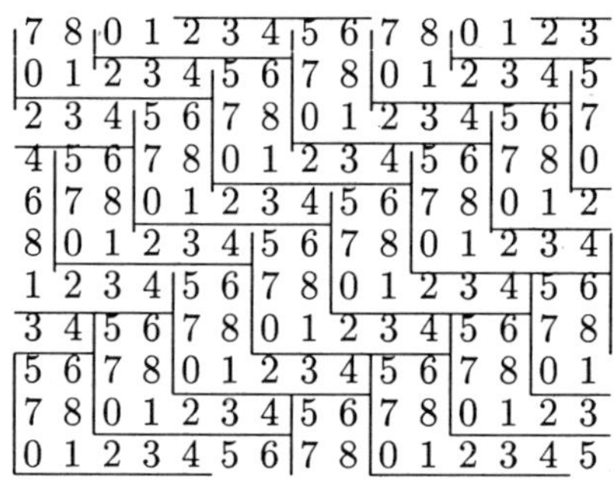


**FIGURE 6.15**
**Tessellation for** $G(9;1,7)$

The approach by Erdos and Hsu [188] is rather different. Since the nodes are labeled by elements from $[0, N-1]$ modulo $N$, these nodes are placed evenly on a circle. In order to find $s$ such that $G(N;1,s)$ has diameter equal to $d(N)$, we travel around the circle at steps of 1 or $s$ counterclockwise (or the other way around as long as it follows a consistent pattern). The problem is how to find $s$ so that all the nodes on the circle can be reached in as few steps as possible. Erdos and Hsu [188] showed that if we pass the node 0 the first time using the jump size $s$ and land at a node which is almost $(3/2)s$ (i.e., $t \cdot s \equiv (3/2)s \bmod N$ for some $t$ and $(t-1) \cdot s < N$), then we only have to travel around the circle at most two times using jump size 1 and $s$. In this case, the lower bound $lb(N) = \lceil\sqrt{3N}\rceil - 2$ can be reached for infinitely many values of $N$. Figure 6.16 gives an example with $N = 3t^2+3t$, $d(G(N;1,3t+2)) = 3t = d(N)$ and $t = 3$. Hence we have $N = 36$ and $d(36) = 9$. It is interesting to note that using the heuristic method as described in Figure 6.14(a), $d(G(36;1,\sqrt{36})) = 10 > d(36) = 9$. All the values of $N < 50$ which do not belong to any infinite class obtained by Erdos and Hsu [188] are $\{35, 38, 39, 42, 45, 46, 49\}$.

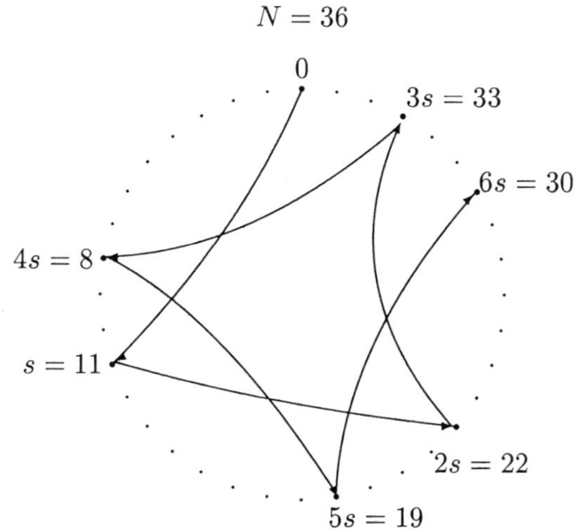


**FIGURE 6.16**
$G(36; 1, 11)$ **with** $d(G) = 9 = d(36)$

The routing problems and fault tolerance properties for $G(N; 1, s)$ and $G(N; s_1, s_2)$ were studied by Erdos and Hsu [188], and Fiol et al. [205]. They also showed that the diameter of $G(N; 1, s)$ after a node failure is at most $d(G(N; s_1, s_2)) + 1$.

As to the connectivity of $G(N; 1, s_2, s_3, \cdots, s_k)$, Hamidoune [249] uses the concept of an atom to prove the following:

***THEOREM 6.13***
*Let $N$ be an integer. Let $1 = s_1 < s_2 < \cdots < s_k < N$ be a sequence of integers such that*

$$s_{i+1} - s_i \geq min\{2, s_i - s_{i-1}\}$$

*for $2 \leq i \leq k-1$. Let $S = \{s_i\}, 1 \leq i \leq k \bmod N$. Then the connectivity of $G(N; 1, s_2, s_3, \cdots, s_k)$ is equal to $k$.*

The above theorem, in fact, gives a sufficient condition that the circulant network $G(N; 1, s_2, s_3, \cdots, s_k)$ has optimal connectivity.

## 6.4 Cayley Graphs

Let $G$ be a group and $S$ a generating set of $G$. The *Cayley digraph* $I(G, S)$ is the digraph with node set $G$ and a set containing an arc from $g$ to $gs$ wherever $g \in G$ and $s \in S$. The *Cayley graph* $I(G, S)$ is the undirected graph with node set $G$ and edge set containing an edge from $g$ to $gs$ and $g$ to $gs^{-1}$ whenever $g \in G$ and $s \in S$. Without confusion, we will state explicitly which of the cases is involved. We sometimes use node and vertex interchangeably.

The study of Cayley graphs, under the name Cayley color-group or Cayley diagrams of groups, can be dated to the 1940s (see [135]). In his study of strongly fixed-vertex-free graphs and later vertex-transitive graphs [596, 595], Sabidussi observed that graphs with vertex-transitive automorphism group are intimately related to Cayley graphs. Let $X$ be a graph and $G(X)$ be its automorphism group. Let $x, y \in V(X)$ and $G'$ be a subgroup of $G(X)$. The vertices $x$ and $y$ are said to be $G'$*-similar* if there is a $f \in G'$ such that $f(x) = y$. We shall say that a subgroup $G'$ of $G(X)$ *acts transitively* on $V(X)$. The graph $X$ is called *vertex-transitive* if there is a $G' \in G(X)$ which is vertex-transitive.

Let $G$ be a group and $S$ be a generating set of $G$ without the identity element. Let $X(G, S)$ be the Cayley graph as defined above. Since the group of left translations of $G$ is a subgroup of $G(X(G, S))$, $X(G, S)$ is vertex-transitive. However, we note that the converse of this is not true. That is, a vertex-transitive graph may not be a Cayley graph, as the Petersen graph of ten vertices shows in Figure 6.17.

Let $H$ be a subgroup of the group $G$. We define an equivalence relation on $G$ by the left cosets of $H$. Consider this equivalence relation $\sim$ on $V(X(G, S))$ and form the graph $X(G, S)/\sim$ as follows (denote $X(G, S)/\sim$ as $X(G, S)/H$):

$$V(X(G,S)/H) = \{aH \mid a \in G\}\,,$$
$$E(X(G,S)/H) = \{\{aH, bH\} \mid aH \neq bH, aH \cap bH\bar{S} \neq \phi\}\,,$$

where $\bar{S} = S \cup S^{-1}$. If $H$ is a normal subgraph in $G$, then $X(G, S)/H = X(G', S')$, where $G' = G/H$ and $S' = \{sH \mid s \in S\}$. The following theorem proved by Sabidussi [596] characterizes the vertex-transitive graphs.

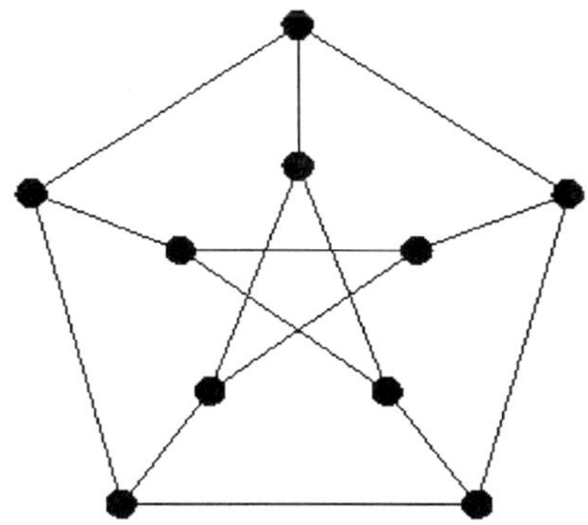

**FIGURE 6.17**
**Petersen Graph** $P(10)$

***THEOREM 6.14***

*Let $X$ be a connected vertex-transitive graph. Let $G'$ be a vertex-transitive subgroup of $G(X)$. There exists a set $S$ of generators of $G(X)$ and a subgroup $H$ of $G$ such that $S \cap H = \emptyset$ and*

$$X \equiv X(G,S)/H.$$

*Conversely, if $G$ is a group, $S$ a set of generators of $G$ without the identity, and $H$ a subgroup of $G$, then $X(G,S)/H$ is a connected vertex-transitive graph.*

One of the important advantage of combining groups and graphs together is that the group structure would allow us to examine the graph more closely in terms of its group characteristics. There are several implications. When Cayley graphs are used as interconnection networks (as proposed by Akers and Krishnamurthy [20]), the inherited properties such as symmetry and scalability can be utilized to perform routing, communication, and computation in a very systematic and efficient way. Moreover, the structure in the group can show us not only the possibility but also the limitation of doing tasks on the networks. Here we mention two examples. The first is that Sabidussi [596] and Chao [100] have shown that there is no vertex-transitive graph with an Abelian automorphism group with more than two vertices. So, we should not spend any time looking for a vertex-transitive graph with an Abelian automorphism group. The second example deals with vertex-transitive graphs with prime number of vertices. It was shown by Turner [671] that a connected graph with a prime number of vertices is vertex-transitive if and only if it is a circulant graph. This gives us indication that networks constructed from a vertex-transitive graph with a prime number of vertices is not only a Cayley graph but also a circulant graph as the latter is a special case of the former. We have included circulant graphs in Section 6.3.

As discussed before, Cayley graphs and vertex-transitive graphs have been studied extensively from the algebraic and in particular, group theoretical, point of view. More recently, there has been tremendous interest in these graphs due to their applications to the construction of interconnection networks for parallel and communication systems. We refer the reader to the following two survey papers: Babai et al. [35] for the diameter of finite groups and Curran and Gallian [138] for Hamiltonian cycles and paths in Cayley graphs and digraphs. Annexstein and Baumslag [30] present lower bound on the diameter of nilpotent groups and upper bounds on the size of bisection of Cayley graphs of solvable groups.

Hamidoune has studied the connectivity of Cayley digraphs [249] and cutsets in Abelian Cayley graphs [250].

We now turn our attention to some special classes of Cayley digraphs and graphs which have been used as interconnection network in parallel computer architectures and other distributed systems.

Let $V$ be a finite set and $T$ be a set of permutations of $V$. Let $|T| = h$. Then we can construct a digraph $X = X(V,T)$ with $V(X) = V$ and $E(X) = \{(a, \alpha(a)) \mid a \in V, \alpha \in T\}$.

When we labeled each arc $(a, \alpha(a))$ in $X$ as $\alpha$, we call $X(V,T)$ a *permutation labeled digraph* (called group action graph in [31] and abbreviated as PL digraph). We note that $T$ may not consist of all permutations of $V$. That is, $T$ may not be equal to the symmetric group $S_v$ where $|V| = v$. Moreover, $T$ may not be a group at all if operation in $T$ is understood to be the composition of permutations. To illustrate this, we give three examples.

***Example 6.1***

Let $V = [1, 10]$ and $T = \{\alpha, \beta, \gamma\}$ where $\alpha, \beta, \gamma$ are defined as follows:

$$\alpha = \begin{pmatrix} 1 & 2 & 3 & 4 & 5 & 6 & 7 & 8 & 9 & 10 \\ 2 & 3 & 4 & 5 & 1 & 9 & 10 & 6 & 7 & 8 \end{pmatrix}$$

$$\beta = \begin{pmatrix} 1 & 2 & 3 & 4 & 5 & 6 & 7 & 8 & 9 & 10 \\ 5 & 1 & 2 & 3 & 4 & 8 & 9 & 10 & 6 & 7 \end{pmatrix}$$

$$\gamma = \begin{pmatrix} 1 & 2 & 3 & 4 & 5 & 6 & 7 & 8 & 9 & 10 \\ 6 & 7 & 8 & 9 & 10 & 1 & 2 & 3 & 4 & 5 \end{pmatrix}$$

□

***Example 6.2***

Let $V = [0, 7]$ and $T = \{\alpha, \beta\}$ where $\alpha, \beta$ are defined as:

$$\alpha = \begin{pmatrix} 0 & 1 & 2 & 3 & 4 & 5 & 6 & 7 \\ 0 & 2 & 4 & 6 & 1 & 3 & 5 & 7 \end{pmatrix}$$

$$\beta = \begin{pmatrix} 0 & 1 & 2 & 3 & 4 & 5 & 6 & 7 \\ 1 & 3 & 5 & 7 & 0 & 2 & 4 & 6 \end{pmatrix}$$

□

***Example 6.3***

Let $V = \{0, 1, 2, 3, 4, 5, 6\}$ and $T = \{\alpha, \beta\}$ where

$$\alpha = \begin{pmatrix} 0 & 1 & 2 & 3 & 4 & 5 & 6 \\ 1 & 2 & 3 & 4 & 5 & 6 & 0 \end{pmatrix}$$

$$\beta = \begin{pmatrix} 0 & 1 & 2 & 3 & 4 & 5 & 6 \\ 2 & 3 & 4 & 5 & 6 & 0 & 1 \end{pmatrix}$$

□

We observe that the PL digraphs are regular with indegree and outdegree equal to the cardinality $|T|$ of $T$. It is interesting to note that the undirected graph version of $X(V, \{\alpha, \beta, \gamma\})$ in Figure 6.18 is really the Peterson graph in Figure 6.17. Examples 6.1 and 6.3 are, in fact, the de Bruijn digraph $D(2, 3)$ and circulant digraph $G(7; 1, 2)$, respectively (see sections 6.1 and 6.3).

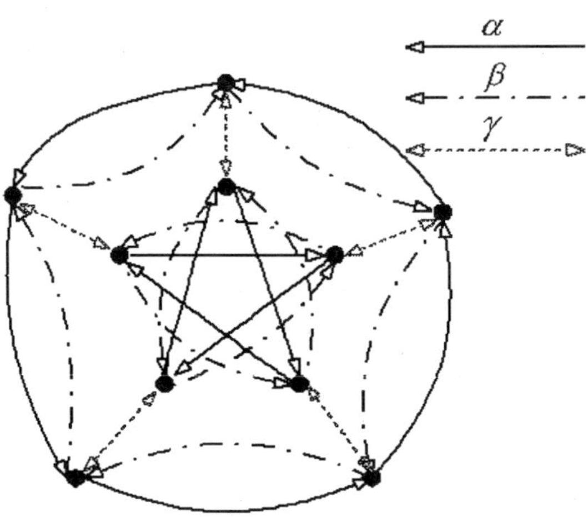


**FIGURE 6.18**
$X(V, \{\alpha, \beta, \gamma\})$

By examining the three examples more closely, we see that the Peterson digraph and the de Bruijn digraph are not Cayley digraphs, but the circulant digraph is. We will show that there is a way to construct a Cayley digraph when given a PL digraph $X(V,T)$. However, when given a Cayley digraph, it is much harder to construct more than one PL digraph.

Given a PL digraph $X(V,T)$, let $G_v(T)$ be the group generated by $T$. Hence $X(G_v(T),T)$ is a Cayley digraph with $V(X) = G_v(T)$ and $E(X) = \{(g, \alpha(g)) \mid \alpha \in T,\ g \in G_v(T)\}$.

Let $H$ be a subgroup of the group $G$. The quotient $G/H$ is defined as $G/H = \{Hg | g \in G\}$. For each $g \in G$, $Hg$ is called the *right coset of $H$ containing $g$*. It is easy to see that the cosets of $H$ partition $G$ into blocks of equal size.

The PL digraph $X(V,T)$ is a connected digraph with both the indegree and outdegree of each vertex equal to $|T|$. Hence $X(V,T)$ contains an Eulerian tour (a closed directed circuit traversing each arc exactly once). Suppose $u$ and $v$ are two nodes in $V(X)$. The sequence of labels on the Eulerian tour from $u$ to $v$ defines a permutation on $G_v(T)$ that maps $u$ to $v$. Hence the group $G_v(T)$ is transitive.

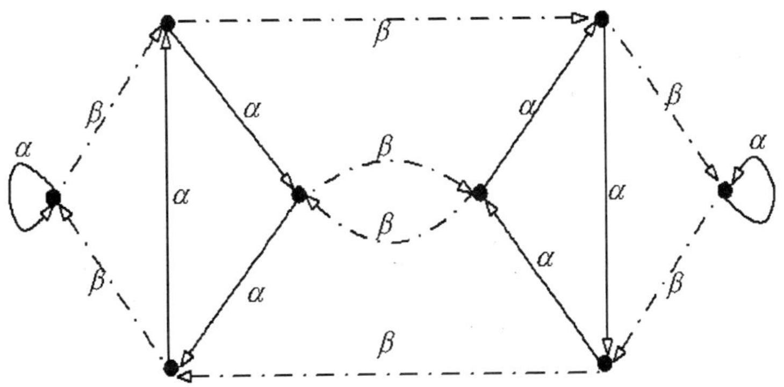


**FIGURE 6.19**
$X(V, \{\alpha, \beta\})$

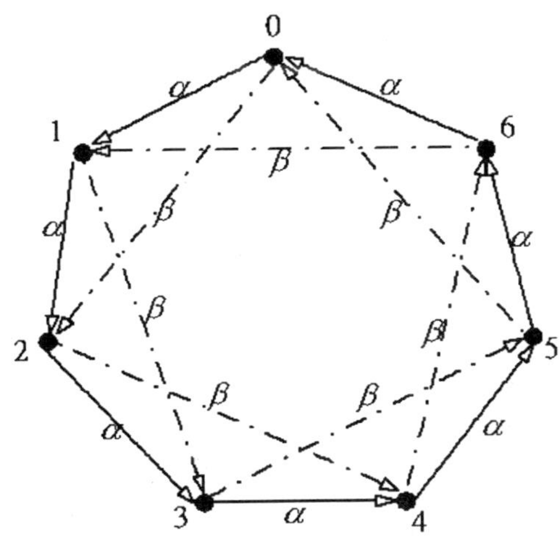


**FIGURE 6.20**
$X(V, \{\alpha, \beta\})$, $\alpha(x) = x + 1$, $\beta(x) = x + 2$

Let $V$ be a set and $G$ be a group of permutations of $V$. For each $v$ in $V$, the *stabilizer* of $v$ in $G$, denoted as $St(G;v)$, is the set of all permutations in $G$ that fix $v$. That is, $St(G;v) = \{\alpha \mid \alpha \in G, \alpha(v) = v\}$. The stabilizer $St(G;v)$, is a subgroup of $G$. Now if $G = G_v(T)$ as defined before, then $G$ is a transitive permutation group. Since in a transitive permutation group all stabilizers are conjugate, they are isomorphic to each other. For this reason, we write $St(G)$ as the *stabilizer group* of $G$.

Now let $H$ be a subgroup of the group $G = G_v(T)$. The *coset digraph* $Cos(G/H, T)$, is the PL digraph $X(G/H, T)$ with $V(X) = G/H$ and

$$E(X) = \{(Hg, Hh) \mid g \in G,\ \alpha \in T,\ \alpha(Hg) = Hh\}\ ,$$

where the permutations in $T$ are extended to act on $G/H$ so that $\alpha(Hg) = H(\alpha(g)) = Hh$. We now formally state the theorem.

***THEOREM 6.15***

*Let $X(V,T)$ be a PL digraph with the set $V$ and a set $T$ of permutations (of elements of $V$). Let $G_v(T)$ be the group generated by $T$. Then $X(G_v(T),T)$ is a Cayley digraph with $V(X) = G_v(T)$ and*

$$E(X) = \{(g, \alpha(g)) \mid \alpha \in T,\ g \in G_v(T)\}\ .$$

*Conversely, suppose we are given a set $V$, a set $T$ of permutations of elements of $V$, and the Cayley digraph $X(G_v(T),T)$. Let $H = St(G_v(T))$. Then the coset digraph $Cos(G_v(T)/H,T)$ is a PL digraph $X(G_v(T)/H,T)$ with $V(X) = G_v(G)/H$ and*

$$E(X) = \{(Hg, Hh) \mid g \in G_v(T), \alpha \in T \text{ such that } \alpha(Hg) = Hh\}\ .$$

It was shown (Theorem 3.2, [31]) that the PL digraph $X(G_v(T)/H,T)$ in Theorem 6.15, where $H = St(G_v(T))$, is in fact isomorphic to $X(V,T)$. We also observe that Theorem 6.15 is similar to Theorem 6.14 where the only difference is the class of PL digraphs vs. the class of vertex-transitive digraphs. It remains to be seen how these two classes of digraphs are related.

The *wreath product* of cyclic group $Z_d$ and cyclic group $Z_n$, denoted by $Z_d \otimes_{wr} Z_n$ and written by $Z_d \otimes Z_n$, is a group of permutations of the set

$$Z_d^n =_{def} [0, d-1]^n\ .$$

Each element in $Z_d \otimes Z_n$ is written as $\pi = (\alpha; a_1, a_2, \cdots, a_n)$, where $\alpha \in Z_n$ and $a_i \in [0, d-1]$. The element $\pi$ in $Z_d \otimes Z_n$ acts on $Z_d^n$ as follows:

$$\pi((b_1, b_2, \cdots, b_n)) = (b_{\alpha-1} + a_{\alpha-1}, b_\alpha + a_\alpha, \cdots, b_0 + a_0, \cdots, b_{\alpha-2} + a_{\alpha-2}),$$

that is, the action consists of a vector addition of $(a_1, a_2, \cdots, a_n)$ to $(b_1, b_2, \cdots, b_n)$ modulo $d$ followed by a sequence of $\alpha$ left cyclic shifts.

Our goal is to define de Bruijn digraph $D(d, n)$ (see Section 6.1) and shuffle-exchange digraph $S(d, n)$ (see Section 6.2) as PL digraphs and then show that butterfly digraph and cube-connected-cycle digraph are their associated Cayley digraphs, respectively.

As in Example 6.2, a de Bruijn digraph can be arc-labeled to become a PL digraph. The de Bruijn digraph $D(d, n)$ has node set $V = Z_d^n = [0, d-1]^n$. For each $t \in Z_d$, we define the permutation $\pi[t; d]$ of $Z_d^n$ as follows:

$$\pi(\alpha, x) = x(a_1 + t \bmod d),$$

where $a_1 \in Z_d$ and $x \in Z_d^{n-1}$. We label each arc of $D(d, n)$ of the form $(a_1 x, xb)$ with permutation $\pi[b - a_1 \bmod d; d]$. With this arc labeling by permutations, we can see that $D(d, n)$ is a PL digraph with two permutations $\pi[0; d]$ and $\pi[1; d]$ equivalent to $\alpha$ and $\beta$ permutations in Example 6.2, respectively. Let

$$T(D) = \{\pi(t; d) \mid t \in Z_d\}.$$

It can be shown that $G_v(T(D))$ is isomorphic to the wreath product $Z_d \otimes Z_n$. Moreover, we have:

***THEOREM 6.16***

*For all $d$ and $n$, the de Bruijn digraph $D(d, n)$ is isomorphic to the coset digraph*

$$Cos(G/H, T(D)),$$

*where $G = Z_d \otimes Z_n$, and $H = \{0\} \otimes Z_n$.*

The *butterfly digraph* $B(d, n)$ is defined as follows. The node set $V(B(d, n)) = Z_n \otimes Z_d^n$. It consists of the following arcs:

$$(l; a_0 a_1 \cdots a_{l-1} u a_{l+1} \cdots a_{n-1}) \to (l + 1 \bmod n; a_0 \cdots a_{l-1} v a_{l+1} \cdots a_{n-1}),$$

where $0 \le l \le n$, $u, a_i \in Z_d$, and for all $v \in Z_d$.

We can turn the butterfly digraph $B(d,n)$ into a PL digraph by using the permutation from $T(D)$ in the following way. We label the arc from node

$$(l; a_0 a_1 \cdots a_{l-1} u a_{l+1} \cdots a_{n-1})$$

to node

$$(l+1 \bmod n; a_0 a_1 \cdots a_{l-1} v a_{l+1} \cdots a_{n-1})$$

with the permutation $\pi(v-u \bmod d; d)$. Hence we can show that $B(d,n)$ is isomorphic to the Cayley digraph $Cay(G_v(T(D)), T(D))$. The following theorem is proved in [31]. It follows from Theorem 6.16 and above discussion.

### THEOREM 6.17

*The de Bruijn digraph $D(d,n)$ is a coset digraph of the butterfly digraph $B(d,n)$.*

The shuffle-exchange digraph $S(d,n)$ is a PL digraph with node set $Z_d^n$ and the arc labels are performed by the permutations $\pi(t;d)$ of $Z_d^n$. Each $a_1, a_n \in Z_d$ and $x \in Z_d^{n-2}$ define a node $a_1 x a_n$ which is adjacent to node $x a_n a_1$ by

$$\pi(a_1 x a_n) = (\pi(0;d))(a_1 x a_n) = (\pi[0;d])(a_1 x a_n) = x a_n a_1$$

and to $d-1$ other nodes by

$$\pi(a_1 x a_n) = (\pi(t;d))(a_1 x a_n) = a_1 x (a_n + t \bmod d) .$$

Let

$$T(S) = \{\pi(t;d) \mid t \in Z_d\} .$$

It can be shown that $G_v(T(S))$ is isomorphic to the wreath product $Z_d \otimes Z_n$. Moreover, we have:

### THEOREM 6.18

*The shuffle-exchange digraph $S(d,n)$ is isomorphic to the coset digraph*

$$Cos(G/H, T(S)),$$

*where $G = Z_d \otimes Z_n$ and $H = \{0\} \otimes Z_n$.*

The cube-connected-cycle digraph (CCC digraph) $C(d,n)$ has node-set $V(C(d,n)) = Z_n \otimes Z_d^n$. There are two kinds of arcs. The first kind connects

$$(l; a_0a_1 \cdots a_{n-1}) \text{ to } (l + 1 \bmod n; a_0a_1 \cdots a_{n-1})$$

and the second connects

$$(l; a_0a_1 \cdots a_{l-1}ua_{l+1} \cdots a_{n-1}) \text{ to } (l; a_0a_1 \cdots a_{l-1}va_{l+1} \cdots a_{n-1})$$

for all $v \in Z_d$.

We now convert the CCC digraph $C(d, n)$ into a PL digraph using the permutations from $T(S)$. The arc of the first kind is labeled with the permutation $\pi(0; d)$ and the arc from node $(l; a_0a_1 \cdots a_{l-1}ua_{l+1} \cdots a_{n-1})$ to the node $(l; a_0a_1 \cdots a_{l-1}va_{l+1} \cdots a_{n-1})$ is labeled with the permutation $\pi(v - u \bmod d; d)$.

It can be shown that the CCC digraph $C(d, n)$ so labeled is isomorphic (as a PL digraph) to the Cayley digraph $Cay(G_v(T(S)), T(S))$. It follows from Theorem 6.18 and above discussions that (also see [31]):

***THEOREM 6.19***

*The shuffle-exchange digraph $S(d, n)$ is a coset digraph of the CCC digraph $C(d, n)$.*

We refer the readers to the survey paper by Annexstein, Baumslag, and Rosenberg [31] for possible applications to parallel architectures, algorithms, and networks as the consequences of the above relationships in Theorems 6.17 and 6.19.

## 6.5 Fat-Tree Networks

A fat tree is not a generic network topology, but it can be interpreted as a concept of network hierarchy design in which the network bandwidth increases as it moves toward a higher level of hierarchy. Based on this definition, many actual interconnection networks can be considered as a fat tree, such as the TMC CM-5 [416], the IBM SP2 system [638], and the Meiko CS-2 [49] transputer system. We briefly discuss here innovative features of an interesting parallel system, namely the CM-5 architecture.

The Thinking Machines Corporation CM-5 is a synchronous MIMD machine which consists of 32 and 16,384 PEs (40 MHz 64-bit Sparcs), each with 32 MB local memory and four 160 MFlops vector-units [416]. PEs, disk arrays, and I/O are supervised by control processors, which uses a special version of the UNIX-system (CMOST). Every node is connected to two user-programmable, interprocessor communication networks, the *Data Network* and the *Control Network*. The Data Network is used for bulk point to point data transfers between individual processors. It uses a scalable 4-ary fat-tree interconnection network with an almost optimal VLSI layout [415]. This multistage network topology increases channel capacities as we ascend from these leaves to the root, and provides multiple shortest paths between leaf nodes. Interior tree nodes are only responsible for routing data between leaves, which consist of PEs, control processors, and I/O channels. The Control Network is a complete binary tree with all system components at the leaves, and user partitions assigned subtrees of the network. The Control Network supports global communication, such as broadcasting, synchronization, and scans. Both networks are designed for low-transmission node-to-neighbor latency and high bandwidth for distant transmissions. A third network, called *Diagnostics Network*, is used for diagnosability and fault recovery. This network is only visible to the system administrator. Distinct CM-5 subnetworks may be allocated to different users, and support for multitasking is also provided. Each allocated partition consists of processing, control and I/O nodes controlled by special processors, called partition managers.

The CM-5 also has SIMD characteristics. C* is a data-parallel extension of ANSI C designed for CM-5, although there is a current effort towards standardization. C* includes primitives for basic communication operations, such as replication, reduction, permutation, and scans, and parallel data structures such that simultaneous application of an operation to all data sets is possible [79]. A parallel variable is defined by its type and shape which determines the number of parallel elements. If a shape exceeds the number of physical processors it is distributed equally to all vector units. For one-dimensional vectors, the allocation is first done on each of the four vector units at the first CM-5 processor, and so on. In C*, actual processor locations within the CM-5 data network are hidden from the programmer; the CM-5 relies on randomized routing to practically avoid worst-case communication delay.

# Chapter 7

## Future Trends in Parallel Communication

A key issue in network design is realizability, which is related to number of PEs, network topology, network latency and bisection bandwidth, memory bandwidth, pin count and wire density constraints, multithreading, and prefetching. Several other important ideas are currently exploited and may affect the design of future parallel and distributed systems. For example,

- One-chip processors [91, 530],
- Superscalar processors with speculative execution (reaching up to 32 instructions per cycle) [434, 677],
- Lightweight processes and microthreading for improving branch prediction accuracy, cache hit rate, and prefetch effectiveness [61, 101, 646],
- Reconfigurable systems [696],
- Holographic memory [560],
- Intelligent RAM integrating processor functions into memory [531],
- Multivalued memory [155],
- Optical fiber communications [42, 489], and
- Evolutionary computing, which complements Nature's massive parallelism with efficient mapping and approximation techniques [38].

With evolutionary computing, fresh new ideas which may actually change our perception of what we consider now as "parallel computing," arise from breakthroughs in exploiting Nature's massive parallelism in *biological* [10, 435], *quantum* [38, 55], or *optical computing* [178]. In this case, we must also radically change parallel algorithm design by either mapping genetic DNA or atomic material energy state changes into solving intractable problems, or, as in the case of more traditional optical computing, relying on efficient intensive communication patterns such as multicasting. As Carver Mead wisely pointed out, "the thing that made the last generation of computers successful may be the thing that holds us back from the next generation." Whether any of these next generation technologies will be ready to be put to practice by the time the limits of silicon-based computing are finally reached [82], remains to be seen.

Since the system bottleneck with existing architectures has moved away from network communication, it is important to focus on the design of efficient operating systems, languages, and compilers that handle efficiently various higher-level software functions related to efficient programmability and fault tolerance, such as

- Automatic parallelization using compiler support,
- Installation, e.g., job scheduling, code allocation, and data placement to PEs,
- Instantiation, e.g., static process creation, and memory allocation,
- Parallel file I/O,
- Dynamic process creation, and termination detection,
- Performance monitoring, and process migration,
- Dynamic garbage collection,
- Dynamic load balancing,
- Efficient context switch,
- User-level communication access bypassing the operating system,
- Fault recovery based on detection, isolation, rollback, and reconfiguration, and
- Visualization using graphical user interfaces.

While the supercomputing industry is moving towards microprocessor-based symmetric multiprocessor systems (SMPs), and large-scale MPP vector systems (which combine features of vector and massively parallel computing), a new increasingly important information technology sector, just below supercomputing, is *high-end computing*. High-end computing arises from a demand for low-cost, high-performance computing.

Although cluster-based distributed shared memory systems, such as networks of workstations configured using scalable routers and processor to network interfaces, belong to this class, with Internet technology it is nowadays possible to think of a few million of the latest microprocessors computing in parallel, achieving an equivalent of a petaflop-distributed computing power. Supercomputers, which require more network bandwidth, may be connected together using dedicated lines to form larger parallel configurations; this is called *metacomputing*.

Of course, the real challenge lies in achieving sustained performance on key applications. This would not only require a low-latency, high-bandwidth interconnection network, but also dealing efficiently with heterogeneous hardware and software environments, e.g., mobile computing. Since no single interface or operating system is right for all users, there is currently an attempt at building a general object oriented communication interface for distributed software written in various programming languages and running under different operating systems. Unfortunately, the approach taken is not a uniform one, resulting in standard "container architectures" like CORBA, JAVA Enterprise Beans, and several other customized interface tools. Application performance on these distributed systems depends on proper interface design, support for multithreading, prefetching, cache coherence, and software fault tolerance, and avoiding memory consistency problems using proper synchronization primitives.

The distributed application performance issue points to further research on (a) *network security*, *dependability*, and *maintainability*, (b) *quality of service* (QoS), and (c) *network performance* [660].

Verifying the strength of network security, with its associated network management, authentication, certification, statistical logging, and interoperability procedures is very important. Without proper security, an intruder may access the system and severely disrupt communications. This can result in improperly used bandwidth, unavailability of the system, and increased costs.

A network dependability and maintainability evaluation should analytically model transient, intermittent, and permanent hardware and software process faults. An ideal environment which would enhance system robustness to random processor node and link faults or process faults, should rely on system rollback, fault tolerant routing, and work redistribution procedures.

QoS relates to latency, throughput, and packet loss requirements for different applications [93]. A definition of the most important QoS performance indicators for a given application follows.

- *Throughput* refers to the average number of packets routed through the network per clock cycle. For some applications the *throughput variance* is also an interesting metric.
- *Latency* refers to the average packet delay, measured from header departure to arrival of last packet bit from the network. For most applications the *delay jitter* (latency variance) is also interesting.
- *Packet loss* refers to the average number of packets lost during transmission, or rejected in order to satisfy QoS constraints.

A network performance study should experimentally evaluate existing protocols for

- Packetization,
- Network flow control, e.g., packet drop and retransmission vs. deflection vs. buffering (congestion tolerance),
- Traffic shaping, e.g., leaky bucket vs. token bucket,
- Buffer management, e.g., FIFO vs. priority queuing vs. weighted fair queuing, and
- Congestion avoidance, i.e., dropping newly injected packets during contention using random early detection vs. weighted random early detection protocols [712].

Any distributed network analysis study must take into account future traffic requirements, e.g., arising from multimedia applications, scaling of existing applications, evolution of networks, as well as cost vs. productivity gain issues. Thus, in order to evaluate new distributed systems, nonnumerical benchmarks are needed, especially for commercial networking applications such as:

- File access and transfer,
- E-mail with attachments,
- Real-time notification, e.g., using (possibly encoded) text,
- Text conference using chat or whiteboard,
- Audio conference,
- Video conference,
- Real-time video monitoring,
- Remote database access, e.g., to digital libraries,
- Data mining,
- Application sharing,
- Web access (text, audio, image, video),
- Virtual reality,
- High-definition digital television, and
- Video on demand.

# Bibliography

[1] A. Abdennadher and T.Y. Feng. On rearrangeability of Omega-Omega networks. In *Proc. IEEE Int. Conf. Parallel Proc.*, volume I, pages 159–165, 1992.

[2] C.M. Ablow and J.L. Brenner. Roots and canonical form for circulant matrices. *Trans. AMS*, 107:360–376, 1963.

[3] F. Abolhassan, J. Keller, and W.J. Paul. On the cost-effectiveness of PRAMs. *Acta Informatica*, 36(6):463–487, 1999.

[4] S. Abraham and K. Padmanabhan. Performance of the direct binary n-cube network for multiprocessors. *IEEE Trans. Computers*, 38(7):1000–1011, 1989.

[5] S. Abraham and K. Padmanabhan. Performance of multicomputer networks under pin-out constraints. *J. Parallel Distrib. Comput.*, 12(3):237–248, 1991.

[6] A. Abu-Sufah, H.E. Hussmann, and D.J. Kuck. On I/O speedup in tightly coupled multiprocessors. *IEEE Trans. Computers*, C-35(6):520–530, 1986.

[7] G. Adams and H. Siegel. The extra stage cube: a fault-tolerant interconnection network for supersystems. *IEEE Trans. Computers*, C-31(5):443–454, 1982.

[8] G.B. Adams, D.P. Agrawal, and H.J. Siegel. Fault tolerant multistage interconnection networks for multiprocessors. *IEEE Computer*, 6:14–27, 1987.

[9] G.B. Adams and H.J. Siegel. On the number of permutations performable by the augmented data manipulator network. *IEEE Trans. Computers*, C-31(4):270–277, 1982.

[10] L. Adleman. Molecular computation of solutions to combinatorial problems. *Science*, 266:1021–1024, 1994.

[11] S.V. Adve and K. Gharachorloo. Shared memory consistency models: a tutorial. *IEEE Trans. Computers*, C-45(12):1145–1155, 1996.

[12] F. Afrati, C.H. Papadimitriou, and G. Papageorgiou. The complexity of cubical graphs. *Information and Control*, 66(1/2):53–60, 1985.

[13] A. Agarwal. Performance tradeoffs in multithreaded processors. *IEEE Trans. Parallel Distrib. Syst.*, C-3(5):525–539, 1992.

[14] D.P. Agrawal, S-C. Kim, and N.K. Swain. Analysis and design of nonequivalent multistage networks. *IEEE Trans. Computers*, C-37(2):232–277, 1988.

[15] D.P. Agrawal and J.S. Leu. Dynamic accessibility testing and path length optimization of multistage interconnection networks. *IEEE Trans. Computers*, 34(3):255–266, 1985.

[16] E. Aharonson and H. Attiya. Counting networks with arbitrary fan-out. *Distr. Comput.*, 8(4):163–169, 1995.

[17] W.A. Aiello, F.T. Leighton, B.M. Maggs, and M. Newman. Fast algorithms for bit serial routing on a hypercube. *Math. Syst. Theory*, 24:253–271, 1991.

[18] M. Ajtai, J. Komlós, and E. Szemerédi. Sorting in $c \log n$ parallel steps. *Combinatorica*, 3:1–19, 1983.

[19] M. Ajtai, J. Komlós, and E. Szemerédi. Halvers and expanders. In *Proc. 33rd IEEE Symp. Foundations Comp. Sci.*, pages 686–692, 1992.

[20] S.B. Akers and B. Krishnamurthy. A group-theoretic model for symmetric interconnection networks. *IEEE Trans. Computers*, C-38:555–565, 1989.

[21] S.G. Akl. *Parallel Sorting Algorithms*. Academic Press, New York, 1985.

[22] M.S. Alam and R.G. Melhem. An efficient modular spare allocation scheme and its application to fault tolerant binary hypercubes. *IEEE Trans. Parallel Distrib. Syst.*, C-2(1):117–126, 1991.

[23] N. Alon. Eigenvalues and expanders. *Combinatorics*, 6:83–96, 1986.

[24] N. Alon, F.R.K. Chung, and R.L. Graham. Routing permutations on graphs via matchings. *SIAM J. Disc. Math.*, 23(7):513–530, 1994.

[25] H. Amano and K. Gaye. A Batcher double-omega network with combining. *IEEE Trans. Fundam. Electronics, Comm. Comput. Sci.*, E75-D(3):307–314, 1992.

[26] G.M. Amdahl. The validity of the single processor approach to achieving large scale computing capabilities. In *AFIPS Conf. Proc.*, volume 30, pages 483–485, 1967.

[27] G.M. Amdahl. Limits of expectation. *Int. J. of Supercomputer Appl.*, 2(1):88–97, 1988.

[28] T. Anderson, S. Owicki, J.B. Saxe, et al. High speed switch scheduling for local area networks. *ACM Trans. Computers Syst.*, 11(4):319–352, 1993.

[29] K.V. Anjan and T.M. Pinkston. An efficient, fully adaptive deadlock recovery scheme: DISHA. In *Proc. 24th IEEE Int. Symp. Comput. Arch.*, pages 210–219, 1995.

[30] F. Annexstein and M. Baumslag. On the diameter and bisector size of Cayley graphs. *Math. System Theory*, 26:271–291, 1993.

[31] F.S. Annexstein, M. Baumslag, and A.L. Rosenberg. Group action graphs and parallel architectures. *SIAM J. Comput.*, 19:544–569, 1990.

[32] S. Arora, F.T. Leighton, and B.M. Maggs. On-line algorithms for path selection in a nonblocking network. *SIAM J. Comput.*, 25(3):600–625, 1996.

[33] J. Aspnes, M. Herlihy, and N. Shavit. Counting networks. *J. ACM*, 41(5):1020–1048, 1994.

[34] S. Assaf and E. Upfal. Fault tolerant sorting networks. *SIAM J. Disc. Math.*, 4:472–480, 1991.

[35] L. Babai, G. Hetyei, W.M. Kantor, and L. Lubotzky. On the diameter of finite groups. In *Proc. 31st IEEE Symp. Foundations Comp. Sci.*, pages 857–865, 1990.

[36] H.G. Badr and S. Podar. An optimal shortest-path routing policy for network computers with regular mesh-connected topologies. *IEEE Trans. Computers*, C-38(10):1362–1371, 1989.

[37] J.L. Baer. *Computer Systems Architecture.* Comp. Sci. Press, 1980.

[38] J. Bailey. *After Thought: The Computer Challenge to Human Intelligence.* Basic Books, 1997.

[39] E.M. Bakker, R.B. Tan, and J. van Leeuwen. Prefix routing schemes in dynamic networks. *Comp. Networks and ISDN Syst.*, 26:403–421, 1993.

[40] P. Banerjee, J.T. Rahmeh, C. Stunkel, V.S. Nair, et al. Algorithm based fault tolerance on a hypercube multiprocessors. *IEEE Trans. Computers*, 39(12):1132–1144, 1990.

[41] P. Bannerjee and A. Dugar. The design, analysis and simulation of a fault-tolerant interconnection network supporting the fetch-and-add primitive. *IEEE Trans. Computers*, C-38(1):30–46, 1989.

[42] J. Bannister, M. Gerla, and M. Kovačević. Routing in optical networks. In *Routing in Comm. Networks*, M. Steenstrup, editor, Prentice-Hall, Englewood Cliffs, NJ, 1995.

[43] A. Bar-Noy, B. Schieber, P. Raghavan, and H. Tamaki. Fast deflection routing for packets and worms. In *Proc. 12th ACM SIGACT-SIGOPS Symp. Princ. Distrib. Comput.*, pages 75–86, 1993.

[44] P. Baran. On distributed communication networks. *IEEE Trans. Comm.*, C-12(1):1–9, 1964.

[45] L.A. Bassalygo and M.S. Pinsker. Complexity of an optimum nonblocking switching network without reconnections. *Problems of Inform. Transm.*, 9:64–66, 1974.

[46] K.E. Batcher. Sorting networks and their applications. In *Proc. AFIPS Spring Joint Comp. Conf.*, volume 32, pages 307–314, 1968.

[47] M. Baumslag and F. Annexstein. A unified framework for off-line permutation routing in parallel networks. *Math. Syst. Theory*, 24(4):233–251, 1991.

[48] R. Becker, D. Nassimi, and J. Perl. The new class of $g$-chain periodic sorters. *J. Parallel Distrib. Comput.*, 54:206–222, 1998.

[49] J. Beecroft, M. Homewood, and M. McLaren. Meiko CS-2 interconnect Elan-Elite design. *Parallel Comput.*, 20(10):1627–1638, 1994.

[50] Y. Ben-Asher and I. Newman. Geometric approach for optimal routing on mesh with buses. In *Proc. 7th IEEE Symp. Parallel Distrib. Comput.*, pages 145–153, 1995.

[51] Y. Ben-Asher, D. Peleg, R. Ramaswami, and A. Schuster. The power of reconfiguration. *J. Parallel Distrib. Comput.*, 13:139–153, 1991.

[52] S. Ben-David, A. Borodin, R. Karp, G. Tardos, and A. Wigderson. On the power of randomization in on-line algorithms. *Algorithmica*, 11(1):2–14, 1994.

[53] V.E. Beneš. On rearrangeable three-stage connecting networks. *Bell Syst. Tech. J.*, 16(5):1481–1491, 1962.

[54] V.E. Beneš. *Mathematical Theory of Connecting Networks and Telephone Traffic*. Academic Press, New York, 1965.

[55] C. Bennett, E. Bernstein, G. Brassard, and U. Vazirani. Strengths and weaknesses of quantum computing. *SIAM J. Comput.*, 26(5):1510–1523, 1997.

[56] P. Berenbrink, F. Meyer auf der Heide, and V. Stemann. Fault-tolerant shared memory simulations. In *Proc. 13th Symp. Theoret. Asp. Comp. Sci.*, volume 1046 of *Lect. Notes Comp. Sci.*, pages 181–192, 1996.

[57] P. Berman, K. Diks, and A. Pelc. Reliable broadcasting in logarithmic time with byzantine link failures. *J. Algorithms*, 22(2):199–211, 1997.

[58] J.C. Bermond, F. Comellas, and D.F. Hsu. Distributed loop networks: a survey. *J. Parallel Distrib. Comput.*, 24:2–10, 1995.

[59] J.C. Bermond and C. Peyrat. Broadcasting in de Bruijn network. In *Proc. 19th SE Conf. on Combinatorics, Graph Theory and Computing, Congress Numeration*, pages 283–292, 1988.

[60] P.J. Bernhard. Bounds on the performance of message routing heuristics. *IEEE Trans. Computers*, C-42(10):1252–1255, 1993.

[61] B. Bershad, T. Anderson, L. Lazowska, and H. Levy. Lightweight remote procedure call. In *Proc. 12th ACM Symp. Oper. Syst. Princ.*, pages 102–122, 1987.

[62] D.P. Bertsekas, C. Ozveren, G.D. Stamoulis, et al. Optimal communication algorithms for hypercubes. *J. Parallel Distrib. Comput.*, 11(4):263–275, 1991.

[63] T. Beth and V. Hatz. Design machines: algebraically well described interconnection networks. *Designs, Codes and Cryptography*, 2(3):287–298, 1992.

[64] D.P. Bhandarkar. Analysis of memory interference in multiprocessors. *IEEE Trans. Computers*, C-24(9):897–908, 1975.

[65] S.N. Bhatt, F.R.K. Chung, J.W. Hong, et al. Optimal emulations by butterfly-like networks. *J. ACM*, 43(2):293–330, 1996.

[66] S.N. Bhatt, F.R.K. Chung, F.T. Leighton, and A.L. Rosenberg. Efficient embeddings of trees in hypercubes. *SIAM J. Comput.*, 21(1):151–162, 1992.

[67] L.N. Bhuyan. An analysis of processor-memory interconnection networks. *IEEE Trans. Computers*, C-34(3):279–283, 1985.

[68] L.N. Bhuyan and C.R. Das. Dependability evaluation of multicomputer networks. In *Proc. IEEE Int. Conf. Parallel Proc.*, pages 576–583, 1986.

[69] R. Bianchini and B.H. Lim. Evaluating the performance of multithreading and prefetching in shared-memory multiprocessors. *J. Parallel Distrib. Comput.*, 37:83–97, 1996.

[70] G. Bilardi and F.P. Preparata. An architecture for bitonic sorting with optimal VLSI performance. *IEEE Trans. Computers*, C-33(7):646–651, 1984.

[71] G. Bilardi and F.P. Preparata. A minimum area VLSI network for $\log N$ time sorting. *IEEE Trans. Computers*, C-34(4):336–343, 1985.

[72] G. Bilardi and F.P. Preparata. Characterization of associative operations with prefix circuits of constant depth and linear size. *SIAM J. Comput.*, 19(2):246–255, 1990.

[73] G. Bilardi and F.P. Preparata. Size-time complexity of boolean networks for prefix computations. *J. ACM*, 36(2):362–382, 1996.

[74] A. Biryukov and S. Even. Cryptanalysis of the Portz interconnection network block cipher. *DIMACS Series in Cryptography and Network Security*, 1997.

[75] C.R. Bisbee and V.P. Nelson. Failure dependency bandwidth on shuffle-exchange networks. *IEEE Trans. Computers*, C-37(7):853–858, 1988.

[76] J.T. Blake and K.S. Trivedi. Multistage interconnection network reliability. *IEEE Trans. Computers*, 38:1600–1603, 1989.

[77] J.T. Blake and K.S. Trivedi. Reliability analysis of interconnection networks using hierarchical composition. *IEEE Trans. Reliability*, 38:111–119, 1989.

[78] G.E. Blelloch. Scans as primitive parallel operations. *IEEE Trans. Computers*, C-38(11):1526–1538, 1989.

[79] G.E. Blelloch. *Vector Model for Data Parallel Computing.* MIT Press, Cambridge, MA, 1990.

[80] N.J. Boden, D. Cohen, R.E. Felderman, A.E. Kulawick, et al. Myrinet: A gigabit-per-second local area network. *IEEE Micro*, 15(1):29–36, 1995.

[81] H.L. Bodlaender, J. van Leeuwen, R. Tan, and D. Thilikos. On interval routing schemes and treewidth. *Information and Computation*, 139(1):92–109, 1997.

[82] M. Bohr. Silicon trends and limits for advanced microprocessors. *Comm. ACM*, 41(3):80–87, 1998.

[83] A. Borodin. Towards a better understanding of pure packet routing. In *Proc. 3rd Workshop Algorithms and Data Struct.*, volume 709 of *Lect. Notes Comp. Sci.*, pages 14–25, 1993.

[84] A. Borodin and J.E. Hopcroft. Routing, merging and sorting on parallel models of computation. *J. Comput. Syst. Sci.*, 30(1):130–145, 1985.

[85] A. Borodin, P. Raghavan, B. Schieber, and E. Upfal. How much can hardware help routing. In *Proc. 25th ACM Symp. Theory Comput.*, pages 573–582, 1993.

[86] C. Botting, S. Rai, and D.P. Agrawal. Reliability computation of multistage interconnection networks. *IEEE Trans. Reliability*, 38:138–145, 1989.

[87] G. Brassard and P. Bratley. *Algorithmics: Theory and Practice.* Prentice-Hall, Englewood Cliffs, NJ, 1988.

[88] A. Broder, M. Fischer, D. Dolev, and B. Simmons. Efficient fault-tolerant routings in networks. *Information and Computation*, 75:52–64, 1987.

[89] J. Bruck, R. Cypher, and C.T. Ho. Fault-tolerant meshes and hypercubes with minimum number of spaces. *IEEE Trans. Computers*, 42:1089–1104, 1993.

[90] J. Bruck, R. Cypher, and C.T. Ho. Wildcard dimensions, coding theory and fault-tolerant meshes and hypercubes. *IEEE Trans. Parallel Distrib. Syst.*, C-44(1):150–155, 1995.

[91] B. Burger and J.R. Goodman. Billion transistor architectures. *IEEE Computer*, 30(9):46–48, 1997.

[92] C. Busch and M. Mavronicolas. A combinatorial treatment of balancing networks. *J. ACM*, 43(5):794–839, 1996.

[93] A. Campbell and K. Nahrstedt, editors. *Building QoS into Distributed Systems.* Chapman and Hall, Boca Raton, FL, 1997.

[94] D.G. Cantor. On nonblocking switching networks. *Networks*, 1:367–377, 1971.

[95] J. Carpinelli and A.Y. Oruc. Applications of edge coloring algorithms to routing in Clos networks. *Networks*, 24:319–326, 1994.

[96] S. Chalasani and A. Varma. Evaluation of two traffic distribution strategies for a dual-network multiprocessor system. *IEEE Trans. Parallel Distrib. Syst.*, C-3(6):375–384, 1992.

[97] M.Y. Chan and F.Y.L. Chin. On embedding rectangular grids in hypercubes. *IEEE Trans. Computers*, C-37(1):1285–1288, 1988.

[98] M.Y. Chan, F.Y.L. Chin, C.N. Chu, et al. Dilation 5 embedding of 3-dimensional grids onto hypercubes. *J. Parallel Distrib. Comput.*, 33(1):98–106, 1995.

[99] C. Chang and P. Mohapatra. Performance improvement of allocation schemes for mesh-connected computers. *J. Parallel Distrib. Comput.*, 52(1):40–68, 1998.

[100] C.Y. Chao. On a theorem of Sabidussi. In *Proc. AMS*, volume 15, pages 291–292, 1964.

[101] R.S. Chappell, J. Stark, S.P. Kim, S.K. Reinhardt, and Y.N Patt. Simultaneous subordinate microthreading (SSMT). In *Proc. 26th IEEE Int. Symp. Comput. Arch.*, pages 186–195, 1999.

[102] M.S. Chen, K.G. Chin, and D.D. Kandlur. Addressing, routing and broadcasting in hexagonal mesh multiprocessors. *IEEE Trans. Computers*, C-39(1):10–18, 1990.

[103] M.S. Chen and K.G. Shin. Adaptive fault-tolerant routing in hypercube multiprocessors. *IEEE Trans. Computers*, C-39(12):1406–1416, 1990.

[104] M.S. Chen and K.G. Shin. Depth-first search approach for fault-tolerant routing in hypercube multicomputers. *IEEE Trans. Parallel Distrib. Syst.*, C-1(2):152–159, 1990.

[105] M.S. Chen and K.G. Shin. Subcube allocation and task migration in hypercube multiprocessors. *IEEE Trans. Computers*, C-39(9):1146–1155, 1990.

[106] V. Cherkassky. Performance evaluation of non-rectangular multistage interconnection networks. *IEEE Trans. Computers*, C-37(10):1269–1272, 1988.

[107] V. Cherkassky and M. Malek. On permuting properties of regular rectangular SW-banyans. *IEEE Trans. Computers*, C-34(6):542–546, 1985.

[108] M.V. Chien and A.Y. Oruc. Optimal nonblocking networks with pin constraints. In *Proc. IEEE Int. Conf. Parallel Proc.*, volume I, pages 422–425, 1991.

[109] M.V. Chien and A.Y. Oruc. High performance concentrators and superconcentrators using multiplexing schemes. *IEEE Trans. Comm.*, 42(11):1045–1051, 1994.

[110] D.D. Chinn, T. Leighton, and M. Tompa. Minimal adaptive routing on the mesh with bounded queue size. In *Proc. 6th ACM Symp. Parallel Algorithms Arch.*, pages 354–363, 1994.

[111] B.S. Chlebus, A. Czumaj, L. Gąsieniec, M. Kowaluk, and W. Plandowski. Parallel alternating-direction access machine. In *Proc. 21st Math. Foundations Comp. Sci.*, volume 1113 of *Lect. Notes Comp. Sci.*, pages 267–278, 1996.

[112] B.S. Chlebus, A. Czumaj, and J.F. Sibeyn. Routing on the PADAM: Degrees of optimality. In *Proc. 3rd Euro-Par Conf.*, volume 1300 of *Lect. Notes Comp. Sci.*, pages 272–279, 1997.

[113] B.S. Chlebus, M. Kaufmann, and J.F. Sibeyn. Deterministic permutation routing on meshes. In *Proc. 5th IEEE Symp. Parallel Distribut. Proc.*, pages 814–821, 1993.

[114] B.S. Chlebus and M. Kukawka. A guide to sorting on the mesh-connected processor array. *Comput. Art. Intel.*, 9:599–610, 1990.

[115] G. Chochia, M. Cole, and T. Heywood. Implementing the hierarchical PRAM on the 2D mesh: analyses and experiments. In *Proc. 7th Symp. on Parallel Distrib. Proc.*, pages 587–595, 1995.

[116] S. Chowdhury and A.M. Holliday. Stability and performance of alternative two-level interconnection networks. In *Proc. IEEE Int. Conf. Parallel Proc.*, volume I, pages 720–721, 1991.

[117] P.J. Chuang and N.F. Tzeng. A fast recognition-complete processor allocation strategy for hypercube computers. *IEEE Trans. Computers*, C-41(4):467–479, 1992.

[118] K.M. Chung and C.K. Wong. Construction of a generalized connector with $5.8n \log n$ edges. *IEEE Trans. Computers*, C-29(11):1029–1032, 1980.

[119] P.S. Chung and K.W. Ross. On nonblocking multirate interconnection networks. *SIAM J. Comput.*, 20(4):726–736, 1991.

[120] V. Chvátal. Lecture notes on the new AKS sorting network. Technical Report 294, Dept. Comp. Sci., Rutgers University, 1992.

[121] K.Y. Chwa and S.L. Hakimi. On fault identification in diagnosable systems. *IEEE Trans. Computers*, C-30(6):414–422, 1981.

[122] C. Clos. A study of nonblocking switching networks. *Bell Technical J.*, pages 407–424, 1953.

[123] C.J. Colbourn. *The Combinatorics of Network Reliability.* Oxford University Press, New York, 1987.

[124] C.J. Colbourn, J.S. Devitt, D.D. Harms, and M. Kraetzl. Assessing reliability of multistage interconnection networks. *IEEE Trans. Computers*, 42(10):1207–1221, 1993.

[125] C.J. Colbourn, L.D. Nel, T.B. Boffey, and D.F. Yates. Probabilistic estimation of damage from fire spread. *Annals of Operations Research*, 50:173–185, 1994.

[126] R. Cole and J. Hopcroft. On edge coloring bipartite graphs. *SIAM J. Comput.*, 11:540–546, 1982.

[127] R. Cole, B. Maggs, and R. Sitaraman. Multi-scale self-simulation: a technique for reconfiguring arrays with faults. *SIAM J. Comput.*, 26(6):1581–1611, 1997.

[128] R. Cole and A. Siegel. Optimal VLSI circuits for sorting. *J. ACM*, 35(4):777–809, 1988.

[129] W.W. Collier. *Reasoning About Parallel Architectures.* Prentice-Hall, Englewood Cliffs, NJ, 1992.

[130] O. Collins, S. Dolinar, R. McEliece, and F. Pollara. VLSI decomposition of the de Bruijn graph. *J. ACM*, 39:931–948, 1992.

[131] R. Conterno and R. Melen. An analytical model for a class of processor-memory interconnection networks. *IEEE Trans. Comput*, C-36(11):1374–1378, 1987.

[132] S.A. Cook. The complexity of theorem proving procedures. In *Proc. 3rd ACM Symp. Theory Comput.*, pages 151–158, 1971.

[133] K. Coolsaet, F. Meyer auf der Heide, and V. Fack. Optimal algorithms for total exchange without buffering on the hypercube. *Bit*, 32:559–569, 1992.

[134] T.H. Cormen and C.E. Leiserson. A hyperconcentrator switch for routing bit-serial messages. *J. Parallel Distrib. Comput.*, 10(3):193–204, 1990.

[135] H.S.M. Coxeter and W.O.J. Moser. *Generators and Relations for Discrete Groups.* Springer-Verlag, New York, 1972.

[136] D.E. Culler, R.M. Karp, D.A. Patterson, A. Sahay, et al. LogP: towards a realistic model of parallel computation. In *Proc. ACM SIGPLAN Notices*, volume 28 (7), pages 1–12, 1993.

[137] D.E. Culler, J.P. Singh, and A. Gupta. *Parallel Computer Architecture: a Hardware/Software Approach.* Academic Press, New York, 1998.

[138] S.J. Curran and J.A. Gallian. Hamiltonian cycles and paths in Cayley graphs and digraphs - survey. *Discr. Math.*, 156:1–18, 1996.

[139] G. Cybenko, D.W. Krumme, and K.N. Venkataraman. Fixed hypercube embedding. *Inf. Proc. Letters*, 25(1):35–39, 1987.

[140] R. Cypher. *Efficient communication in massively parallel computers.* PhD thesis, Dept. Comp. Sci., University of Washington, Seattle, 1989.

[141] R Cypher. Theoretical aspects of VLSI pin limitations. *SIAM J. Comput.*, 22(2):356–378, 1993.

[142] R. Cypher and L. Gravano. Storage-efficient, deadlock-free packet routing algorithms for torus networks. *IEEE Trans. Computers*, 43(12):1376–1385, 1994.

[143] R. Cypher, F. Meyer auf der Heide, C. Scheideler, and B. Vöking. Universal algorithms for store-and-forward and wormhole routing. In *Proc. 28th ACM Symp. Theory Comput.*, pages 356–365, 1996.

[144] R. Cypher and C.G. Plaxton. Deterministic sorting in nearly logarithmic time on the hypercube and related computers. *J. Comput. Syst. Sci.*, 47:501–548, 1993.

[145] R. Cypher and J.L.C. Sanz. Cubesort: a parallel algorithm for sorting $n$ data items with $s$-sorters. *J. Algorithms*, 13(2):211–234, 1992.

[146] A. Czumaj, F. Meyer auf der Heide, and V. Stemann. Shared memory simulations with triple-logarithmic delay. In *Proc. 3rd European Symp. Algorithms*, volume 979 of *Lect. Notes Comp. Sci.*, pages 46–59, 1995.

[147] A. Czumaj, F. Meyer auf der Heide, and V. Stemann. Simulating shared memory in real time: on the computation power of reconfigurable architectures. *Information and Computation*, 137(2):103–120, 1997.

[148] A.T. Dabhura and G.M. Masson. An $\mathcal{O}(N^{2.5})$ fault identification algorithm for diagnosable systems. *IEEE Trans. Computers*, C-33(6):486–492, 1984.

[149] A.T. Dahbura, K.K. Sabnani, and L. King. The comparison approach to multiprocessor diagnosis. *IEEE Trans. Computers*, C-36(3):373–378, 1987.

[150] W.J. Dally. Performance analysis of k-ary n-cube interconnection networks. *IEEE Trans. Computers*, C-39(6):775–785, 1990.

[151] W.J. Dally. Virtual-channel flow control. *IEEE Trans. Parallel Distrib. Syst.*, C-2(2):194–205, 1992.

[152] W.J. Dally and H. Aoki. Deadlock-free adaptive routing in multicomputer networks using virtual channels. *IEEE Trans. Parallel Distrib. Syst.*, C-4(4):466–474, 1993.

[153] W.J. Dally and C. Seitz. Deadlock free message routing in multiprocessor interconnection networks. *IEEE Trans. Computers*, C-36(5):547–553, 1987.

[154] S.P. Dandamudi and D.L. Eager. Hierarchical interconnection networks for multicomputer systems. *IEEE Trans. Computers*, C-39(6):786–797, 1990.

[155] T.T. Dao, E.J. McCluskey, and L.K. Russell. Multivalued integrated injection logic. *IEEE Trans. Computers*, C-26(12):1233–1241, 1977.

[156] C.R. Das, J.T. Kreulen, M.J. Thazhuthaveetil, and L.N. Bhuyan. Dependability modeling for multiprocessor computers. *The Computer J.*, 23(10):7–19, 1990.

[157] C.R. Das, P. Mohapatra, L. Tien, and L.N. Bhuyan. An availability model for MIN-based multiprocessors. *IEEE Trans. Parallel Distrib. Syst.*, 4(10):1118–1129, 1993.

[158] B. Dasgupta, F.K. Hwang, and Y.C. Yao. Permutation routing in $t$-nary hypercubes. *J. Parallel Distrib. Comput.*, 1995, submitted.

[159] M. Davio and C. Ronse. Insertion networks. *IEEE Trans. Computers*, 34(6):565–570, 1985.

[160] N.G. de Bruijn. A combinatorial problem. In *Proc. Abademe Van Waterschappen*, volume 49, pages 758–764, 1946.

[161] O. Delmas and S. Perennes. Circuit-switched gossiping in 3-dimensional torus networks. In *Proc. 2nd Euro-Par Conf.*, volume 1123 of *Lect. Notes Comp. Sci.*, pages 370–373, 1996.

[162] F. Desprez, E. Fleury, and Loi M. T9000 et C104: la nouvelle generation de transputers. Technical Report 93–01, ENS Lyon, Lip, Lyon, France, 1993.

[163] D.M. Dias and M. Kumar. Preventing congestion in multistage networks in the presence of hotspots. In *Proc. 2nd IEEE Symp. Parallel Distrib. Comput.*, pages 548–551, 1990.

[164] S.R. Dickey and R. Kenner. Design of components for a low-cost combining network. *VLSI Design*, 2(4):287–303, 1995.

[165] S.R. Dickey and O. Percus. Performance differences among combining switch architectures. In *Proc. IEEE Conf. Parallel Proc.*, volume I, pages 110–117, 1992.

[166] M. Dietzfelbinger and F. Meyer auf der Heide. Simple, efficient shared memory simulations. In *Proc. 5th ACM Symp. Parallel Algorithms Arch.*, pages 110–118, 1993.

[167] J. Ding and L.N. Bhuyan. Performance evaluation of multistage networks with finite buffers. In *Proc. IEEE Int. Conf. Parallel Proc.*, volume 1, pages 592–599, 1991.

[168] J. Ding and L.N. Bhuyan. An adaptive submesh allocation strategy for two-dimensional mesh connected systems. In *Proc. IEEE Int. Conf. Parallel Proc.*, volume II, pages 193–200, 1993.

[169] J. Ding and L.N. Bhuyan. Analysis of multi-queue buffer allocation schemes in multistage interconnection networks. Technical Report 53/93, Dept. Comput. Sci., Texas A & M University, 1993.

[170] J. Ding and L.N. Bhuyan. Finite buffer analysis of multistage interconnection networks. *IEEE Trans. Computers*, C-43(2):243–247, 1994.

[171] K.S. Ding, C.T. Ho, and J.J. Tsay. Matrix transpose on meshes with wormhole and XY-routing. In *Proc. 6th IEEE Symp. Parallel Distrib. Comput.*, pages 656–663, 1994.

[172] D. Dolev, C. Dwork, N. Pippenger, and A. Wigderson. Superconcentrators, generalizers and generalized connectors with limited depth. In *Proc. 15th ACM Symp. Theory Comput.*, pages 42–51, 1983.

[173] D. Dolev, J. Halpern, B. Simmons, and R. Strong. A new look at fault tolerant network routing. *Information and Computation*, 72:180–196, 1987.

[174] J. Dongarra, J.L. Martin, and J. Worlton. Computer benchmarking: paths and pitfalls. *IEEE Spectrum*, 24(7):38–43, 1987.

[175] J.J. Dongarra, H.W. Meuer, and E. Strohmaier. Top 500 supercomputer sites. http://www.top500.org/, 2000. Twelfth edition.

[176] B.G. Douglas and A.Y. Oruc. Self-routing and route balancing in connection networks. In *Proc. IEEE Int. Conf. Parallel Proc.*, volume I, pages 331–337, 1990.

[177] M. Dowd, Y. Perl, L. Rudolph, and M. Saks. The periodic balanced sorting network. *J. ACM*, 36(4):738–757, 1989.

[178] T.J. Drabik et al. 2D silicon/ferroelectric liquid crystal spatial light modulators. *IEEE Micro*, 15(4):67–76, 1995.

[179] D.Z. Du, B. Gao, F.K. Hwang, and J.H. Kim. On multirate rearrangeable Clos networks. *SIAM J. Comput.*, 28(2):463–470, 1999.

[180] D.Z. Du and F.K. Hwang. Generalized de Bruijn digraphs. *Networks*, 18:27–38, 1988.

[181] J. Duato. A necessary and sufficient condition for deadlock-free adaptive routing in wormhole networks. *IEEE Trans. Parallel Distrib. Syst.*, C-6(10):1055–1067, 1995.

[182] J. Duato. Theory of deadlock-free adaptive multicast routing in wormhole networks. *IEEE Trans. Parallel Distrib. Syst.*, C-6(9):976–987, 1995.

[183] J. Duato. A necessary and sufficient condition for deadlock-free routing in cut-through and store-and-forward networks. *IEEE Trans. Parallel Distrib. Syst.*, C-7(8):841–854, 1996.

[184] J. Duato, S. Yalamanchili, and L. Ni. *Interconnection Networks: An Engineering Approach.* IEEE Computer Society Press, 1997.

[185] S. Dutt and J.P. Hayes. Subcube allocation in hypercube computers. *IEEE Trans. Computers*, C-40(3):341–351, 1991.

[186] A.C. Elster, M.Ü. Uycar, and A.P. Reeves. Fault tolerant matrix operations on hypercube multiprocessors. In *Proc. IEEE Int. Conf. Parallel Proc.*, volume III, pages 169–176, 1989.

[187] C. Engelmann and J. Keller. Simulation-based comparison of hash functions for emulated shared memory. In *Proc. Parallel Lang. Arch. Eur.*, volume 694 of *Lect. Notes Comp. Sci.*, pages 1–11, 1993.

[188] P. Erdos and D.F. Hsu. Distributed loop network with minimum transmission delay. *Theoretical Comp. Sci.*, 100:223–241, 1992.

[189] S. Even, S. Muthukrishnan, M.S. Paterson, and S.C. Sahinalp. Layout of the Batcher bitonic sorter. In *Proc. Int. Symp. Parallel Arch. Algorithms and Networks*, pages 172–181, 1998.

[190] V. Faber, O.M. Lubeck, and A.B. White. Superlinear speedup of an efficient sequential algorithm is not possible. *Parallel Comput.*, 3:259–260, 1986.

[191] U. Feige and P. Raghavan. Exact analysis of hot-potato routing. In *Proc. 33rd IEEE Symp. Foundations Comp. Sci.*, pages 553–562, 1992.

[192] D.G. Feitelson. Job scheduling in multiprogrammed parallel systems. IBM Research Report RC-19970, Dept. Electr. Eng., University of Maryland at College Park, 1997.

[193] P. Feldman. Fault tolerance of minimal path routing in a network. In *Proc. 17th ACM Symp. Theory Comput.*, pages 327–334, 1985.

[194] R. Feldmann and P. Mysliwietz. The shuffle exchange network has a Hamiltonian path. *Math. Syst. Theory*, 29(5):471–486, 1997.

[195] S. Felperin, P. Raghavan, and E. Upfal. A theory of wormhole routing in parallel computers. *IEEE Trans. Comm.*, C-45(6):704–713, 1996.

[196] E. Felten, S. Karlin, and Otto S. The traveling salesman problem on a hypercubic, MIMD computer. In *Proc. IEEE Int. Conf. Parallel Proc.*, pages 6–10, 1985.

[197] T.Y. Feng. Data manipulating functions in parallel processors and their implementation. *IEEE Trans. Computers*, C-23(3):309–318, 1974.

[198] T.Y. Feng and C.L. Wu. Fault diagnosis for a class of multistage shuffle/exchange networks. *IEEE Trans. Computers*, 30(10):743–758, 1981.

[199] T.Y. Feng and Q. Zhang. On fault diagnosis of multiprocessor networks with for valid states. *JPDC*, 9:237–251, 1990.

[200] A. Ferreira. Hypercubes. In *Parallel and Distributed Computer Handbook*, A. Zomaya, editor, McGraw-Hill, New York, 1996.

[201] A. Ferreira and M.D. Grammatikakis. Improved probabilistic routing on generalized hypercubes. *Theoret. Comput. Sci., North-Holland*, 158(1-2):53–64, 1995.

[202] A.G. Ferreira, E. Fleury, and M.D. Grammatikakis. Multicasting control and communications on multihop stack-ring OPS net-

works. In *Proc. IEEE Int. Conf. Mass. Parallel Proc. using Optical Interconnections*, pages 39–44, 1997.

[203] C.M. Fiduccia, R.W. Forcade, and J.S. Zito. Geometry and diameter bounds of directed Cayley graphs of Abelian groups. *SIAM J. Disc. Math.*, 11(1):157–167, 1998.

[204] M.A. Fiol, J.L.A. Yebra, and I. Alegre. Line digraph iteration and the $(d, k)$ digraph problem. *IEEE Trans. Computers*, C-33:400–403, 1984.

[205] M.A. Fiol, L.A. Yebra, I. Alegre, and M. Valero. A discrete optimization problem in local networks and data alignment. *IEEE Trans. Computers*, C-36:702–713, 1987.

[206] H.P. Flatt and K. Kennedy. Performance of parallel processors. *Parallel Comput.*, 12(1):1–20, 1989.

[207] P. Flocchini, B. Mans, and N. Santoro, editors. *Proc. 1st Colloquium Struct. Info Comm. Complexity*, volume 1. Carleton Univ. Press, 1994.

[208] M.J. Flynn. Very high speed computing. In *Proc. IEEE*, volume C-54 (12), pages 1901–1909, 1966.

[209] P. Fraigniaud. Asymptotically optimal broadcasting and gossiping in faulty hypercubes multicomputers. *IEEE Trans. Computers*, C-41(11):1410–1419, 1992.

[210] P. Fraigniaud. Performance analysis of broadcasting in hypercubes with restricted communication capabilities. *J. Parallel Distrib. Comput.*, 16(1):15–26, 1992.

[211] P. Fraigniaud and C. Gavoille. Universal routing schemes. *Distr. Comput.*, 10(2):65–78, 1997.

[212] P. Fraigniaud and E. Lazard. Methods and problems of communication in usual networks. *Disc. Appl. Math. - Special Issue on Broadcasting and Gossiping*, 53:79–133, 1994.

[213] P. Fraigniaud and J.G. Peters. Structured communication in torus networks. In *Proc. 28th Hawaii Int. Conf. Syst. Sci.*, volume II, pages 584–593, 1995.

[214] P. Fraigniaud and C. Peyrat. Broadcasting in a hypercube when some calls fail. *Inf. Proc. Letters*, 39(3):115–119, 1991.

[215] M.A. Franklin, D. Wann, and W. Thomas. Pin limitations and partitioning of VLSI interconnection networks. *IEEE Trans. Computers*, C-31(11):1109–1116, 1982.

[216] G.N. Frederickson and R. Janardan. Separator-based strategies for efficient message routing. In *Proc. 27th IEEE Symp. Foundations Comp. Sci.*, pages 428–437, 1986.

[217] H. Fredricksen. A survey of full length non-linear shift register cycle algorithms. *SIAM Review*, 24:195–221, 1982.

[218] K.A. Frenkel. Evaluating two massively parallel machines. *Comm. ACM*, 29:752–758, 1986.

[219] O. Gabber and Z. Galil. Explicit construction of linear size superconcentrators. *J. Computer Syst. Sci.*, 22(3):407–420, 1981.

[220] B. Gao and F.K. Hwang. Wide-sense nonblocking for multirate 3-stage Clos networks. *Theoret. Comput. Sci., North-Holland*, 182(1-2):171–182, 1997.

[221] D. Gautier de Lahaut and C. Germain. Static communications in parallel scientific programs. In *Proc. Parallel Arch. Lang., Europe Conf.*, volume 817 of *Lect. Notes Comp. Sci.*, pages 262–276, 1994.

[222] I. Gazit and M. Malek. On the number of permutations performable by extra-stage multistage interconnection networks. *IEEE Trans. Computers*, C-38(2):297–301, 1989.

[223] A. Ghafoor and P. Solë. Performance of fault tolerant diagnostic on the hypercube systems. *IEEE Trans. Computers*, 38(8):1164–1171, 1989.

[224] C.J. Glass and L.M. Ni. The turn model for adaptive routing. *J. ACM*, 41(5):874–902, 1994.

[225] L.R. Goke and G.J. Lipovski. Banyan networks for partitioning multiprocessor systems. In *Proc. 1st Annual Symp. on Comp. Architecture*, pages 21–28, 1973.

[226] A.V. Goldberg, B.M. Maggs, and S.A. Plotkin. A parallel algorithm for reconfiguring a multibutterfly network with faulty switches. *IEEE Trans. Computers*, 43(3):321–326, 1994.

[227] L.A. Goldberg, Y. Matias, and S. Rao. An optical simulation of shared memory. In *Proc. 6th ACM Symp. Parallel Algorithms Arch.*, pages 257–267, 1994.

[228] I.J. Good. Normally recurring decimals. *J. London Math. Soc.*, 21:167–169, 1946.

[229] J.M. Gordon and Q.F. Stout. Fault tolerant message routing on large parallel systems. In *Proc. 2nd Symp. Frontiers Massively Parallel Comput.*, pages 155–158, 1988.

[230] J.M. Gordon and Q.F. Stout. Hypercube message routing in the presence of faults. In *Proc. 3rd Conf. Hypercube Concurrent Comput. Appl.*, pages 318–327, 1988.

[231] A. Gottlieb and C.P. Kruskal. Complexity results for permuting data and other computations on parallel processors. *J. ACM*, 31(2):193–209, 1984.

[232] N. Graham, F. Harary, M. Livingston, et al. Subcube fault tolerance in hypercubes. *Information and Computation*, 102(1):280–314, 1993.

[233] M.D. Grammatikakis, H. Dollani, and S. Liesche. Synchronization on Cray-T3E virtual shared memory. In *Proc. Conf. Cray Users Group*, Stuttgart, Germany, 1998.

[234] M.D. Grammatikakis, D.F. Hsu, and F.K. Hwang. Adaptive and oblivious routing algorithms on the d-cube. In *Proc. 4th ACM Int. Symp. Algorithms Comput.*, volume 762 of *Lect. Notes Comp. Sci.*, pages 167–176, 1993.

[235] M.D. Grammatikakis, D.F. Hsu, M. Kraetzl, and J.F. Sibeyn. Packet routing in fixed-connection networks: A survey. *JPDC*, 54:77–132, 1998.

[236] M.D. Grammatikakis, J.S. Jwo, M. Kraetzl, and S.H. Wang. Dynamic and static packet routing on symmetric communication networks. In *Proc. IEEE Conf. Global Comm.*, pages 1501–1505, San Francisco, Calif., 1994.

[237] M.D. Grammatikakis, M. Kraetzl, and E. Fleury. Shortest-path and hot-potato routing on unbuffered toroidal networks. In *Proc. IEEE Conf. Global Comm.*, pages 165–169, 1997.

[238] M.D. Grammatikakis, M. Kraetzl, and E. Fleury. Continuous routing on packet switches (C104 - Telegraphos). *Int. J. Foundations Comp. Sci. - Special Issue on Interc. Networks*, 9(2):121–138, 1999.

[239] M.D. Grammatikakis and S. Liesche. Parallel priority queues and sorting for parallel simulation. *IEEE Trans. Software Engin.*, SE-26(5):401–422, 2000.

[240] A.G. Greenberg and J. Goodman. Sharp approximate models of deflection routing in mesh networks. *IEEE Trans. Comm.*, C-41(1):210–223, 1993.

[241] A.G. Greenberg and B. Hajek. Deflection routing in hypercube networks. *IEEE Trans. Comm.*, C-40(6):1070–1081, 1992.

[242] E. Gunduzhan and A.Y. Oruc. Complexity of buffered concentrators and superconcentrators. Technical Report INRL TR 9701, Dept. Electr. Eng., University of Maryland at College Park, 1997.

[243] K.D. Günther. Prevention of deadlocks in packet switched data transport systems. *IEEE Trans. Comm.*, 29(4):512–524, 1981.

[244] W. Guo and A.Y. Oruc. Regular sparse crossbar concentrators. *IEEE Trans. Computers*, C-47(3):363–368, 1998.

[245] J. Gustafson, G. Montry, and R. Benner. Development of parallel methods for a 1024-processor hypercube. *SIAM J. Sci. Stat. Comput.*, 9(4):609–638, 1988.

[246] S.C. Hakimi and A.T. Amin. Characterization of connection assignment of diagnosable systems. *IEEE Trans. Computers*, C-23(1):86–88, 1974.

[247] P. Hall. On representatives of subsets. *J. London Math. Soc.*, 10:26–30, 1935.

[248] S.E. Hambrusch, F. Hameed, and A. Khokhar. Communication operations on coarse-grained architectures. *Parallel Comput.*, 21:731–751, 1995.

[249] Y.O. Hamidoune. On the connectivity of Cayley digraphs. *Europ. J. Combinatorics*, 5:309–312, 1984.

[250] Y.O. Hamidoune and O. Serra. On small cuts separating an abelian Cayley graph into two equal parts. *Math. Syst. Theory*, 29(4):407–409, 1996.

[251] T. Han and D.F. Stanat. "Move and smooth" routing algorithms on mesh connected computers. In *Proc. 28th Allerton Conf. Comm. Control Comput.*, pages 236–245, 1990.

[252] Y. Han, Y. Igarashi, and M. Truszczynski. Indexing functions and time lower bounds for sorting on a mesh-connected computer. *Disc. Appl. Math.*, 36:141–152, 1992.

[253] M. Harchol-Balter and P.E. Black. Queuing analysis of oblivious packet-routing networks. In *Proc. 5th ACM-SIAM Symp. Disc. Algorithms*, pages 583–592, 1994.

[254] M. Harchol-Balter and D. Wolfe. Bounding delays in packet-routing networks. In *Proc. 27th ACM Symp. Theory Comput.*, pages 248–257, 1995.

[255] D.D. Harms, M. Kraetzl, C.J. Colbourn, and J.S. Devitt. *Network Reliability: Experiments with a Symbolic Algebra Environment.* CRC Press, Boca Raton, FL, 1995.

[256] D.T. Harper and J.R. Jump. Evaluation of reduced bandwidth multistage networks. *J. Parallel Distrib. Comput.*, C-9:304–311, 1990.

[257] J. Hastad, F.T. Leighton, and M. Newman. Reconfiguring a hypercube in the presence of faults. In *Proc. 19th ACM Symp. Theory Comput.*, pages 274–284, 1987.

[258] J. Hastad, F.T. Leighton, and M. Newman. Fast computation using faulty hypercubes. In *Proc. 21st ACM Symp. Theory Comput.*, pages 251–263, 1989.

[259] S.M. Hedetniemi, S.T. Hedetniemi, and A.L. Liestman. A survey of gossiping and broadcasting in communication networks. *Networks*, 18:319–349, 1987.

[260] H. Hellerman. On the average speed of a multiple-module storage system. *IEEE Trans. Electronic Comput.*, EC-15(4):670, 1966.

[261] H. Hellwagner and A. Reinefeld, editors. *Proc. SCI: Scalable Coherent Interface: Architecture and Software for High Performance Computation Clusters.* Springer-Verlag, New York, 1999.

[262] D.P. Helmbold and C.E. McDowell. Modeling speedup greater than $n$. In *Proc. IEEE Int. Conf. Parallel Proc.*, volume III, pages 219–224, 1989.

[263] K.T. Herley and G. Bilardi. Deterministic simulations of PRAMs on bounded-degree networks. *SIAM J. Comput.*, 23(2):276–292, 1994.

[264] K.T. Herley, A. Pietracaprina, and G. Pucci. Implementing shared memory on multi-dimensional meshes and on the fat-tree. In *Proc. 3rd European Symp. Algorithms*, volume 979 of *Lect. Notes Comp. Sci.*, pages 60–74, 1995.

[265] M. Herlihy, B.H. Lim, and N. Shavit. Scalable concurrent counting. *ACM Trans. Computers Syst.*, C-13(4):343–364, 1995.

[266] M.C. Heydemann, J. Opatrny, and D. Sotteau. Broadcasting and spanning trees in de Bruijn and Kautz networks. *Discr. Appl. Math.*, 37/38:297–317, 1992.

[267] M.C. Heydemann, J. Opatrny, and D. Sotteau. Embeddings of hypercubes and grids into de Bruijn graphs. *J. Parallel Distrib. Comput.*, 23:104–111, 1994.

[268] M. Hill. Multiprocessors should support simple memory-consistency models. *IEEE Computer*, 31(8):28–34, 1998.

[269] M.D. Hill and J.R. Larus. Cache considerations for multiprocessor programmers. *Comm. ACM*, 33(8):97–102, 1990.

[270] D.S. Hirschberg and D.A. Lelewer. Efficient decoding of prefix codes. *Comm. ACM*, 33(4):449–459, 1990.

[271] M.G. Hluchyj and M.J. Karol. Queuing in high-performance packet switching. *IEEE Trans. Sel. Areas Comm.*, C-6(9):1587–1597, 1988.

[272] C.T. Ho and S.L. Johnsson. Embedding meshes in boolean cubes by graph decomposition. *J. Parallel Distrib. Comput.*, 8(4):325–339, 1990.

[273] C.T. Ho and M.Y. Kao. Efficient broadcast on hypercubes with wormhole and *e*-cube routings. *Parallel Proc. Letters*, 5(2):213–222, 1995.

[274] C.T. Ho and M.Y. Kao. Optimal broadcast in all-port wormhole-routed hypercubes. *IEEE Trans. Parallel Distrib. Syst.*, C-6(2):200–204, 1995.

[275] R.C. Ho, C.H. Yang, M.A. Horowitz, and D.L. Dill. Architecture validation for processors. In *Proc. 22nd IEEE Int. Symp. Comput. Arch.*, pages 404–413, 1995.

[276] R.W. Hockney. The communication challenge for MPP: Intel Paragon and Meiko CS-2. *Parallel Comput.*, 20(3):389–398, 1994.

[277] J. Hromkovič, R. Klasing, B. Monien, and R. Peine. Dissemination of information in interconnection networks (broadcasting and gossiping). In *Comb. Network Theory*, D.F. Hsu and D.Z. Du, editors, Kluwer Academic, Norwell, MA, 1995.

[278] D.F. Hsu. On container width and length in graphs, groups and networks. *IEEE Trans. Fundamentals of Electronics, Comm. and Comput. Sci.*, E77-A(4):668–680, 1994.

[279] D.F. Hsu and X.D. Jia. Extremal problems in the construction of distributed loop networks. *SIAM J. Disc. Math.*, 7:57–71, 1994.

[280] D.F. Hsu and H.F. Liu. Generalized shuffle-exchange digraphs. Technical Report 98-01, Dept. Comp. Info. Science, Fordham University, 1998.

[281] D.F. Hsu and L. Tomasz. Note on the k-diameter of k-regular k-connected graphs. *Disc. Math.*, 132:291–296, 1994.

[282] D.F. Hsu and D.S. Wei. Efficient routing and sorting schemes for de Bruijn networks. *IEEE Trans. Parallel Distrib. Syst.*, 8:1–14, 1997.

[283] J.M. Hsu and P. Banerjee. Performance evaluation of hardware support for message passing in distributed memory multicomputers. In *Proc. IEEE Int. Conf. Parallel Proc.*, volume I, pages 604–607, 1991.

[284] W.T.Y. Hsu and P.C. Yew. The impact of wiring constraints on hierarchical network performance. In *Proc. 6th IEEE Int. Parallel Proc. Symp.*, pages 580–588, 1992.

[285] W.T.Y. Hsu and P.C. Yew. Performance evaluation of wire-limited hierarchical networks. *J. Parallel Distrib. Comput.*, 41(2):156–172, 1997.

[286] A.J. Huang and Z.G. Mou. Parallel partition expansion for the solution of arbitrary recurrences. In *Proc. IEEE Int. Conf. Parallel Proc.*, volume III, pages 109–114, 1992.

[287] K.H. Huang and J.A. Abraham. Algorithm based fault tolerance for matrix operations. In *Proc. IEEE Int. Conf. Parallel Proc.*, pages 518–528, 1984.

[288] S.T. Huang and S.K. Tripathi. Finite state model and computability theory: new analysis tools for permutation networks. *IEEE Trans. Computers*, C-35(7):591–601, 1986.

[289] W.K Huang and F. Lombardi. On the constant diagnosability of baseline interconnection networks. *IEEE Trans. Computers*, 39(12):1485–1488, 1990.

[290] F.K. Hwang. Control algorithms for rearrangeable Clos networks. *IEEE Trans. Comm.*, 31:952–954, 1983.

[291] F.K. Hwang and Y.H. Xu. Double loop networks with minimum delay. *Discr. Math.*, 66:109–118, 1987.

[292] F.K. Hwang and Y.C. Yao. Comments on the oblivious routing algorithm of Kaklamanis, Krizanc and Tsantilas in the hypercube. *Theory Comput. Syst.*, 31:63–66, 1998.

[293] F.K. Hwang, Y.C. Yao, and M.D. Grammatikakis. d-move local permutation routing for the d-cube. *Disc. Appl. Math.*, 72(3):199–207, 1997.

[294] K. Hwang. *Advanced Computer Architecture: Parallelism, Scalability, Programmability.* McGraw-Hill, New York, 1993.

[295] K. Hwang and T.P. Chang. Combinatorial reliability analysis of multiprocessor computers. *IEEE Trans. Reliability*, 31:469–473, 1982.

[296] R.A. Iannucci, G.R. Gao, Halstead R.H. Jr., and B. (Eds) Smith. *Proc. Workshop Multithreaded Comput.* Kluwer Academic, Norwell, MA, 1994.

[297] D. Ierardi. 2D-bubblesorting in average time $\mathcal{O}(\sqrt{n \lg n})$. In *Proc. 6th ACM Symp. Parallel Algorithms Arch.*, pages 36–45, 1994.

[298] M. Imase and M. Itoh. Design to minimize diameters on building-block network. *IEEE Trans. Computers*, C-30:439–442, 1981.

[299] M. Imase, T. Soneoka, and K. Okada. Connectivity of regular directed graphs with small diameter. *IEEE Trans. Computers*, C-34:267–273, 1985.

[300] J. JáJá. *An Introduction to Parallel Algorithms.* Addison-Wesley, Reading, MA, 1992.

[301] J. JáJá and K.W. Ryu. Load balancing and routing on the hypercube and related networks. *J. Parallel Distrib. Comput.*, 14:431–435, 1992.

[302] C. Jan and A.Y. Oruc. Fast self-routing permutation switching on an asymptotically minimum cost network. *IEEE Trans. Computers*, C-42(12):1369–1379, 1993.

[303] J.F. Jenq and S. Sahni. All pairs shortest paths on a hypercube multiprocessor. In *Proc. IEEE Int. Conf. Parallel Proc.*, pages 713–716, 1987.

[304] C.R. Jesshope, P.R. Miller, and Y.S. Yantchev. High performance communications in processor networks. In *Proc. 16th IEEE Int. Symp. Comput. Arch.*, pages 150–157, 1989.

[305] C.R. Jesshope, P.R. Miller, and Y.S. Yantchev. The mad-postman network chip. In *in Proc. Transputing Conf., IOS Press*, volume II, pages 517–536, 1991.

[306] H. Jiang, L.N. Bhuyan, and J.K. Muppala. MVAMIN: mean value analysis algorithms for multistage interconnection networks. *J. Parallel Distrib. Comput.*, 12(3):189–201, 1991.

[307] S.L. Johnsson and C.T. Ho. Algorithms for matrix transposition for boolean $n$-cube configured ensemble architectures. *SIAM J. Matrix Appl.*, 9(3):419–454, 1988.

[308] S.L. Johnsson and C.T. Ho. Optimum broadcasting and personalized communication in hypercubes. *IEEE Trans. Computers*, C-38(12):1249–1267, 1989.

[309] S.L. Johnsson and C.T. Ho. Generalized shuffle permutations on boolean cubes. *J. Parallel Distrib. Comput.*, 16:1–14, 1992.

[310] S.L. Johnsson and C.T. Ho. Optimal communication channel utilization for matrix transpose and related permutations on boolean cubes. *Disc. Appl. Math.*, 53(1-3):251–274, 1994.

[311] J.M. Joy and R.D. Bergeron. The hypercube as a dynamically reconfigurable processor mesh. *J. Parallel Distrib. Comput.*, 48(1):130–142, 1998.

[312] B. Juurlink, P.S. Rao, and J.F. Sibeyn. Worm-hole gossiping on meshes and tori. In *Proc. 2nd Euro-Par Conf.*, volume 1123 of *Lect. Notes Comp. Sci.*, pages 361–369, 1996.

[313] B. Juurlink, J.F. Sibeyn, and P.S. Rao. Gossiping on meshes and tori. *IEEE Trans. Parallel Distrib. Syst.*, C-9(6):513–525, 1998.

[314] N. Kahale. Eigenvalues and expansion of regular graphs. *J. ACM*, 42(5):1091–1106, 1995.

[315] N. Kahale, F.T. Leighton, Y. Ma, C.G. Plaxton, T. Suel, and E. Szemerédi. Lower bounds for sorting networks. In *Proc. 27th ACM Symp. Theory Comput.*, pages 548–558, 1995.

[316] N. Kahale and T. Leighton. Greedy dynamic routing on arrays. In *Proc. 6th ACM-SIAM Symp. Disc. Algorithms*, pages 558–566, 1995.

[317] C. Kaklamanis, A.R. Karlin, F.T. Leighton, et al. Asymptotically tight bounds for computing with faulty arrays of processors. In *Proc. 31st IEEE Symp. Foundations Comp. Sci.*, pages 285–296, 1990.

[318] C. Kaklamanis and D. Krizanc. Optimal sorting on mesh-connected processor arrays. In *Proc. 4th Symp. on Parallel Algorithms Arch*, pages 50–59. ACM, 1992.

[319] C. Kaklamanis, D. Krizanc, L. Narayanan, and T. Tsantilas. Randomized sorting and selection on mesh-connected processor arrays. In *Proc. 3rd ACM Symp. Parallel Algorithms Arch.*, pages 17–28, 1991.

[320] C. Kaklamanis, D. Krizanc, and S. Rao. Simple path selection for optimal routing on processor arrays. In *Proc. 4th ACM Symp. Parallel Algorithms Arch.*, pages 23–30, 1992.

[321] C. Kaklamanis, D. Krizanc, and A. Tsantilas. Tight bounds for oblivious routing on the hypercube. *Math. Syst. Theory.*, 24(4):223–232, 1991.

[322] H. Karloff and P. Raghavan. Randomized algorithms and pseudorandom numbers. *J. ACM*, 40(3):454–476, 1993.

[323] R. Karp, M. Luby, and F. Meyer auf der Heide. Efficient PRAM simulation on a distributed memory machine. *Algorithmica*, 16(4/5):517–542, 1996.

[324] M. Kaufmann, H. Lauer, and H. Schröder. Fast deterministic hot-potato routing on meshes. In *Proc. 5th ACM Int. Symp. Algorithms Comput.*, volume 834 of *Lect. Notes Comp. Sci.*, pages 333–341, 1994.

[325] M. Kaufmann, U. Meyer, and J.F. Sibeyn. Towards practical permutation routing on meshes. In *Proc. 6th IEEE Symp. Parallel Distrib. Comput.*, pages 664–671, 1994.

[326] M. Kaufmann, U. Meyer, and J.F. Sibeyn. Matrix transpose on meshes: theory and practice. *Computers and Artificial Intelligence*, 16(2):107–140, 1997.

[327] M. Kaufmann, S. Rajasekaran, and J.F. Sibeyn. Matching the bisection bound for routing and sorting on the mesh. In *Proc. 4th ACM Symp. Parallel Algorithms Arch.*, pages 31–40, 1992.

[328] M. Kaufmann, R. Raman, and J.F. Sibeyn. Randomized routing on meshes with buses. *Algorithmica*, 18:417–444, 1997.

[329] M. Kaufmann, H. Schröder, and J.F. Sibeyn. Asymptotically optimal and practical routing on the reconfigurable mesh. *Parallel Proc. Letters*, 5:81–95, 1995.

[330] M. Kaufmann and J.F. Sibeyn. Deterministic routing on circular arrays. In *Proc. 4th IEEE Symp. Parallel Distrib. Comput.*, pages 376–383, 1992.

[331] M. Kaufmann and J.F. Sibeyn. Randomized multipacket routing and sorting on meshes. *Algorithmica*, 17:224–244, 1997.

[332] M. Kaufmann, J.F. Sibeyn, and T. Suel. Derandomizing routing and sorting algorithms for meshes. In *Proc. 5th ACM-SIAM Symp. Disc. Algorithms*, pages 669–679, 1994.

[333] M. Kaufmann, J.F. Sibeyn, and T. Suel. Beyond the bisection bound: fast ranking and counting on meshes. In *Proc. 3rd European Symp. Algorithms*, volume 979 of *Lect. Notes Comp. Sci.*, pages 75–88, 1995.

[334] A. Kavianpour and K.H. Kim. Diagnosabilities of hypercubes under the pessimistic one-step diagnosis strategy. *IEEE Trans. Comput*, 40(2):231–237, 1991.

[335] J. Keller. Fast rehashing in PRAM emulations. *Theoret. Comput. Sci., North-Holland*, 155:349–363, 1996.

[336] P. Kermani and L. Kleinrock. Virtual cut through: a new computer communication switching technique. *Computer Networks*, 3(4):267–286, 1979.

[337] G. Kim and H. Yoon. On submesh allocation for mesh multicomputers: a best-fit allocation and a virtual submesh allocation for faulty meshes. *IEEE Trans. Parallel Distrib. Syst.*, C-9(2):175–185, 1998.

[338] J. Kim, C.R. Das, and W. Lin. A top-down processor allocation scheme for hypercube computers. *IEEE Trans. Parallel Distrib. Syst.*, C-2(1):20–30, 1991.

[339] J.H. Kim, Z. Liu, and A.A. Chien. Compressionless routing: a framework for adaptive and fault-tolerant routing. *IEEE Trans. Parallel Distrib. Syst.*, C-8(3):229–244, 1997.

[340] J.U. Kim, K.H. Shim, and Park K.H. A link-disjoint subcube for processor allocation in hypercube computers. *Parallel Comput.*, 22(12):1579–1595, 1997.

[341] N.M. Kini, A. Kumar, and D.P. Agrawal. Quantitative reliability analysis of redundant multistage interconnection networks. *Reliability of Computer and Comm. Networks, DIMACS Series in Discr. Math. Theoret. Comp. Sci.*, 5:153–170, 1991.

[342] M. Klugerman and C.G. Plaxton. Small-depth counting networks. In *Proc. 24th ACM Symp. Theory Comput.*, pages 417–428, 1992.

[343] D.E. Knuth. *The Art of Computer Programming: Sorting and Searching.* Addison-Wesley, Reading, MA, 1973.

[344] P.M. Kogge. *The Architecture of Pipelined Computers.* McGraw-Hill, New York, 1981.

[345] S. Konstantinidou and L. Snyder. The Chaos router. *IEEE Trans. Computers*, C-43(12):1386–1397, 1994.

[346] D.M. Koppelman and A.Y. Oruc. A self-routing permutation network. *J. Parallel Distrib. Comput.*, 10(2):140–151, 1990.

[347] S.C. Kothari and S. Lakshmivarahan. A note on rearrangeable networks. Technical report, Dept. Electr. Engin. Comp. Sci., University of Oklahoma, 1983.

[348] M. Kraetzl and C.J. Colbourn. Threshold channel graphs. *Combinatorics, Probability and Comput.*, 2:337–349, 1993.

[349] M. Kraetzl and C.J. Colbourn. Transformations on channel graphs. *IEEE Trans. Comm.*, 41:664–666, 1993.

[350] E. Kranakis and L. Kirousis, editors. *Proc. 2nd Colloq. Struct. Inf. & Comm. Compl.*, volume 2. Carleton Univ. Press, 1995.

[351] D. Krizanc. A note on off-line permutation routing on a mesh-connected processor array. *Parallel Proc. Letters*, 1:67–70, 1991.

[352] D. Krizanc. Oblivious routing with limited buffer capacity. *J. Computer Syst. Sci.*, 43:317–327, 1991.

[353] D. Krizanc. Integer sorting on a mesh-connected array of processors. *Inf. Proc. Letters*, 47:283–289, 1993.

[354] D. Krizanc and L. Narayanan. Zero-one sorting on the mesh. *Parallel Proc. Letters*, 5(2):149–155, 1995.

[355] D. Krizanc, S. Rajasekaran, and T. Tsantilas. Optimal routing algorithms for mesh connected processor arrays. In *Proc. VLSI Algorithms and Architectures*, volume 319 of *Lect. Notes Comp. Sci.*, pages 411–422, Springer-Verlag, New York, 1988.

[356] L. Kronsjo. PRAM models. In *Handbook Parallel Distrib. Comput.*, A. Zomaya, editor, McGraw-Hill, New York, 1996.

[357] P. Krueger and D. Babbar. The effects of precedence and priority constraints on the performance of scan scheduling for hypercube multiprocessors. *J. Parallel Distrib. Comput.*, 39(2):95–104, 1996.

[358] P. Krueger, T.H. Lai, and V.A. Dixit-Radiya. Job scheduling is more important than processor allocation for hypercube computers. *IEEE Trans. Parallel Distrib. Syst.*, C-5(5):488–497, 1994.

[359] C.P. Kruskal, L. Rudolph, and M. Snir. The power of parallel prefix. *IEEE Trans. Computers*, C-34(10):965–968, 1985.

[360] C.P. Kruskal and M. Snir. The performance of multistage interconnection networks for multiprocessors. *IEEE Trans. Computers*, C-32(12):1091–1098, 1983.

[361] C.P. Kruskal and M. Snir. A unified theory of interconnection network structure. *Theoretical Comp. Sci.*, 48(1):75–94, 1986.

[362] H. Kumamoto, K. Tanaka, K. Inone, and E.J. Henley. Dagger sampling Monte Carlo evaluation of system reliability. *IEEE Trans. Reliability*, 29:122–125, 1980.

[363] M. Kumar and D.S. Hirschberg. An efficient implementation of Batcher's odd-even merge algorithm and its application in parallel sorting schemes. *IEEE Trans. Computers*, C-32(3):254–264, 1983.

[364] V. Kumar, A. Grama, A. Gupta, and G. Karypis. *Introduction to Parallel Computing: Design and Analysis of Algorithms.* Benjamin/Cummings, Menlo Park, CA, 1994.

[365] V.P Kumar and A.L. Reibman. Failure dependent performance analysis of a fault tolerant multistage interconnection network. *IEEE Trans. Computers*, 38(12):1703–1713, 1989.

[366] M. Kunde. Lower bounds for sorting on mesh-connected architectures. *Acta Informatica*, 24:121–130, 1987.

[367] M. Kunde. Routing and sorting on mesh connected processor arrays. In *Proc. VLSI Algorithms Arch.*, volume 319 of *Lect. Notes Comp. Sci.*, pages 423–433, 1988.

[368] M. Kunde. Balanced routing: towards the distance bound on grids. In *Proc. 3rd ACM Symp. Parallel Algorithms Arch.*, pages 260–271, 1991.

[369] M. Kunde. Concentrated regular data streams on grids: sorting and routing near to the bisection bound. In *Proc. 32nd IEEE Symp. Foundations Comp. Sci.*, pages 141–150, 1991.

[370] M. Kunde. Block gossiping on grids and tori: deterministic sorting and routing match the bisection bound. In *Proc. European Symp. Algorithms*, volume 726 of *Lect. Notes Comp. Sci.*, pages 272–283, 1993.

[371] M. Kunde, R. Niedermeier, K. Reinhardt, and P. Rossmanith. Optimal average case sorting on arrays. In *Proc. 12th Symp. Theoret. Asp. Comput. Sci.*, Lect. Notes Comp. Sci., pages 503–513, 1995.

[372] M. Kunde, R. Niedermeier, and P. Rossmanith. Faster sorting and routing on grids with diagonals. In *Proc. 11th Symp. Theoret. Asp. Comput. Sci.*, volume 775 of *Lect. Notes Comp. Sci.*, pages 225–236, 1994.

[373] M. Kunde and T. Tensi. Multi-packet routing on mesh connected processor arrays. In *Proc. 1st ACM Symp. Parallel Algorithms Arch.*, pages 336–343, 1989.

[374] M. Kunde and T. Tensi. *k-k* routing on multidimensional mesh-connected arrays. *J. Parallel Distrib. Comput.*, 11:146–155, 1991.

[375] M. Kutyłowski, K. Loryś, and B. Oesterdiekhoff. Periodic merging networks. *Theory Comput. Syst.*, 31(5):551–578, 1998.

[376] M. Kutyłowski, K. Loryś, B. Oesterdiekhoff, and R. Wanka. Fast and feasible periodic sorting networks of constant depth. In *Proc. 35th IEEE Symp. Foundations Comp. Sci.*, pages 369–380, 1994.

[377] M. Kutyłowski and R. Wanka. Periodic sorting on two-dimensional meshes. *Parallel Proc. Letters*, 2:213–220, 1992.

[378] B.C. Kuzmaul. Fast deterministic routing on hypercubes using small buffers. *IEEE Trans. Computers*, C-39(11):1388–1393, 1990.

[379] Q.H. Kwon and K.Y. Chwa. An algorithm for scheduling jobs in hypercube systems. *IEEE Trans. Parallel Distrib. Syst.*, C-9(9):856–866, 1998.

[380] R.E. Ladner and M.J. Fischer. Parallel prefix computation. *J. ACM*, 27(4):831–838, 1980.

[381] T.H. Lai and S. Sahni. Anomalies in parallel branch-and-bound algorithms. *Comm. ACM*, 27(6):594–602, 1984.

[382] S. Lakshmivarahan and S.K. Dhall. *Analysis and Design of Parallel Algorithms.* McGraw-Hill, New York, 1990.

[383] S. Lakshmivarahan and S.K. Dhall. *Parallel Computing using the Prefix Problem.* Oxford University Press, New York, 1994.

[384] S. Lakshmivarahan, S.K. Dhall, and Miller L. Parallel sorting algorithms. *Adv. Comput.*, pages 295–354, 1984.

[385] A. LaMarca and R.E. Ladner. The influence of caches on the performance of sorting. In *Proc.* $8^{th}$ *ACM-SIAM Symp. Disc. Algorithms*, pages 370–379, 1997.

[386] L. Lamport. How to make a multiprocessor computer that correctly executes multiprocess programs. *IEEE Trans. Computers*, C-28(9):690–691, 1979.

[387] H.W. Lang, M. Schimmler, H. Schneck, and H. Schröder. Systolic sorting on a mesh connected network. *IEEE Trans. Computers*, C-34(7):652–658, 1985.

[388] T. Lang and L. Kurisaki. Nonuniform traffic spots (NUTS) in multistage interconnection networks. *J. Parallel Distrib. Comput.*, 10:55–67, 1990.

[389] D.H. Lawrie. Access and alignment of data in an array processor. *IEEE Trans. Computers*, C-24(12):1145–1155, 1975.

[390] C.T. Lea and D.J. Shyy. Tradeoff of horizontal decomposition versus vertical stacking in rearrangeable nonblocking networks. *IEEE Trans. Comm.*, 39(6):899–904, 1991.

[391] C.Y. Lee. Analysis of switching networks. *Bell System Technical Journal*, 34(6):1287–1315, 1955.

[392] C.Y. Lee and A.Y. Oruc. Design of efficient and easily routable generalized connectors. *IEEE Trans. Comm.*, 43(2):646–650, 1995.

[393] G. Lee, B.C. Kang, and R.Y. Kain. Analysis of finite buffered multistage combining networks. *IEEE Trans. Parallel Distrib. Syst.*, C-6(7):760–766, 1995.

[394] G.H. Lee, C.P. Kruskal, and D.J. Kuch. On the effectiveness of combining in resolving "hot spot" contention. *J. Parallel Distrib. Comput.*, 20(2):136–144, 1994.

[395] K.Y. Lee. On the rearrangeability of $(2\log_2 n - 1)$ stage permutation networks. *IEEE Trans. Computers*, C-34(5):412–425, 1985.

[396] K.Y. Lee. A new Beneš network control algorithm. *IEEE Trans. Computers*, C-36(6):768–772, 1987.

[397] S. Lee and K.G. Shin. Interleaved all-to-all reliable broadcast on meshes and hypercubes. *IEEE Trans. Parallel Distrib. Syst.*, C-5(5):449–458, 1994.

[398] T.C. Lee and J.P. Hayes. A fault tolerant communication scheme for hypercube computers. *IEEE Trans. Computers*, C-41(10):1242–1255, 1992.

[399] F.T. Leighton. Tight bounds on the complexity of parallel sorting. *IEEE Trans. Computers*, C-34(4):344–354, 1985.

[400] F.T. Leighton. Average case analysis of greedy routing algorithms. In *Proc. 2nd ACM Symp. Parallel Algorithms Arch.*, pages 2–10, 1990.

[401] F.T. Leighton. *Introduction to Parallel Algorithms and Architectures: Arrays • Trees • Hypercubes.* Academic Press, New York, 1992.

[402] F.T. Leighton. Methods for message routing in parallel machines. *Theoret. Comput. Sci., North-Holland*, 128(1-2):31–62, 1994.

[403] F.T. Leighton and Y. Ma. Tight lower bounds on the size of fault-tolerant merging and sorting networks with destructive faults. In

*Proc. 5th ACM Symp. Parallel Algorithms Arch.*, pages 30–41, 1993.

[404] F.T. Leighton and Y. Ma. Breaking the $\Theta(nlog^2n)$ barrier for sorting with faults. *J. Comput. Syst. Sci.*, 54(2):265–304, 1997.

[405] F.T. Leighton, Y. Ma, and C.G. Plaxton. Highly fault-tolerant sorting circuits. In *Proc. 32nd IEEE Symp. Foundations Comp. Sci.*, pages 458–469, 1991.

[406] F.T. Leighton, Y. Ma, and T. Suel. On probabilistic networks for selection, merging and sorting. *Theory Comput. Syst.*, 30(6):559–582, 1997.

[407] F.T. Leighton and B.M. Maggs. Fast algorithms for routing around faults in multibutterflies and randomly-wired splitter networks. *IEEE Trans. Computers*, C-41(5):578–587, 1992.

[408] F.T. Leighton, B.M. Maggs, A. Ranade, and S. Rao. Randomized routing and sorting on fixed-connection networks. *J. Algorithms*, 17:157–205, 1994.

[409] F.T. Leighton, B.M. Maggs, and S. Rao. Universal packet routing algorithms. *Combinatorica*, 14(2):167–186, 1994.

[410] F.T. Leighton, B.M. Maggs, and A.W. Richa. Fast algorithms for finding $\mathcal{O}$(congestion+dilation) packet routing schedules. *Combinatorica*, 19(3):375–401, 1999.

[411] F.T. Leighton, B.M. Maggs, and R. Sitaraman. On the fault tolerance of some popular bounded-degree networks. *SIAM J. Comput.*, 27(5):1303–1333, 1998.

[412] F.T. Leighton, F. Makedon, and I.G. Tollis. A $2n - 2$ step algorithm for routing in an $n \times n$ array with constant size queues. *Algorithmica*, 14(4):561–572, 1995.

[413] F.T. Leighton, M.J. Newman, A.G. Ranade, et al. Dynamic tree embeddings in butterflies and hypercubes. *SIAM J. Comput.*, 21(4):639–654, 1992.

[414] F.T. Leighton and C.G. Plaxton. Hypercubic sorting networks. *SIAM J. Comput.*, 27(1):1–47, 1998.

[415] C.E. Leiserson. Fat-trees: universal networks for hardware-efficient supercomputing. *IEEE Trans. Computers*, C-34(10):892–901, 1985.

[416] C.E. Leiserson, Z.S. Abuhamdeh, D.C. Douglas, et al. The network architecture of the connection machine CM-5. *J. Parallel Distrib. Comput.*, 33(2):145–158, 1996.

[417] A. Lempel. $M$-ary closed sequences. *J. Comb. Theory.*, 10:253–258, 1971.

[418] J. Lenfant. Permutations of data: a Beneš network control algorithm for frequently used permutations. *IEEE Trans. Computers*, C-27(7):637–647, 1978.

[419] J. Lengauer. VLSI theory. In *Handbook Theoret. Comput. Sci.*, J. van Leeuwen, editor, pages 805–833. North-Holland, Amsterdam, 1990.

[420] V. Leppänen and M. Penttonen. Work-optimal simulation of PRAM models on meshes. *Nordic J. Comput.*, 2:51–69, 1995.

[421] J. Leung and S.M. Shende. On multi-dimensional packet routing on meshes with buses. *J. Parallel Distrib. Comput.*, 20:187–197, 1994.

[422] Y.W. Leung. Online fault identification in multistage interconnection networks. *Parallel Comput.*, 19:693–702, 1993.

[423] G. Lev, N. Pippenger, and L.G. Valiant. A fast parallel algorithm for routing in permutation networks. *IEEE Trans. Computers*, C-30(2):93–100, 1981.

[424] G.J. Li and B.W. Wah. Coping with anomalies in parallel branch-and-bound algorithms. *IEEE Trans. Computers*, C-35(6):568–573, 1986.

[425] K. Li. IVY: a shared virtual memory system for parallel computing. In *Proc. IEEE Conf. Parallel Proc.*, pages 94–101, 1988.

[426] K. Li and K.H. Cheng. A two-dimensional buddy system for dynamic resource allocation in a partitionable mesh connected system. *J. Parallel Distrib. Comput.*, 12(1):79–83, 1991.

[427] K. Li and P.R. Hudak. Memory coherence in shared virtual memory systems. *ACM Trans. Computers Syst.*, C-7(4):321–359, 1989.

[428] S.T. Liang and W.T. Tsai. Multi-failure fault tolerance of embedded loops on hypercubes. *IEEE Conf Parallel Distr. Proc., Dallas, TX*, pages 511–516, 1999.

[429] Y.C. Lim and C. Wang. Performance of crossbar multiprocessor system. *J. Parallel Distrib. Comput.*, 7(2):321–334, 1989.

[430] G. Lin and N. Pippenger. Parallel algorithms for routing in non-blocking networks. *Math. Syst. Theory.*, 27(1):29–40, 1994.

[431] W. Lin. Manipulating general vectors on synchronous binary $n$-cube. *IEEE Trans. Computers*, C-42(7):863–870, 1993.

[432] X. Lin, P.K. McKinley, and L.M. Ni. Deadlock-free multicast wormhole routing in 2D mesh multicomputers. *IEEE Trans. Parallel Distrib. Syst.*, C-5(8):793–804, 1994.

[433] X. Lin and L.M. Ni. Multicast communication in multicomputer networks. *IEEE Trans. Parallel Distrib. Syst.*, C-4(10):1105–1117, 1993.

[434] M.H. Lipasti and J.P. Shen. Superspeculative microarchitecture for beyond AD 2000. *IEEE Computer*, 30(9):59–67, 1997.

[435] R.J. Lipton. DNA solution of hard computational problems. *Science*, 268:542–545, 1995.

[436] D. Lisinski, F.T. Leighton, and B.M. Maggs. Empirical evaluation of randomly-wired multistage networks. In *Proc. IEEE Int. Conf. Comput. Design.*, pages 380–385, 1990.

[437] J.C. Liu and K. Shin. Polynomial testing of packet switching networks. *IEEE Trans. Computers*, 38(2):202–217, 1989.

[438] T. Liu, W.K. Huang, F. Lombardi, and L.N. Bhuyan. A submesh allocation scheme for mesh-connected multiprocessor systems. In *Proc. IEEE Int. Conf. Parallel Proc.*, volume II, pages 159–163, 1995.

[439] Y.S. Liu and S.R. Dickey. Simulation and analysis of enhanced switch architectures for interconnection networks in massively parallel shared machines. In *Proc. 2nd Symp. Frontiers Massively Parallel Comput.*, pages 487–490, 1988.

[440] Z. Liu and T.Y Sung. Routing and transmitting problems in de Bruijn networks. *IEEE Trans. Computers*, 45:1056–1062, 1996.

[441] M. Livingston and Q.F. Stout. Fault tolerance of allocation schemes in massively parallel computers. In *Proc. 2nd Symp. Frontiers Massively Parallel Comput.*, pages 491–494, 1988.

[442] M.L. Livingston and Q.F. Stout. Embeddings in hypercubes. *Math. and Comput. Modeling*, 11:222–227, 1988.

[443] V. Lo, K.J. Windisch, W. Liu, and B. Nitzberg. Noncontiguous processor allocation algorithms for mesh-connected multicomputers. *IEEE Trans. Parallel Distrib. Syst.*, C-8(7):712–726, 1997.

[444] B.N. Locanthi. *Homogeneous machine.* PhD thesis, Dept. Comp. Sci., California Institute of Technology, 1989.

[445] F. Lombardi, C. Feng, and W.K. Huang. Detection and location of multiple faults in baseline interconnection networks. *IEEE Trans. Computers*, 41(10):1340–1344, 1992.

[446] A. Lubiw. Counterexample to a conjecture of Szymanski on hypercube routing. *Inf. Proc. Letters*, 35:57–61, 1990.

[447] Y.D. Lyuu. Fast fault-tolerant parallel communication and on-line maintenance for hypercubes using information dispersal. *Math. Syst. Theory.*, 24(4):273–294, 1991.

[448] Y. Ma. An $\mathcal{O}(N \log N)$ size fault tolerant sorting network. In *Proc. 28th ACM Symp. Theory Comput.*, pages 266–275, 1996.

[449] Y. Ma, S. Sen, and I.D. Scherson. The distance bound for sorting on mesh-connected processor arrays is tight. In *Proc. 27th IEEE Symp. Foundations Comp. Sci.*, pages 255–263, 1986.

[450] B. Maggs and R. Sitaraman. Simple algorithms for routing on butterfly networks with bounded queues. In *Proc. 24th ACM Symp. Theory Comput.*, pages 150–161, 1992.

[451] B.M. Maggs. Randomly wired multistage networks. *Statistical Sci.*, 8(1):70–75, 1993.

[452] B.M. Maggs, L.R. Matheson, and R.E. Tarjan. Models of parallel computation: a survey and synthesis. In *Proc. 28th Hawaii Int. Conf. Syst. Sci.*, volume II, pages 61–70, 1995.

[453] B.M. Maggs and B. Vöcking. Improved routing and sorting on multibutterflies. In *Proc. 29th ACM Symp. Theory Comput.*, pages 517–530, 1997.

[454] F. Makedon and A. Symvonis. Optimal algorithms for the many-to-one routing problem on two-dimensional meshes. *Microprocessors and Microsystems*, 17:361–367, 1993.

[455] F. Makedon and A. Symvonis. Optimal algorithms for multipacket routing problems on rings. *J. Parallel Distrib. Comput.*, 22:37–43, 1994.

[456] F. Makedon and A. Symvonis. Flit-serial packet routing on meshes and tori. *Math. Syst. Theory.*, 29(4):311–330, 1996.

[457] M. Malek and E. Opper. Multiple fault diagnosis of SW-banyan networks. In *13th Int. Symp. Fault Tolerant Comput.*, pages 446–449, 1983.

[458] Y. Mansour and B. Patt-Shamir. Many-to-one packet routing on grids. In *Proc. 27th ACM Symp. Theory Comput.*, pages 258–267, 1995.

[459] Y. Mansour and L. Schulman. Sorting on a ring of processors. *J. Algorithms*, 11:622–630, 1990.

[460] J.M. Marberg and E. Gafni. Sorting in constant number of row and column phases on a mesh. *Algorithmica.*, 3:561–572, 1988.

[461] G.M. Masson. Binomial switch networks for concentrators and distributors. *IEEE Trans. Computers*, C-25(9):873–883, 1977.

[462] G.M. Masson. Expected capacity of $(m, 2)$ networks. *IEEE Trans. Computers*, C-32(7):649–656, 1983.

[463] G.M. Masson and B.W. Jordan. Generalized multistage connection networks. *Networks*, 2:191–209, 1972.

[464] T.R. Mathies. Percolation theory and computing with faulty arrays of processors. In *Proc. 3rd ACM-SIAM Symp. Disc. Algorithms*, pages 100–103, 1992.

[465] E.W. Mayr and C.G. Plaxton. Pipelined parallel prefix computations and sorting on a pipelined hypercube. *J. Parallel Distrib. Comput.*, 17:374–380, 1993.

[466] R.J. McMillen and H.J. Siegel. Routing schemes for the augmented data manipulator network in an MIMD system. *IEEE Trans. Computers*, C-31(12):1202–1214, 1982.

[467] K. Mehlhorn and U. Vishkin. Randomized and deterministic simulations of PRAMs by parallel machines with restricted granularity of parallel memories. *Acta Informatica*, 9(1):29–59, 1984.

[468] R. Mehrotra and E.F. Gehringer. Superlinear speedup through randomized algorithms. In *Proc. IEEE Int. Conf. Parallel Proc.*, volume I, pages 291–300, 1985.

[469] R. Melen and J.S. Turner. Nonblocking multirate networks. *SIAM J. Comput.*, 18(2):301–313, 1989.

[470] R. Melen and J.S. Turner. Nonblocking multirate distribution networks. *IEEE Trans. Comm.*, 41(2):362–369, 1993.

[471] J.M. Mellor-Crummey and M.L. Scott. Algorithms for scalable synchronization on shared memory multiprocessors. *ACM Trans. Computers Syst.*, C-9(1):21–65, 1991.

[472] F. Meyer auf der Heide. Efficient simulations among several models of parallel computers. *SIAM J. Comput.*, 15(1):106–119, 1986.

[473] F. Meyer auf der Heide, B. Oesterdiekhoff, and R. Wanka. Strongly adaptive token distribution. *Algorithmica*, 15(5):413–427, 1996.

[474] F. Meyer auf der Heide and C. Scheideler. Routing with bounded buffers and hot-potato routing in vertex-symmetric networks. In *Proc. 3rd European Symp. Algorithms*, volume 979 of *Lect. Notes Comp. Sci.*, pages 341–354, 1995.

[475] F. Meyer auf der Heide and C. Scheideler. Space-efficient routing in vertex-symmetric networks. In *Proc. 7th ACM Symp. Parallel Algorithms Arch.*, pages 137–146, 1995.

[476] F. Meyer auf der Heide, C. Scheideler, and V. Stemann. Exploiting storage redundancy to speed up randomized shared memory simulations. *Theoret. Comput. Sci., North-Holland*, 162(2):245–281, 1996.

[477] M. Michael and M.L. Scott. Simple, fast and practical non-blocking and blocking concurrent queue algorithms. In *Proc. ACM SIGACT-SIGOPS Symp. Princ. Distrib. Comput.*, pages 267–275, 1996.

[478] M. Mitzenmacher. Bounds on the greedy routing algorithm for array networks. In *Proc. 6th ACM Symp. Parallel Algorithms Arch.*, pages 346–353, 1994.

[479] P. Mohapatra. Dynamic real-time task scheduling on hypercubes. *J. Parallel Distrib. Comput.*, 14(2):91–100, 1997.

[480] P. Mohapatra and C.R. Das. On dependability evaluation of mesh connected multicomputers. *IEEE Trans. Computers*, C-44(9):1073–1084, 1995.

[481] P. Mohapatra and C.R. Das. Performance analysis of finite-buffered asynchronous multistage interconnection networks. *IEEE Trans. Parallel Distrib. Syst.*, 7(1):18–25, 1996.

[482] P. Mohapatra, S. Wong, and C.R. Das. Performance analysis of combining multistage interconnection networks. In *Proc. IEEE Int. Conf. Parallel Proc.*, volume I, pages 13–16, 1994.

[483] P. Mohapatra, C. Yu, and C.R. Das. A lazy scheduling scheme for hypercube computers. *J. Parallel Distrib. Comput.*, 27(1):26–37, 1995.

[484] B. Monien, R. Diekmann, R. Feldmann, et al. Efficient use of parallel and distributed systems: from theory to practice. In *Proc. Comp. Sci. Today: Recent Trends & Devel.*, volume 1000 of *Lect. Notes Comp. Sci.*, pages 62–77, 1995.

[485] Z.G. Mou, C. Constantinescu, and T.J. Hickey. Optimal mappings of divide and conquer algorithms to mesh connected parallel architectures. In *Proc. Int. Comput. Symp., Taichung, Taiwan*, pages 273–274, 1992.

[486] Z.G. Mou and M. Goodman. A comparison of communication costs for three parallel program paradigms on hypercube and mesh architectures. In *Proc. 5th SIAM Conf. Parallel Proc. Sci. Comput.*, pages 491–500, 1992.

[487] Z.G. Mou and X. Wang. A divide and conquer method of solving tridiagonal systems on hypercube massively parallel computers. In *Proc. 3rd IEEE Symp. Parallel Distrib. Comput.*, pages 810–817, 1991.

[488] A. Mourad, B. Oezden, and M. Malek. Comprehensive testing of multistage interconnection networks. *IEEE Trans. Computers*, 40(8):935–951, 1991.

[489] B. Mukherjee. *Optical Communication Networks.* McGraw-Hill, New York, 1997.

[490] K. Mulmuley, U.V. Vazarani, and V.V. Vazarani. Matching is as easy as matrix inversion. In *19th Annual ACM Symp. Theory Comput.*, pages 345–354, 1987.

[491] S. Muthukrishnan, M.S. Paterson, and T. Suel. Compact grid layouts of multilevel networks. In *Proc. 31st ACM Symp. Theory Comput.*, pages 455–463, 1999.

[492] V.S.S. Nair and J.A. Abraham. Average checksum codes for fault-tolerant matrix operations on processor arrays. In *Proc. Int. Conf. Supercomput.*, pages 284–290, 1987.

[493] V.S.S. Nair, J.A. Abraham, and P. Banerjee. Efficient techniques for the analysis of algorithm-based fault tolerance schemes. *IEEE Trans. Computers*, C-40(9):499–503, 1996.

[494] K. Nakano. A bibliography of published papers on dynamically reconfigurable architectures. *Parallel Proc. Letters*, 5:111–124, 1995.

[495] D. Nassimi. Parallel algorithms for the classes of $2^b$ descend and ascend computations on a simd hypercube. *IEEE Trans. Parallel Distrib. Syst.*, C-4(12):1372–1381, 1993.

[496] D. Nassimi. Nearly logarithmic-time parallel algorithm for the class of $2^b$ ascend computations on a SIMD hypercube. *J. Parallel Distrib. Comput.*, 20:289–302, 1994.

[497] D. Nassimi and S. Sahni. Bitonic sorting on a mesh-connected parallel computer. *IEEE Trans. Computers*, C-28(1):2–7, 1979.

[498] D. Nassimi and S. Sahni. An optimal routing algorithm for mesh-connected parallel computers. *J. the ACM*, 27(1):6–29, 1980.

[499] D. Nassimi and S. Sahni. Data broadcasting in SIMD computers. *IEEE Trans. Computers*, C-30(2):101–106, 1981.

[500] D. Nassimi and S. Sahni. A self-routing Beneš network and parallel permutation algorithms. *IEEE Trans. Computers*, C-30(4):332–340, 1981.

[501] D. Nassimi and S. Sahni. Optimal BPC permutations on a cube connected SIMD computer. *IEEE Trans. Computers*, C-31(4):338–341, 1982.

[502] D. Nassimi and S. Sahni. Parallel algorithms to set-up the Beneš permutation network. *IEEE Trans. Computers*, C-31(2):148–154, 1982.

[503] D. Nassimi and S. Sahni. Parallel permutation and sorting algorithms and a new generalized connection network. *J. ACM*, 29(3):642–667, 1982.

[504] I. Newman and A. Schuster. Hot-potato algorithms for permutation routing. *IEEE Trans. Parallel Distrib. Syst.*, C-6(11):1168–1176, 1995.

[505] I. Newman and A. Schuster. Hot-potato worm routing via store-and-forward packet routing. *J. Parallel Distrib. Comput.*, 30:76–84, 1995.

[506] L.M. Ni, Y. Gui, and S. Moore. Performance evaluation of switch-based wormhole networks. In *Proc. IEEE Conf. Parallel Proc.*, volume 1, pages 32–40, 1995.

[507] D. Nicol. Inflated speedups in parallel simulation via malloc( ). *Int. J. Simulation*, 2(4):413–426, 1992.

[508] R. Niedermeier, K. Reinhardt, and P. Sanders. Towards optimal locality in mesh-indexings. In *Proc. 11th Symp. Fundamentals Comput. Theory*, volume 1279 of *Lect. Notes Comp. Sci.*, pages 364–375, 1997.

[509] M. Nigam and S. Sahni. Sorting $n$ numbers on $n \times n$ reconfigurable meshes with buses. *J. Parallel Distrib. Comput.*, 23:37–48, 1994.

[510] D. Nussbaum and A. Agarwal. Scalability of parallel machines. *Comm. ACM*, 34(3):56–61, 1991.

[511] J.P. Ofman. A universal automaton. *Trans. Moscow Math Soc.*, pages 200–215, 1965.

[512] A.D. Oh and H.A. Choi. Generalized measure of fault tolerance in $n$-cube networks. *IEEE Trans. Parallel Distrib. Syst.*, C-4(6):702–703, 1993.

[513] L. Oliker and R. Biswas. Parallelization of a dynamic unstructured application using three leading paradigms. In *Proc. Int. Conf. Supercomput.*, pages 11–18, 1999.

[514] A.R.O. Omondi and J.D. Brock. Implementing a dictionary on hypercube machines. In *Proc. IEEE Int. Conf. Parallel Proc.*, pages 707–710, 1987.

[515] D.C. Opferman and N.T. Tsao-Wu. On a class of rearrangeable switching networks, part i: control algorithms, part ii: enumeration studies and fault diagnosis. *Bell Technical J.*, 50(5):1579–1600, 1971.

[516] A.Y. Oruc and H.M. Huang. Crosspoint complexity of sparse crossbar concentrators. *IEEE Trans. Inform. Theory*, 35(9):1466–1471, 1996.

[517] A.Y. Oruc and M.Y. Oruc. On testing isomorphisms of permutation networks. *IEEE Trans. Computers*, C-34(10):958–961, 1985.

[518] P. Ouyang and Palem K.V. Very efficient cyclic shifts on hypercubes. *J. Parallel Distrib. Comput.*, 39(1):79–86, 1996.

[519] L.B. Page and J.E. Perry. An algorithm for exact fault-tree probability without cut sets. *IEEE Trans. Reliability*, 35:544–588, 1986.

[520] M.A. Palis, S. Rajasekaran, and D.S.L. Wei. Packet routing and PRAM emulations on star graphs and leveled networks. *J. Parallel Distrib. Comput.*, 20:145–157, 1994.

[521] P. Panaite. Hypercube permutations routable under dimension orderings. *Inf. Proc. Letters*, 59(4):185–189, 1996.

[522] C. Papadimitriou and K. Steiglitz. *Combinatorial Optimization.* Prentice-Hall, Englewood Cliffs, NJ, 1982.

[523] I. Parberry. A computer assisted optimal depth lower bound for nine-input sorting networks. *Math. Syst. Theory*, 24:101–116, 1991.

[524] D.S. Parker. Notes on shuffle/exchange-type switching networks. *IEEE Trans. Computers*, 29(4):213–222, 1980.

[525] D.S. Parker and C.S. Raghavendra. The Gamma network. *IEEE Trans. Computers*, C-31(4):367–373, 1984.

[526] D. Parkinson. Parallel efficiency can be greater than unity. *Parallel Comput.*, 3:261–262, 1986.

[527] J.H. Patel. Processor-memory interconnections for multiprocessors. In *Proc. 6th Annual Symp. on Computer Architecture*, pages 168–177, 1979.

[528] J.H. Patel. Performance of processor-memory interconnections for multiprocessors. *IEEE Trans. Computers*, C-30:771–780, 1981.

[529] M.S. Paterson. Improved sorting networks with $\mathcal{O}(\log N)$ depth. *Algorithmica*, 5:65–92, 1990.

[530] Y.N. Patt, S.J. Patel, M. Evers, D.H. Friendly, et al. One billion transistors, one uniprocessor, one chip. *IEEE Computer*, 30(9):51–57, 1997.

[531] D. Patterson. A case for intelligent DRAM: IRAM. *IEEE Micro*, 17(2):34–44, 1997.

[532] M.C. Pease. The indirect binary $n$-cube microprocessor array. *IEEE Trans. Computers*, C-26:458–473, 1977.

[533] D. Peleg. A note on optimal time broadcast in faulty hypercubes. *J. Parallel Distrib. Comput.*, 26(1):132–135, 1995.

[534] D. Peleg and B. Simmons. On fault tolerant routings in general networks. *Information and Computation*, 74(1):33–44, 1987.

[535] D. Peleg and E. Upfal. The token distribution problem. *SIAM J. Comput.*, 18(2):229–243, 1989.

[536] D. Peleg and E. Upfal. A time randomness tradeoff for oblivious routing. *SIAM J. Comput.*, 19(2):256–266, 1990.

[537] J.G. Peters and M. Syska. Circuit-switched broadcasting in torus networks. *IEEE Trans. Parallel Distrib. Syst.*, C-7(3):246–255, 1996.

[538] J.C. Peterson, J.O. Tuazon, D. Lieberman, and M. Pniel. The Mark III hypercube-ensemble concurrent computer. In *Proc. IEEE Int. Conf. Parallel Proc.*, volume I, pages 71–73, 1985.

[539] G.F. Pfister and V.A. Norton. Hot-spot contention and combining in multistage interconnection networks. *IEEE Trans. Computers*, C-34(10):943–948, 1985.

[540] J. Pieprzyk. Non-linearity of exponent permutations. In *Proc. Advances in Cryptology-Eurocrypt*, pages 80–92, 1989.

[541] J. Pieprzyk and G. Finkelstein. Permutations that maximize non-linearity and their cryptographic significance. In *Proc. IFIP Conf. on Comput. Security in Age of Info.*, pages 63–74, 1989.

[542] A. Pietracaprina and G. Pucci. Tight bounds on deterministic PRAM emulations with constant redundancy. In *Proc. 2nd European Symp. Algorithms*, volume 855 of *Lect. Notes Comp. Sci.*, pages 391–400, 1994.

[543] A. Pietracaprina and G. Pucci. Improved deterministic PRAM simulation on the mesh. In *Proc. 22nd Int. Colloquium on Automata Lang. and Programming*, volume 944 of *Lect. Notes Comp. Sci.*, pages 372–383, 1995.

[544] A. Pietracaprina, G. Pucci, and J.F. Sibeyn. Constructive deterministic PRAM simulation on a mesh-connected computer. In *Proc. 6th ACM Symp. Parallel Algorithms Arch.*, pages 248–256, 1994.

[545] T.M. Pinkston and S. Warnakulasuriya. On deadlocks in interconnection networks. In *Proc. 24th IEEE Int. Symp. Comput. Arch.*, pages 38–49, 1997.

[546] M.S. Pinsker. On the complexity of a concentrator. In *Proc. 7th Int. Teletraffic Congr.*, pages 3181–3184, 1973.

[547] N. Pippenger. Superconcentrators. *SIAM J. Comput.*, 6:298–304, 1972.

[548] N. Pippenger. Telephone switching networks. In *Proc. Symp. Appl. Math.*, volume 26, pages 101–133, 1982.

[549] N. Pippenger. Parallel communication with limited buffers. In *Proc. 25th IEEE Symp. Foundations Comp. Sci.*, pages 127–136, 1984.

[550] N. Pippenger. Communication networks. In *Handbook Theoret. Comput. Sci.*, J. van Leeuwen, editor, pages 807–833. North-Holland, Amsterdam, 1990.

[551] N. Pippenger. The expected capacity of concentrators. *SIAM J. Math.*, 4(1):121–129, 1991.

[552] C.G. Plaxton. Load balancing, selection and sorting on the hypercube. In *Proc. 1st ACM Symp. Parallel Algorithms Arch.*, pages 64–73, 1989.

[553] C.G. Plaxton. A hypercubic sorting network with nearly logarithmic depth. In *Proc. 24th ACM Symp. Theory Comput.*, pages 405–416, 1992.

[554] C.G. Plaxton and T. Suel. A lower bound for sorting networks based on the shuffle permutation. *Math. Syst. Theory*, 27:491–508, 1994.

[555] C.G. Plaxton and T. Suel. A super-logarithmic lower bound for hypercubic sorting networks. In *Proc. 21st EATCS Int. Coll. Aut. Lang. Progr.*, pages 618–629, 1994.

[556] C.D. Polychronopoulos. *Parallel Programming and Compilers*. Kluwer Academic, Norwell, MA, 1988.

[557] M. Portz. On the use of interconnection networks in cryptography. In *Proc. Advances in Cryptology-Eurocrypt*, volume 547 of *Lect. Notes Comp. Sci.*, pages 302–315, 1991.

[558] F.P. Preparata, G. Metze, and R.T. Chien. On the connection assignment problem of diagnosable systems. *IEEE Trans. Electr. Comput.*, EC:848–854, 1967.

[559] F.P. Preparata and J. Vuillemin. The cube-connected cycle: A versatile network for parallel computation. *Comm. ACM*, 24(5):300–309, 1981.

[560] D. Psaltis, G. Waltis, and G.W. Burr. Holographic data storage. *IEEE Computer*, 31(2):52–60, 1998.

[561] M.J. Quinn. *Designing Efficient Algorithms for Parallel Computers.* McGraw-Hill, New York, 1987.

[562] Y. Rabani and E. Tardos. Distributed packet switching in arbitrary networks. In *Proc. 28th ACM Symp. Theory Comput.*, pages 366–375, 1996.

[563] M. Rabin. Efficient dispersal of information for security, load balancing and fault tolerance. *J. ACM*, 36(2):335–348, 1989.

[564] P. Raghavan. Probabilistic construction of deterministic algorithms; Approximating packing integer programs. In *Proc. 27th IEEE Symp. Foundations Comp. Sci.*, pages 10–18, 1986.

[565] P. Raghavan. Robust algorithms for packet routing in a mesh. *Math. Syst. Theory*, 28(1):1–11, 1995.

[566] C.S. Raghavendra and R.V. Boppana. On self-routing in Beneš and shuffle-exchange networks. *IEEE Trans. Computers*, C-40(9):1057–1064, 1991.

[567] C.S. Raghavendra and A. Varma. Rearrangeability of the five-stage shuffle-exchange network for $n = 8$. *IEEE Trans. Comm.*, 35(8):808–812, 1986.

[568] M.T. Raghunath and A. Ranade. Customizing interconnection networks to suit packaging hierarchies. Technical Report 92-725, Comput. Sci. Division, University of California, Berkeley, 1992.

[569] M.T. Raghunath and A. Ranade. Designing interconnection networks for multilevel packaging. In *Proc. Int. Conf. Supercomput.*, pages 772–781, 1993.

[570] S. Rai, J.L. Trahan, and Smailus T. Processor allocation in hypercube multiprocessors. *IEEE Trans. Parallel Distrib. Syst.*, C-6(6):606–616, 1995.

[571] S. Rajasekaran. $k$-$k$ routing, $k$-$k$ sorting and cut-through routing on the mesh. *J. Algorithms*, 19(3):361–382, 1995.

[572] S. Rajasekaran. Mesh connected computers with fixed and reconfigurable buses: packet routing and sorting. *IEEE Trans. Computers*, C-45(5):529–539, 1996.

[573] S. Rajasekaran and R. Overholt. Constant queue routing on a mesh. *J. Parallel Distrib. Comput.*, 15:160–166, 1992.

[574] S. Rajasekaran and M. Raghavachari. Optimal randomized algorithms for multipacket and cut through routing on the mesh. *J. Parallel Distrib. Comput.*, 26:257–260, 1995.

[575] S. Rajasekaran and T. Tsantilas. Optimal routing algorithms for the mesh-connected processor arrays. *Algorithmica*, 8:21–38, 1992.

[576] P. Ramanathan and K.G. Shin. Reliable broadcast in hypercube multicomputers. *IEEE Trans. Computers*, C-37(12):1654–1657, 1988.

[577] A. Ranade. How to emulate shared memory. *J. Comput. Syst. Sci.*, 42(3):307–326, 1991.

[578] A. Ranade, S. Schleimer, and D.S. Wilkerson. Nearly tight bounds for wormhole routing. In *Proc. 35th Symp. Foundations Comp. Sci.*, pages 346–255, 1994.

[579] A.G. Ranade and S.L. Johnsson. The communication efficiency for meshes, boolean cubes and cube connected cycles for wafer scale integration. In *Proc. IEEE Int. Conf. Parallel Proc.*, pages 479–482, 1987.

[580] R. Ranka and S. Sahni. *Hypercube Algorithms with Application to Image Processing and Pattern Recognition.* Springer-Verlag, New York, 1990.

[581] V.N. Rao and V. Kumar. On the efficiency of parallel backtracking. *IEEE Trans. Parallel Distrib. Syst.*, C-4(4):427–437, 1993.

[582] D. Rau, J.A.B. Fortes, and H.J. Siegel. Destination tag routing techniques based on a state model for the IADM network. In *Proc. 15th IEEE Int. Symp. Comput. Arch.*, pages 318–324, 1988.

[583] C.V. Ravi. On the bandwidth and interference in interleaved memory systems. *IEEE Trans. Computers*, C-21:899–901, 1972.

[584] S.M. Reddy, D.K. Pradhan, and J.G. Kuhl. Directed graphs with minimum diameter and maximal connectivity. Technical report, School Engin., Oakland Univ., Rochester, MI, 1980.

[585] J. Reif and L.G. Valiant. A logarithmic time sort for linear size networks. *J. the ACM*, 34(1):68–76, 1987.

[586] R. Reischuk. Probabilistic parallel algorithms for sorting and selection. *SIAM J. Comput.*, 14:396–411, 1985.

[587] D. Richards. Parallel sorting – a bibliography. *SIGACT News*, 18(1):28–48, 1986.

[588] A. Roberts and A. Symvonis. A general method for deflection worm routing on meshes based on packet routing algorithms. Technical Report 490, Dept. Comp. Sci., University of Sydney, Sydney, 1994.

[589] R.S. Roberts and S.C. Kothari. On computing the combinatorial power of SW banyan networks. *IEEE Trans. Computers*, C-38(5):761–765, 1989.

[590] A.L. Rosenberg. Product-shuffle networks: toward reconciling shuffles and butterflies. *Discr. Appl. Math.*, 37/38:465–488, 1992.

[591] M. Rosenblum, J. Chapin, D. Teodosiu, et al. Implementing efficient fault containment for shared-memory multiprocessors. *Comm. ACM*, 39(9):52–61, 1996.

[592] P. Ružička. A note on the efficiency of an interval routing algorithm. *The Computer J.*, 34(5):475–476, 1991.

[593] Y. Saad and M.H. Schultz. Topological properties of hypercubes. *IEEE Trans. Computers*, 37:867–872, 1988.

[594] Y. Saad and M.H. Schultz. Data communication in hypercubes. *J. Parallel Distrib. Comput.*, 6:115–135, 1989.

[595] G. Sabidussi. On a class of fixed-point-free graphs. In *Proc. Amer. Math. Soc.*, volume 9, pages 800–804, 1958.

[596] G. Sabidussi. Vertex-transitive graphs. *Monatsh. Math.*, 68:426–438, 1969.

[597] K. Sado and Y. Igarashi. Some parallel sorts on a mesh-connected processor array and their time efficiency. *J. Parallel Distrib. Comput.*, 3:398–410, 1986.

[598] N. Santoro and R. Khatib. Labelling and implicit routing in networks. *The Computer J.*, 28(1):5–8, 1985.

[599] S. Savage, M. Burrows, G. Nelson, P. Sobalvarro, and T. Anderson. Eraser: A dynamic data race detector for multi-threaded programs. In *Proc. 16th ACM Symp. on OS Princ.*, pages 26–37, Saint-Malo, France, 1997.

[600] S.A. Savari. Average case analysis of five two-dimensional bubble-sorting algorithms. In *Proc. 5th ACM Symp. Parallel Algorithms Arch.*, pages 336–345, 1993.

[601] E.R. Scheinerman. On the expected capacity of binomial and random concentrators. *SIAM J. Comput.*, 19(1):156–163, 1990.

[602] I.D. Scherson and S. Sen. Parallel sorting in two-dimensional VLSI models of computation. *IEEE Trans. Computers*, C-38(2):238–249, 1989.

[603] C.P. Schnorr and A. Shamir. An optimal sorting algorithm for mesh connected computers. In *Proc. 18th Symp. Theory Comput.*, pages 255–263. ACM, 1986.

[604] U. Schwiegelsohn. A shortperiodic two-dimensional sorting technique for VLSI networks. In *Proc. Conf. Systolic Arrays*, pages 257–264, 1988.

[605] D.S. Scott and J. Brandenburg. Minimal mesh embedding in binary hypercubes. *IEEE Trans. Computers*, C-37(10):1284–1285, 1988.

[606] M.L. Scott and Mellor-Crummey J.M. Fast contention-free combining tree barriers. *Int. J. Parallel Progr.*, C-22(4):449–481, 1994.

[607] S.L. Scott. Synchronization and communication in the T3E multiprocessor. In *Proc. 7th ACM Conf. Arch. Supp. Prog. Lang. Oper. Syst.*, pages 26–36, 1996.

[608] S.L. Scott and G.S. Sohi. The use of feedback in multiprocessors and its application to tree saturation control. *IEEE Trans. Parallel Distrib. Syst.*, C-1(4):385–398, 1990.

[609] S.L. Scott and G. Thorson. The Cray T3E network: adaptive routing in a high performance 3D torus. In *Proc. 4th Symp. Hot. Interc.*, pages 147–156, 1996.

[610] C.L. Seitz. The cosmic cube. *Comm. ACM*, 28(1):22–33, 1985.

[611] A. Sengupta and S. Viswanathan. On fault-tolerant fixed routing in hypercube. *Inf. Proc. Letters*, 51:93–99, 1994.

[612] A. Sengupta, K. Zemoudeh, and S. Bandyopadhyay. Self-routing algorithms for strongly regular multistage interconnection networks. *J. Parallel Distrib. Comput.*, 14(2):187–192, 1992.

[613] S.W. Seo and T.Y. Feng. A new routing scheme for concatenating two omega networks. In *Proc. Parallel Lang. Arch. Eur.*, volume 817 of *Lect. Notes Comp. Sci.*, pages 831–834, 1994.

[614] A.S. Sethi and N. Deo. Interference in multiprocessor system with local memory access probability. *IEEE Trans. Computers*, C-28(2):157–163, 1979.

[615] C.E. Shannon. A mathematical theory of communication. *Bell System Technical Journal*, 27:379–423 and 623–656, 1948.

[616] A. Shaout and D. Smyth. Data flow models for a hypercube multiprocessor network. In *Proc. IEEE Int. Conf. Parallel Proc.*, volume I, pages 351–354, 1991.

[617] D.D. Sharma and D.K. Pradhan. Processor allocation in hypercube multicomputers: fast and efficient strategies for cubic and noncubic allocation. *IEEE Trans. Parallel Distrib. Syst.*, C-6(10):1108–1123, 1995.

[618] X. Shen, Q. Hu, and W. Liang. Realization of an arbitrary permutation on a hypercube. *Inf. Proc. Letters*, 51:237–243, 1994.

[619] T.L. Sheng, W. Lin, C.R. Das, and M.J. Irwin. Distributed fault diagnosis in the butterfly parallel processor. In *Proc. IEEE Int. Conf. Parallel Proc.*, volume I, pages 172–175, 1989.

[620] D.J. Shyy and C.T. Lea. $\log_2^d(n, m, p)$ strictly nonblocking networks. *IEEE Trans. Comm.*, 39(10):1502–1510, 1991.

[621] D.J. Shyy and C.T. Lea. Rearrangeable nonblocking $\log_2^d(n, m, p)$ networks. *IEEE Trans. Comm.*, 42(5):2084–2086, 1994.

[622] J.F. Sibeyn. Matrix techniques for faster routing of affine permutations on a mesh interconnection network. Technical Report 90-17, Dept. Comp. Sci., Utrecht University, the Netherlands, 1990.

[623] J.F. Sibeyn. Routing permutations on mesh interconnection networks. In *Proc. 2nd IEEE Symp. Parallel Distrib. Comput.*, pages 94–97, 1990.

[624] J.F. Sibeyn. *Algorithms for routing on meshes.* PhD thesis, Utrecht University, the Netherlands, 1992.

[625] J.F. Sibeyn. Desnakification of mesh sorting algorithms. In *Proc. 2nd Euro Symp. Algorithms*, volume 855 of *Lect. Notes Comp. Sci.*, pages 377–390, 1994.

[626] J.F. Sibeyn. Deterministic routing and sorting on rings. In *Proc. 8th IEEE Int. Parallel Proc. Symp.*, pages 406–410, 1994.

[627] J.F. Sibeyn. Routing on triangles, tori and honeycombs. *Int. J. Foundations of Comput. Sci.*, 8(3):269–287, 1997.

[628] J.F. Sibeyn. Routing with finite speeds of memory and network. In *Proc. 22nd Symp. Math. Foundations Comp. Sci.*, volume 1295 of *Lect. Notes Comp. Sci.*, pages 488–497, 1997.

[629] J.F. Sibeyn. Sample sort on meshes. In *Proc. 3rd Euro-Par Conf.*, volume 1300 of *Lect. Notes Comp. Sci.*, pages 389–398, 1997.

[630] J.F. Sibeyn, B.S. Chlebus, and M. Kaufmann. Deterministic permutation routing on meshes. *J. Algorithms*, 22:111–141, 1997.

[631] J.F. Sibeyn and M. Kaufmann. Deterministic 1-$k$ routing on meshes. In *Proc. 11th Symp. Theoret. Asp. Comput. Sci.*, volume 775 of *Lect. Notes Comp. Sci.*, pages 237–248, 1994.

[632] J.F. Sibeyn and M. Kaufmann. Solving cheap graph problems on meshes. In *Proc. 20th Symp. Math. Foundations Comp. Sci.*, volume 969 of *Lect. Notes Comp. Sci.*, pages 412–422, 1995.

[633] H.J. Siegel. The theory underlying the partitioning of permutation networks. *IEEE Trans. Computers*, C-29(9):791–800, 1980.

[634] H.J. Siegel and R.J. McMillen. Using the augmented data manipulator network in PASM. *IEEE Comput.*, 14:25–33, 1981.

[635] A. Silberschatz, J. Peterson, and P. Galvin. *Operating System Concepts.* Addison-Wesley, Reading, MA, third edition, 1991.

[636] L. Simoncini, F. Saheban, and A.D. Friedman. Design of self-diagnosable multiprocessor systems with concurrent computation and diagnosis. *IEEE Trans. Comput*, C-29(6):540–546, 1980.

[637] M. Snir. Depth-size trade-offs for parallel prefix computation. *J. Algorithms*, 7(2):185–201, 1986.

[638] M. Snir, P. Hochschild, D.D. Frye, and K.J. Gildea. The communication software and parallel environment of the IBM SP2. *IBM Syst. J.*, 34(2):205–221, 1995.

[639] P.G. Sobalvarro. Analytical modeling of multistage multipath networks. *IEEE Trans. Parallel Distrib. Syst.*, 7(10):1059–1064, 1996.

[640] M. Šoch and P. Tvrdík. Optimal gossip in store-and-forward noncombining 2-D tori. In *Proc. 3rd Int. Euro-Par Conf.*, volume 1300 of *Lect. Notes Comp. Sci.*, pages 234–241, 1997.

[641] M.A. Sridhar. A fast algorithm for testing isomorphism of permutation networks. *IEEE Trans. Computers*, C-38(6):903–909, 1989.

[642] M.A. Sridhar and C.S. Raghavendra. Minimal full-access networks: enumeration and characterization. *J. Parallel Distrib. Comput.*, 9(4):347–356, 1990.

[643] G.D. Stamoulis and J.N. Tsitsiklis. An efficient algorithm for multiple simultaneous broadcasts in the hypercube. *Inf. Proc. Letters*, 46(5):219–224, 1993.

[644] G.D. Stamoulis and J.N. Tsitsiklis. Efficient routing schemes for multiple broadcasts in hypercubes. *IEEE Trans. Parallel Distrib. Syst.*, C-4(7):725–739, 1993.

[645] G.D. Stamoulis and J.N. Tsitsiklis. Greedy routing in hypercubes and butterflies. *IEEE Trans. Comm.*, 42(11):3051–3061, 1994.

[646] D. Stein and D. Shah. Implementing lightweight threads. In *Proc. USENIX Conf.*, pages 1–9, 1992.

[647] D. Steinberg. Notes on invariant properties of the shuffle/exchange and a simplified cost-effective version of the Omega network. *IEEE Trans. Computers*, C-32(5):444–450, 1983.

[648] I. Stojmenovic. Computational geometry on a hypercube. In *Proc. IEEE Int. Conf. Parallel Proc.*, pages 100–103, 1988.

[649] H.S. Stone. Parallel processing with the perfect shuffle. *IEEE Trans. Computers*, C-20:153–161, 1971.

[650] Q.F. Stout and B. Wagar. Intensive hypercube communication. *J. Parallel Distrib. Comput.*, 10:167–181, 1990.

[651] T. Suel. Improved bounds for routing and sorting on multi-dimensional meshes. In *Proc. 6th ACM Symp. Parallel Algorithms Arch.*, pages 26–35, 1994.

[652] T. Suel. Routing and sorting on meshes with row and column buses. In *Proc. 8th IEEE Int. Parallel Proc. Symp.*, pages 411–417, 1994.

[653] X.H. Sun and J. Zhu. Performance considerations of shared virtual memory machines. *IEEE Trans. Parallel Distrib. Syst.*, C-6(11):1185–1194, 1995.

[654] S. Sur and P.K. Srimani. Self-diagnosability properties of hypercubes and star graphs and diagnosis algorithms. *J. Comput. Syst. Sci.*, 8(4):217–223, 1993.

[655] T. Szymanski. On the permutation capability of a circuit-switched hypercube. In *Proc. IEEE Int. Conf. Parallel Proc.*, volume I, pages 103–110, 1989.

[656] T. Szymanski and C. Fang. Randomized routing of virtual connections in extended banyans. *IEEE Trans. Comm.*, 3(9):2521–2531, 1995.

[657] T.H. Szymanski and V.C. Hamacher. On the permutation capability of multistage interconnection networks. *IEEE Trans. Computers*, 36:810–822, 1987.

[658] T.H. Szymanski and V.C. Hamacher. On the universality of multipath multistage interconnection networks. *J. Parallel Distrib. Comput.*, 7:541–549, 1989.

[659] H. Tamaki. Efficient self-embedding of butterfly networks with random faults. In *Proc. 33rd IEEE Symp. Foundations Comp. Sci.*, pages 533–541, 1992.

[660] A. Tanenbaum. *Computer Networks.* Prentice-Hall, Englewood Cliffs, NJ, 1999.

[661] S. Thanawastien, V.P. Nelson, and S. Tantaratana. Optimal fault detection and testing sequence for shuffle/exchange networks. In *Proc. 13th Int. Symp. Fault Tolerant Comput.*, pages 442–445, 1983.

[662] C.D. Thompson. Generalized connection networks for parallel processor interconnection. *IEEE Trans. Computers*, C-27(2):1119–1125, 1978.

[663] C.D. Thompson. Area-time complexity for VLSI. In *Proc. 11th Symp. Theory Comput.*, pages 81–88, 1979.

[664] C.D. Thompson. The VLSI complexity of sorting. *IEEE Trans. Computers*, C-31(12):1171–1184, 1983.

[665] C.D. Thompson and H.T. Kung. Sorting on a mesh-connected parallel computer. *Comm. ACM*, 20(4):263–270, 1977.

[666] P.W. Thompson and J.D. Lewis. The STC104 packet routing chip. *VLSI Design*, 2(4):305–314, 1995.

[667] S.B. Tien and C.S. Raghavendra. Algorithms and bounds for shortest paths and diameter in faulty hypercubes. *IEEE Trans. Parallel Distrib. Syst.*, C-4(6):713–718, 1993.

[668] J. Torrellas and Z. Zhang. The performance of the Cedar multistage switching network. *IEEE Trans. Parallel Distrib. Syst.*, C-8(4):321–336, 1997.

[669] Y.C. Tseng, T.H. Lin, S.K.S. Gupta, and D.K. Panda. Bandwidth-optimal complete exchange on wormhole-routed 2D/3D torus networks. *IEEE Trans. Parallel Distrib. Syst.*, C-8(4):380–396, 1997.

[670] J.O. Tuazon, J.C. Peterson, M. Pniel, and D. Lieberman. Caltech/JPL MARK II hypercube concurrent processor. In *Proc. IEEE Int. Conf. Parallel Proc.*, volume I, pages 666–673, 1985.

[671] J. Turner. Point-symmetric graphs with a prime number of point. *J. Comb. Theory*, 3:136–145, 1967.

[672] N.F. Tzeng. Design of a novel combining structure for shared-memory multiprocessors. In *Proc. IEEE Int. Conf. Parallel Proc.*, volume I, pages 1–8, 1989.

[673] N.F. Tzeng. Alleviating the impact of tree saturation on multistage interconnection network performance. *J. Parallel Distrib. Comput.*, 12(2):107–117, 1991.

[674] N.F. Tzeng. An approach to the performance improvement of multistage interconnection networks with nonuniform traffic spots. In *Proc. IEEE Int. Conf. Parallel Proc.*, volume I, pages 542–545, 1991.

[675] N.F. Tzeng. A reliable butterfly network for distributed memory multiprocessors. In *Proc. IEEE Int. Conf. Parallel Proc.*, volume 1, pages 323–327, 1991.

[676] N.F. Tzeng. An effective fault tolerant technique for circular butterfly parallel systems. *IEEE Trans. Computers*, 41(10):223–226, 1992.

[677] A.K. Uht, V. Sindagi, and S. Somanathan. Effect reduction techniques. *IEEE Computer*, 30(5):71–81, 1997.

[678] E. Upfal. Efficient schemes for parallel communication. *J. ACM*, 31(2):507–517, 1984.

[679] E. Upfal. An $\mathcal{O}(\log N)$ deterministic packet-routing scheme. *J. ACM*, 39(1):55–70, 1992.

[680] E. Upfal, S. Felperin, and M. Snir. Randomized routing with shorter paths. In *Proc. 5th ACM Symp. Parallel Algorithms Arch.*, pages 283–292, 1993.

[681] L.G. Valiant. Graph-theoretic properties in computational complexity. *J. Computer Syst. Sci.*, 13(3):278–285, 1976.

[682] L.G. Valiant. A scheme for fast parallel communication. *SIAM J. Comput.*, 11(2):350–361, 1982.

[683] L.G. Valiant. Optimality of a two phase strategy for routing in interconnection networks. *IEEE Trans. Computers*, C-32(9):861–863, 1983.

[684] L.G. Valiant. General purpose parallel architectures. In *Handbook Theoret. Comp. Sci.*, J. van Leeuwen, editor, pages 943–971. North-Holland, Amsterdam, 1990.

[685] L.G. Valiant and G.J. Brebner. Universal schemes for parallel communication. In *Proc. 13th ACM Symp. Theory Comput.*, pages 263–277, 1981.

[686] J. van Leeuwen and R.B. Tan. Interval routing. *The Computer J.*, 30(4):298–307, 1987.

[687] A. Varma and C.S. Raghavendra. Rearrangeability of multistage shuffle/exchange networks. *IEEE Trans. Comm.*, 36(10):1138–1147, 1988.

[688] A. Varma and C.S. Raghavendra. Reliability analysis of redundant-path interconnection networks. *IEEE Trans. Reliability*, 38:130–137, 1989.

[689] P.J. Varman and K. Doshi. Sorting with linear speedup on a pipelined hypercube. *IEEE Trans. Computers*, C-41(1):97–105, 1992.

[690] E.A. Varvarigos and A. Banerjee. Routing schemes for multiple random broadcasts in arbitrary network topologies. *IEEE Trans. Parallel Distrib. Syst.*, C-7(8):886–895, 1996.

[691] E.A. Varvarigos and D.P. Bertsekas. Communication algorithms for isotropic tasks in hypercubes and wraparound meshes. *Parallel Comput.*, 18:1233–1257, 1992.

[692] E.A. Varvarigos and D.P. Bertsekas. Multinode broadcast in hypercubes and rings with randomly distributed length of packets. *IEEE Trans. Parallel Distrib. Syst.*, C-4(2):144–154, 1993.

[693] E.A. Varvarigos and D.P. Bertsekas. Dynamic broadcasting in parallel computing. *IEEE Trans. Parallel Distrib. Syst.*, C-6(2):120–131, 1995.

[694] E.A. Varvarigos and D.P. Bertsekas. Transposition of banded matrices: a nearly isotropic task. *Parallel Comput.*, 21:243–264, 1995.

[695] J. Vounckx, G. Deconinck, R. Lauwereins, et al. The FTMPS-project: design and implementation of fault-tolerance techniques for massively parallel systems. In *Proc. High Performance Comput. and Netw. Conf.*, volume 797 of *Lect. Notes Comp. Sci.*, pages 401–406, 1994.

[696] J. Vuillemin, B. Patrice, S. Didier, R. Mark, et al. Programmable active memories: reconfigurable systems come of age. *IEEE Trans. VLSI Syst.*, 4(1):56–69, 1996.

[697] A. Wagner and D.G. Corneil. Embedding trees in a hypercube is NP-complete. *SIAM J. Comput.*, 19(3):570–590, 1990.

[698] A. Waksman. A permutation network. *J. ACM*, 15(1):159–163, 1968.

[699] B. Wang, G. Chen, and F. Lin. Constant time sorting on a processor array with a reconfigurable bus system. *Inf. Proc. Letters*, 34:187–192, 1990.

[700] R. Wanka. Fast general sorting on meshes of arbitrary dimension without routing. Technical Report 87, Fachbereich Informatik, Universität-GH Paderborn, Paderborn, Germany, 1991.

[701] D.S. Wei. Quicksort and permutation routing on the hypercube and de Bruijn networks. In *Proc. Int. Symp. Parallel Arch. Algorithms and Networks.*, pages 238–245, 1994.

[702] R.P. Weicker. A detailed look at some popular benchmarks. *Parallel Comput.*, 17:1153–1172, 1991.

[703] B. Wilkinson. Cascaded rhombic crossbar interconnection networks. *J. Parallel Distrib. Comput.*, 10:96–101, 1990.

[704] B. Wilkinson. *Computer Architecture; Design and Performance.* Prentice-Hall, Englewood Cliffs, NJ, 1996.

[705] D.S. Wise. Compact layouts of banyan/FFT networks. *VLSI Syst. and Comput.*, pages 186–195, 1981.

[706] C.K. Wong and D. Coppersmith. A combinatorial problem related to multimodule memory organization. *J. ACM*, 21(3):392–402, 1974.

[707] C. Wu and T.Y. Feng. The universality of the shuffle-exchange network. *IEEE Trans. Computers*, C-30(5):324–332, 1981.

[708] C. Wu and M. Lee. Performance analysis of multistage interconnection network configurations and operations. *IEEE Trans. Computers*, 41(1):18–27, 1992.

[709] C.-l. Wu and T.Y. Feng. On a class of multistage interconnection networks. *IEEE Trans. Computers*, C-29(8):694–702, 1980.

[710] C.L. Wu and T.Y. Feng. The reverse-exchange interconnection network. *IEEE Trans. Computers*, C-29(9):801–811, 1980.

[711] J. Wu. Safety levels - an efficient mechanism for achieving reliable broadcasting in hypercubes. *IEEE Trans. Parallel Distrib. Syst.*, C-44(5):702–706, 1995.

[712] X. Xiao and L.M. Ni. Internet QoS: A big picture. *IEEE Network*, 13(2):8–18, 1999.

[713] C.B. Yang, R.C.T. Lee, and W.T. Chen. Parallel graph algorithms based upon broadcast communications. *IEEE Trans. Computers*, C-39(12):1468–1471, 1990.

[714] C.L. Yang, G.M. Masson, and R.A. Leonetti. On fault isolation and identification in $t$-diagnosable systems. *IEEE Trans. Computers*, C-35(7):639–643, 1986.

[715] Y. Yang and G.M. Masson. Nonblocking broadcast switching networks. *IEEE Trans. Computers*, C-40(9):1005–1015, 1991.

[716] D.W.L. Yen, J.H. Patel, and E.S. Davidson. Memory interference in synchronous multiprocessor systems. *IEEE Trans. Computers*, C-31(11):1117–1121, 1982.

[717] A. Youssef. Cartesian product networks. In *Proc. IEEE Int. Conf. Parallel Proc.*, volume I, pages 684–685, 1991.

[718] A. Youssef and B. Arden. Topology of efficiently controllable banyan multistage networks. *Int. J. Microprocessors and Microsystems*, 16(1):3–13, 1992.

[719] M. Zaki and M.M. Elboraey. Analysis of reliability models for interconnecting MIMD systems. *The Computer J.*, 31(4):304–312, 1988.

[720] E.W. Zegura. Evaluating blocking probability in generalized connectors. *IEEE/ACM Trans. Netw.*, 3(4):387–398, 1995.

[721] Y. Zhu. Efficient processor allocation strategies for mesh-connected parallel computers. *J. Parallel Distrib. Comput.*, 16(4):328–337, 1992.

# Index